Polyester Films

# Polyester Films

## Materials, Processes and Applications

*Edited by*

*Miko Cakmak*
*Schools of Materials and Mechanical Engineering*
*Purdue University*
*West Lafayette, IN, USA*

*Jehuda Greener*
*J. Greener Consulting*
*Rochester, NY, USA*

WILEY

Published by John Wiley & Sons, Inc., Hoboken, New Jersey.
Published simultaneously in Canada.



*Library of Congress Cataloging-in-Publication Data*
Names: Cakmak, Miko, author. | Greener, Jehuda, author. | John Wiley & Sons, publisher.
Title: Polyester films : materials, processes and applications / Miko Cakmak, Jehuda Greener.
Description: Hoboken, NJ : Wiley, 2023. | Includes index.
Identifiers: LCCN 2023000322 (print) | LCCN 2023000323 (ebook) | ISBN 9781119535751 (cloth) | ISBN 9781119535799 (adobe pdf) | ISBN 9781119535744 (epub)
Subjects: LCSH: Polyester films.
Classification: LCC TP1180.P6 C35 2023 (print) | LCC TP1180.P6 (ebook) | DDC 668.4/225–dc23/eng/20230113
LC record available at https://lccn.loc.gov/2023000322
LC ebook record available at https://lccn.loc.gov/2023000323

Cover Design: Wiley
Cover Image: © MamiGibbs/Getty Images

Set in 9.5/12.5pt STIXTwoText by Straive, Pondicherry, India

*For Kaia, Gulgun, Emre, Karen, Helena, Maya, Eli and Ari*

# Contents

# List of Contributors

*Abdellah Ajji*
Chemical Engineering Department
Polytechnique Montreal
Montreal, Quebec, Canada

*Miko Cakmak*
Schools of Materials and Mechanical Engineering
Purdue University
West Lafayette, IN, USA

*Ebrahim Jalali Dil*
PolyExpert Inc.,
Laval, Quebec, Canada

*Abdullah Al Faysal*
Multifunctional Composites Manufacturing Laboratory (MCML)
Department of Mechanical and Industrial Engineering,
University of Toronto
Toronto, Ontario, Canada

*Kazuki Fukushima*
Department of Chemistry and Biotechnology School of Engineering
The University of Tokyo
Tokyo, Japan

Japan Science and Technology Agency (JST)
Precursory Research for Embryonic Science and Technology (PRESTO)
Saitama, Japan

*Elena Gabirondo*
POLYMAT
University of the Basque Country UPV/EHU
Donostia-San Sebastián
Spain

*Jehuda Greener*
J. Greener Consulting,
Rochester, NY, USA

*Coralie Jehanno*
POLYMAT
University of the Basque Country UPV/EHU
Donostia-San Sebastián
Spain

POLYKEY
Joxe Mari Korta Center
Donostia-San Sebastian
Spain

***Karnav Kanuga***
Medtronic Inc.,
Dublin, Ireland

***Patrick C. Lee***
Multifunctional Composites
Manufacturing Laboratory (MCML)
Department of Mechanical and
Industrial Engineering
University of Toronto
Toronto, Ontario, Canada

***Dennis J. Massa***
Massa Science and Technology
Consulting
Pittsford, NY, USA

***Ion Olazabal***
POLYMAT
University of the Basque
Country UPV/EHU
Donostia-San Sebastián
Spain

***Amir Saffar***
ProAmpac Inc.,
Terrebonne, Quebec, Canada

***Haritz Sardon***
POLYMAT, University of the Basque
Country UPV/EHU
Donostia-San Sebastián
Spain

***Troy Su***
Multifunctional Composites
Manufacturing Laboratory (MCML)
Department of Mechanical
and Industrial Engineering
University of Toronto
Toronto, Ontario, Canada

***Andy H. Tsou***
National Taiwan University
Houston, TX, USA

***Robert A. Weiss***
Department of Chemical Engineering
University of Connecticut
Storrs, CT, USA

***Baris Yalcin***
Global Packaging R&D
The Coca-Cola Company
Atlanta, GA, USA

***Jianxiang Zhao***
Multifunctional Composites
Manufacturing Laboratory (MCML)
Department of Mechanical and
Industrial Engineering,
University of Toronto
Toronto, Ontario, Canada

# Preface

Polyester film serves as a foundation, literally and figuratively, to many important technologies that define this technological age. In fact, many advances in the digital and information revolution over the past four decades owe a great deal to developments in several key material technologies such as polyester films, which comprise and enable many critical components in various flat panel displays, photovoltaics, microelectronics systems, and biomedical applications among others. Polyester films often serve as either substrates (bottom or carrier layers) or superstrates (top or protective layers) in a particular two-dimensional film structure in many of these application categories and are the materials of choice owing to a combination of remarkable physical and mechanical properties and a relatively low cost. These materials are especially well suited for roll-to-roll operations, which provide significant cost benefits and operational flexibility in the manufacture of many critical film components in the various application areas.

Through contributions of recognized experts, this volume presents the state of the art of many aspects of polyester film technology covering materials, processing, and applications and identifies areas with potential opportunities for further progress. The overriding objective of this manuscript is to identify the multitude of options (tool-kit) available to material and product designers in manipulating the properties of polyester films to meet specific performance and product requirements. These include synthetic modifications (copolymerization), physical enhancements (blending), and process upgrades (tenter-frame changes, coextrusion, and coating).

A brief historical review of polyester film technology and overview of properties and applications of key polyester films, PET and PEN, is presented in Chapter 1. The vast synthetic options available for manipulating the structure and properties of polyesters are surveyed in Chapter 2, with a special focus on polyester ionomers given in Chapter 4. Chapter 3 identifies the main blending options available to improve the properties of commodity polyesters. The unique rheo-optical properties and testing methodology of polyester films are discussed in Chapter 5.

Chapter 6 presents the micro-layer coextrusion technology, which provides a unique and not fully explored methodology for preparing multi-layered films. Heat-setting is one of the critical steps in the tentering process commonly used for manufacturing polyester films. Key aspects of this step are reviewed in Chapter 7. The use of polyester films in biomedical applications is surveyed in Chapter 8, and finally, the important topic of polyester recycling is covered in Chapter 9 with a special focus on upcycling.

The editors would like to thank all the contributors to this volume for their excellent contributions, patience, and understanding through the difficult Covid-19 pandemic while this book project was undertaken.

*Miko Cakmak and Jehuda Greener*

# 1

# Polyester Films: An Overview

*Jehuda Greener*

*J. Greener Consulting, Rochester, NY, USA*

## 1.1 Introduction

Many advances in the digital and information revolution over the past four decades owe a great deal to developments in several key material technologies. A case in point is polyester films, which comprise and enable many critical components in various flat panel displays (FPDs) [1–3], photovoltaics [4, 5], microelectronics systems [6], and biomedical applications [7], among others. Polyester films often serve as either substrates (bottom or carrier layers) or superstrates (top or protective layers) [8] in a particular two-dimensional film structure in many of these application categories and are the materials of choice owing to a combination of remarkable physical and mechanical properties and a relatively low cost. These materials are especially well suited for roll-to-roll (R2R) operations [9], which provide significant cost benefits and operational flexibility in the manufacture of many critical film components in the various application areas.

Polyesters are polymers produced via a polycondensation reaction from diols and diacids or other related precursors (e.g. diesters), and as such, cover a vast material space with a wide range of molecular structures and physical properties [10]. However, only a small fraction of the known polyesters are good film formers suitable for high-end film applications. The best-known and most common film polyester is poly(ethylene terephthalate) (PET), and somewhat less known but equally important is poly(ethylene-2,6-naphthalene dicarboxylate) (PEN), see Figure 1.1. These polyesters and many variations thereof are

*Polyester Films: Materials, Processes and Applications*, First Edition.
Edited by Miko Cakmak and Jehuda Greener.

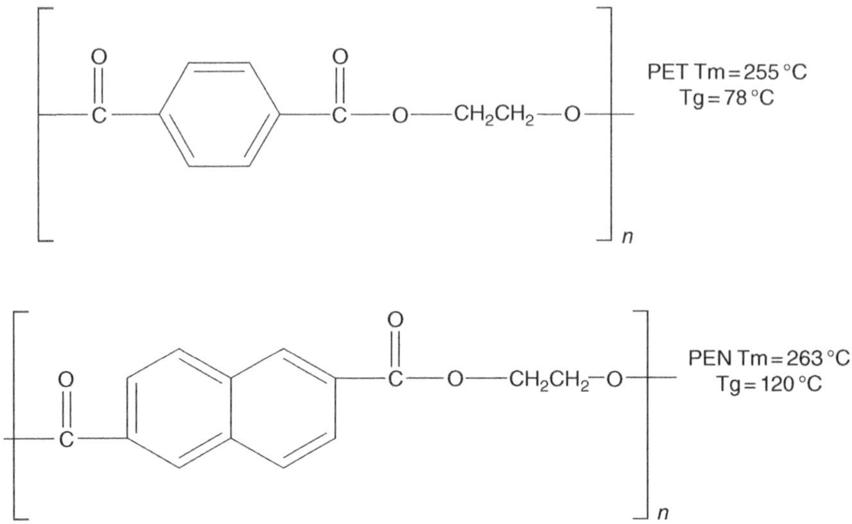


**Figure 1.1** Molecular structures of PET and PEN.

transparent, semi-crystalline, possess relatively high glass transition temperatures ($T_g$s) and are used in a variety of film products spanning a wide range of applications.

We note that the PET film market is a relatively small segment of the global PET resin market, which is dominated by the textile (fiber) industry and by packaging applications, driven mainly by growth in the beverage sector [11, 12]. Yet, as noted above, this segment of the market is critical to many mature and emerging technologies.

The physical and chemical properties of these and related film products can be manipulated and enhanced through a variety of well-established material design strategies. PET, a polycondensation product of ethylene glycol (EG) and terephthalic acid (TA) or dimethyl terephthalate (DMT), can be modified by changing the comonomer composition to impart certain desired properties. For example, by substituting some of the TA with an ionic diacid or diester such as (sodiosulfo) isophthalate (SIP), it is possible to increase the hydrophilicity of the base polymer while still retaining its film-forming ability [13, 14]. Similarly, by substituting some of the EG with a poly(ethylene glycol) (PEG) segment the polymer is "softened" and its $T_g$ can be suppressed by controlling the length and substitution level of the PEG moiety [15]. By contrast, by fully substituting TA with the bulkier naphthalate diacid moiety, to form PEN, the $T_g$ and stiffness of the polyester film are significantly enhanced, as will be discussed in Section 1.3. Many other comonomers can be used to tweak certain properties of the base polyester films to achieve a desired performance [10]. However, a synthetic modification of the base polyester is often not feasible from an economic or operational standpoint.

Another common approach to boosting and manipulating the properties of polyester films is to blend the base polyester with another miscible polymer possessing some desirable attributes. For example, blending PET with fully miscible polyether-imide (PEI) can raise the $T_g$ of the blended resin relative to PET and enhance its overall thermal stability [16]. Along these lines, addition of various immiscible fillers, solid, liquid, or gas, at a certain loading level to create a polyester composite can also enhance or modify certain bulk or surface properties of the polyester film [16]. In order for the polyester composite to retain its essential film-forming ability, the polyester phase must be the major ("continuous") phase of the composite material, while the added filler must be well dispersed and confined to the minor (discrete) phase of the composite. If the characteristic domain size of the minor component is less than the wavelength of visible light, then the composite material system is defined as a *nanocomposite*, which carries some implications for the physical and optical properties of the corresponding material.

All of the aforementioned material design approaches are potentially useful so long as the modified polyester resin is processable using suitable film processing methodologies and existing film-making infrastructure. In fact, one of the most common and generally highly effective approaches to boosting or modifying the properties of polyester films is through judicious film processing methods. The physical and mechanical properties of melt-cast PET or PEN films are generally inferior and do not meet the requirements of most film applications. A significant boost in properties and overall physical performance is achieved by converting the polyester resin into film using the so-called tenter-frame (*tentering*) process [17,18] commonly used for the manufacture of polyester films. This process, depicted schematically in Figure 1.2, generally comprises five main steps: (1) extrusion, (2) melt casting, (3) machine direction (MD) stretching, (4) transverse direction (TD) stretching, and (5) heat-setting (constrained high-temperature annealing). The melt-cast film following Step 2 is largely amorphous and structureless and, as noted above, has inferior mechanical properties. The stretching steps applied within the tenter-frame (Steps 3 and 4 above), as shown in Figure 1.2, induce the formation of a desired crystalline morphology and biaxial molecular alignment along both principal in-plane directions, while the heat-setting step helps refine and modify the crystalline microstructure of the biaxially oriented film, thereby enhancing its dimensional stability and mechanical properties [19–21]. Further improvement in dimensional stability often requires an additional annealing step, called heat relaxation [22], whereby the biaxially oriented and heat-set film is exposed to high temperatures ($T_g < T < T_m$) under low tension. Heat relaxation is often done offline (i.e. outside the tenter-frame machine) because of the low conveyance tension requirements. By judicious choice of the tenter-frame process

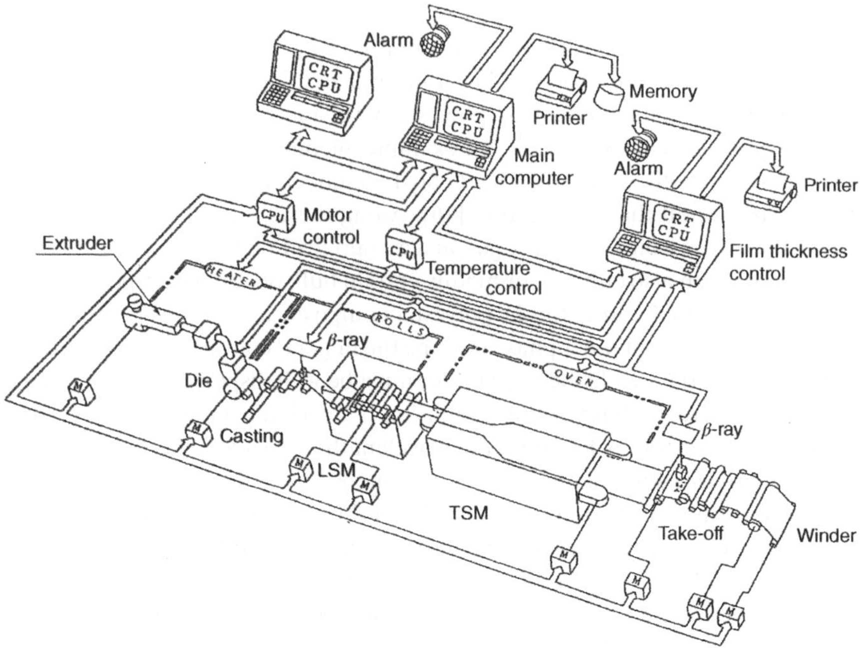


**Figure 1.2** Schematic of the tenter-frame process.

conditions, especially casting temperature, stretch ratios, stretch temperatures, stretching rates, and heat-set temperature, the properties and overall quality and uniformity of the film can be manipulated and optimized for its intended application. It is noted that the order of the stretching steps as listed above, although most common in conventional film-making machines, can be reversed or both stretching steps can be applied simultaneously ("simul-stretching"), which would require changes in machine configuration and process conditions [23]. Simul-stretching is especially useful when a small machine size is desirable. But fundamentally the film must be biaxially oriented to insure uniform and balanced mechanical and physical properties in the plane of the film, so that the exact sequence of the stretching steps is not inherently important. The applied stretch ratios are selected to maximize molecular orientation and mechanical properties and attain a desirable *thickness uniformity* based on the so-called *Considere Construction* [24], which is a strain-hardening condition.[1] Typically, optimal stretch ratios for both PET and PEN vary in the range of 3–4X with the actual

1 Strain-hardening is required to achieve thickness uniformity.

value depending mainly on the stretch temperature, $T_s$, the stretch rate (related to line speed and dimensions of the stretching stations), and the target thickness of the film product. $T_s$ usually falls in the range $\{T_g, T_{cc}\}$, where $T_{cc}$ is the so-called cold-crystallization temperature of the processed resin. The choice of the optimal heat-set temperature, $T_{HS}$, depends on several process and product considerations discussed in detail elsewhere [21]. If the base polymer (PET, PEN, or other similar polyesters) is modified by copolymerization or by blending, the optimal process conditions must be adjusted accordingly, assuming that a viable process window exists.

The tenter-frame process requires specialized equipment described in detail by Tobita et al. [17]. This process is especially well suited for relatively slow crystallizing polymers such as PET and PEN and is designed to insure that the crystalline morphology develops in a controlled and measured way to produce a film with the desired "microstructure" and properties as well as acceptable thickness and structural uniformity. Thus, any of the changes in polymer composition noted above (comonomer type and substitution level, polymer blend composition, filler loading, etc.) must take into consideration the corresponding impact on film processability in the tenter-frame process. If a given material change adversely affects film processability and cost, then the corresponding approach may be deemed ineffective and/or impractical.

Film properties can also be enhanced by coating the polyester film product with a variety of functional and ancillary layers that provide a desired functionality and effective control of the surface properties of the film [25]. A coating step may be incorporated within the tenter-frame process before or after the stretching and heat-setting steps, or it can be added offline, i.e. in a separate R2R line. The types of surface properties that can be imparted by coating include adhesion [26], electric charge dissipation (anti-static control) [27], barrier [28], friction coefficient, scratch resistance, color, reflectivity, planarity, roughness, light management (to be discussed in Section 1.4), etc.

Polyester films can also be functionalized and their *bulk* properties modified via coextrusion [29, 30]. Addition of layer(s) to the base polyester film by coextrusion provides wide latitude in boosting certain surface and bulk properties that are otherwise not attainable with a single extruded polyester layer or with a coated layer. Designing an effective coextruded film structure, however, requires that the selected coextruded polymers can withstand the tenter-frame process conditions and that they adhere well to the base polyester layer while providing the desired functionality. The main advantages of coextrusion over coating are the ability to add relatively thick functional layers anywhere within the film structure (i.e. not only on the external surfaces as in the case of coatings) and the relative ease and operational flexibility in adding a coextrusion capability to an existing tenter-frame machine. Of special note here is the micro-layer coextrusion technology, a subset of film coextrusion, developed originally at Dow Chemical in the 1960s. This process methodology, discussed in detail in Chapter 6 of this volume, makes it possible to create a multilayered structure that offers considerable design flexibility in manipulating the physical and mechanical properties of the film through a wide variety of process and material options [31].

In the next section, we provide a brief historical overview of polyester film technology since its inception in the mid-twentieth century. Some important properties of polyester films are surveyed in Section 1.3 and contrasted with those of other high-performance polymer film types, highlighting some key advantages of polyester films. Although these films are used in a host of application areas, as noted above, we review one particular area – optical films for FPDs – in Section 1.4 to illustrate the great utility of polyester films in this critically important emerging technology space. Finally, a brief summary of this important materials technology is given in Section 1.5.

## 1.2 Historical Overview

The first synthetic polyester ever made is attributed to the work of Berzelius in 1847, who reacted tartaric acid with glycerol [32]. But the earliest synthetic polyester produced via a controlled step-growth polymerization reaction was prepared by Carothers in his pioneering work on synthetic polymers in the late 1920s [33]. The earliest patent on biaxially stretched PET was filed by Swallow and Baird [34] of ICI in 1950, followed by a patent filed by Scarlett [35] of DuPont in 1958 on the biaxial stretching (tenter-frame) process, which was further refined and extended by C. J. Heffelfinger [36] in 1966. A drawing from Scarlett's patent shown in Figure 1.3 represents one of the earliest renditions of the tenter-frame process. One of the earliest applications of PET film was as a film base (substrate) for X-ray films. Consumer photographic films (at the time, one of the largest polymer film markets) have been in use with some type of a solvent-cast cellulosic film base (cellulose nitrate or cellulose triacetate) since the late nineteenth century. Early attempts in the 1950s to apply PET film as a film base to replace cellulosics in consumer photographic roll films were unsuccessful, primarily because of the high strength (toughness) and the tendency of PET-based films to retain high curl during photographic processing, making it difficult to finish and process using conventional photo-processing equipment. However, PET film base was found to be much better suited for X-ray film applications, mainly because of its low moisture sensitivity, high-dimensional stability, and high stiffness, especially compared to cellulosic films. With the development of biaxially oriented PET film technology at DuPont in the late 1950s, all X-ray film products migrated to PET film base, followed soon after by motion picture print films and various graphic

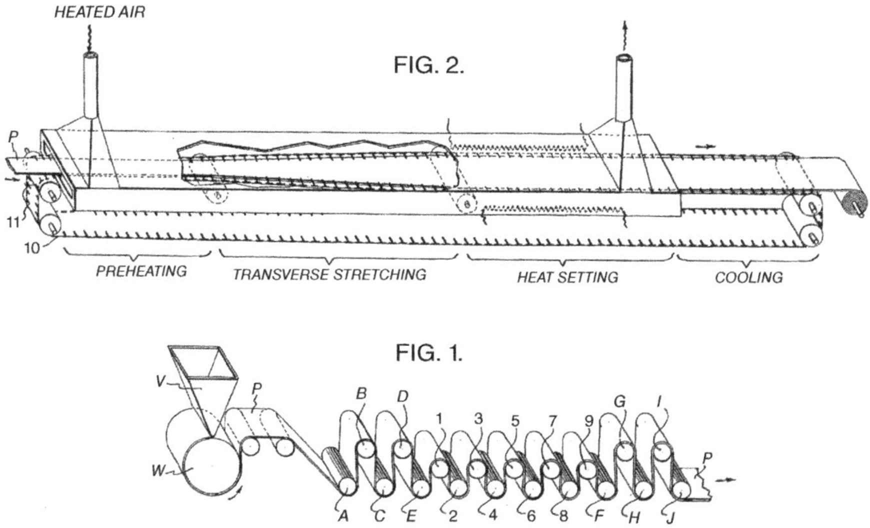


**Figure 1.3** Drawing of the tenter-frame process taken from Scarlett's 1958 patent. *Source:* Scarlett [35]/Arthur C Scarlett/Public Domain.

arts film products. Since then PET films have been produced at a wide range of gauges (1–400 μm) serving many industries and covering many diverse application areas [37] including imaging, packaging, electronics, display, and biomedical. Although PEN was discovered and patented around the same time as PET [38], it was somewhat slower in making inroads into the polyester film industry despite clear advantages over PET in thermal stability and stiffness (see Section 1.3). The slow adaptation of PEN had been mostly due to the high cost and scarcity of one of its precursors, 2,6-naphthalene dicarboxylate (2,6-NDC). PEN production started to gain traction in the 1990s with the growing demand for specialty consumer photographic films (APS format) and the growing markets in the display and microelectronics sectors where the unique attributes of PEN films (very high stiffness and high-temperature stability. See Section 1.3) are especially attractive.

## 1.3 Physical Properties

As noted in Section 1.1, there are many types of polyester films in use today in many segments of industry, covering a wide variety of applications and gauges and possessing a range of properties, but in this brief survey, we will focus on the two key polyester film types, PET and PEN, and contrast these films with other high-performance polymer films used in several high-end application areas.

**Table 1.1** Selected properties of high-performance films.

| Property\Polymer | PET | PEN | PC | PES | PAR | PCO | PI |
|---|---|---|---|---|---|---|---|
| $T_g$ (°C) | 78 | 120 | 150 | 220 | 340 | 330 | >350 |
| $T_m$ (°C) | 255 | 263 | na | na | na | na | na |
| CLTE ppm/°C[a] | 15 | 13 | 60–70 | 54 | 53 | 74 | 30–60 |
| % Transmission[b] | >85 | 85 | >90 | 90 | 90 | 91.6 | Yellow |
| Water absorption (%) | 0.14 | 0.14 | 0.4 | 1.4 | 0.40 | 0.03 | 1.8 |
| Young's modulus (GPa) | 5.3 | 6.1 | 1.7 | 2.2 | 2.9 | 1.9 | 2.5 |
| Tensile strength (MPa) | 225 | 275 | na | 83 | 100 | 50 | 231 |
| WVTR (g/($m^2$ 24 h))[c] | 21.3 | 6.7 | | | | | |
| OTR ($cm^3$/($m^2$ 24 h atm))[d] | 55 | 21 | | | | | |

[a] Linear thermal expansion coefficient over −55–85 °C.
[b] Light transmission over 400–700 nm.
[c] Water vapor transmission rate.
[d] Oxygen transmission rate.
*Source:* MacDonald [37, 39].

Several important physical properties of PET and PEN films are listed in Table 1.1 and compared to those of polycarbonate (PC), polyethersulfone (PES), polycyclo-olefin (polynorbornene, PCO), polyarylate (PAR), and polyimide (PI) films, all of which are considered high-performance films. All the materials listed are commercially available film products. Although the relevant properties and optimal property ranges depend on the particular application category and (film making) process conditions as discussed in Section 1.1, the properties listed are fairly representative of the corresponding materials and are especially important in various high-end applications such as display, microelectronics, and photovoltaics. Of the films listed, only PET and PEN are semi-crystalline. All the other film materials in this table are high-$T_g$ amorphous polymers used mostly in niche applications. High-temperature stability is especially important in various microelectronic and photovoltaic applications that involve high-temperature device fabrication and printing steps with tight registration tolerances. Although the $T_g$s of PET and PEN are generally lower, these films can be fabricated and used at temperatures above their corresponding $T_g$s because of their respective crystalline networks that boost the rigidity, stiffness, and dimensional stability of the film well above $T_g$. In fact, if properly heat-set and heat stabilized, these films can be fabricated and used at temperatures up to 150 and 200 °C for PET and PEN, respectively. (By film *fabrication* we mean the steps taken to convert the film product into a functional component or device in display, microelectronics, photovoltaics, etc., as opposed to the film-making process or film *processing*, i.e. the process of converting the raw polymer resin into the film.) The lower $T_g$s of PET and PEN also facilitate conversion of the corresponding polymer resins into film via the tenter-frame melt process discussed briefly in Section 1.1. Film processing of some of the high $T_g$ amorphous resins listed in Table 1.1 requires a very high-temperature melt-casting process or solvent casting process, both of which are costly and undesirable from operational and environmental considerations.

Other results in Table 1.1 show, with a few exceptions, significant advantages for PET and PEN films in dimensional stability (lowest thermal expansion coefficient, CLTE, and water absorption), stiffness (highest Young's modulus), and toughness/tear resistance (highest tensile strength). Of the two polyester films, PEN is clearly stiffer, stronger, and has a wider operating temperature range owing mainly to its higher $T_g$, but, of course, these advantages must be weighed against a significantly higher price point relative to PET. PEN also has significantly better moisture and oxygen barrier properties than PET [37], which is critically important in some display, microelectronic and photovoltaic applications. It must be noted, though, that this improvement in barrier performance for PEN film is well below the "super barrier" performance required in some display applications (e.g., OLED display), which is usually attained by vacuum deposition of multiple inorganic–organic barrier layers on a PET substrate [40,41].

Despite the semi-crystalline nature of the polyester films listed in Table 1.1, both PET and PEN films are optically clear and highly transmissive. The high optical transmission is possible for these semi-crystalline films because the characteristic sizes of the crystalline domains generated in the biaxially oriented polyester films by the tenter-frame process are smaller than the wavelength of visible light. We note, however, that both the PET and PEN films as well as other polyester films are highly optically anisotropic because of the high molecular order induced by the tenter-frame process. This optical anisotropy is typically manifested by high out-of-plane birefringence. The in-plane birefringence for both of these films, however, is expected to be low if the biaxial stretching process is well balanced, i.e. if the MD and TD stretching steps induce similar levels of molecular alignment in both principal in-plane directions. This point will be discussed in more detail in Chapter 5 of this volume.

We note that while the properties listed in Table 1.1 are fairly representative of the respective polymers, many physical properties of polyester films as well as other polymer film types are not absolute but are rather dependent on the conditions applied during the film-making process. It was noted in Section 1.1 that the dimensional stability of polyester films could be significantly enhanced through an added heat relaxation (stabilization) step. The mechanical properties and thermal stability can likewise be improved and optimized through changes in the stretching and heat-setting conditions in the tenter-frame process. To illustrate

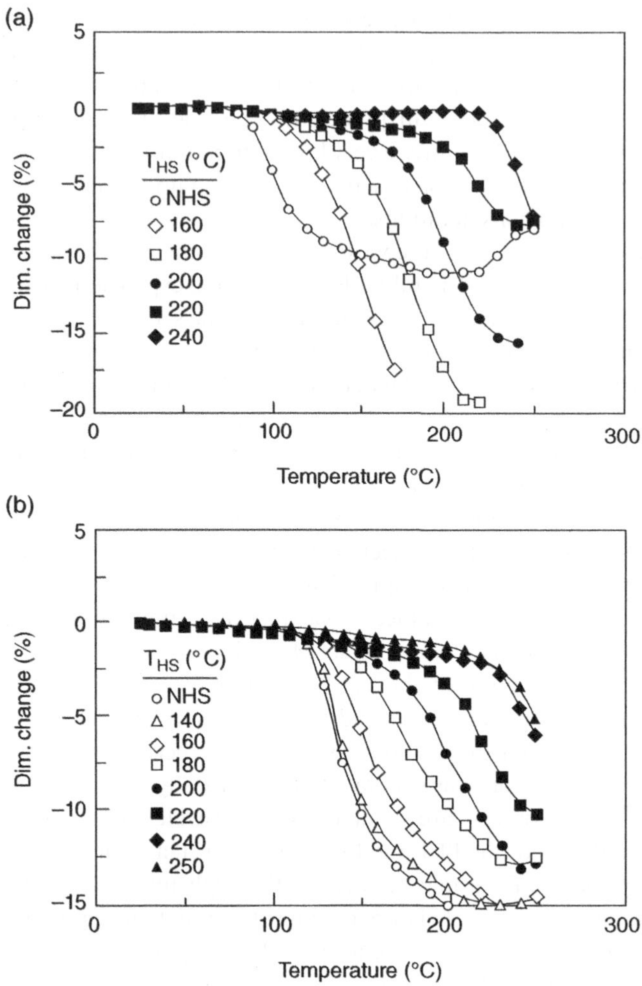


**Figure 1.4** TMA scans for (a) PET and (b) PEN films: % change in sample dimension (along MD) vs. temperature: Effect of heat-set temperature. (NHS: non-heat-set sample.) *Source:* Greener et al. [21]/with permission from John Wiley & Sons.

this point we examine the effect of heat-set temperature on the dimensional stability of PET and PEN films as expressed by thermo-mechanical analysis (TMA) scans shown in Figure 1.4 [21]. These results demonstrate the dramatic effect of heat-set temperature in suppressing the shrinkage of the film sample by shifting the onset of shrinkage up to temperatures well above the corresponding $T_g$s, close to the applied heat-set temperature for both film types, thus extending their corresponding temperature fabrication and application ranges. These results also underscore the considerable leverage we have over the physical performance of the film sample through judicious film processing. Though the effect of heat-set temperature on dimensional stability, as illustrated in Figure 1.4, is clearly evident, other variations in process conditions can have a more subtle effect on the physical and mechanical performances such that the properties listed in Table 1.1 would depend to some extent on film grade, thickness, and vendor, and the values shown should be considered only as a rough guide.

The viscoelastic properties of solid polymer films also play an important role in controlling their dimensional stability, and in particular, their propensity to acquire curl (curvature or non-planarity) in R2R and other rolling operations [42,43]. The property most relevant in this regard is the time-dependent relaxation modulus, $E(t)$; generally the faster the film relaxes when subjected to a constant strain, the higher its propensity to acquire curl in rolling operations. Figure 1.5 compares the relaxation moduli of several commercial film samples, including PET, cellulose acetate (CA), polypropylene (PP), and nylon (MXD). Of the films shown, PET and MXD are the slowest to relax, thus these films are likely to curl less in R2R operations. This is critically important in applications with tight planarity specifications. The relaxation properties of PET or PEN films can be modified by physical aging in a way that the relaxation time is increased (relaxation process is slowed down) with more extensive aging [44]. Thus, long storage times at high temperatures ($<T_g$) (annealing storage) can moderate somewhat the propensity of the film to acquire curl and thereby improve the dimensional integrity of the film product.

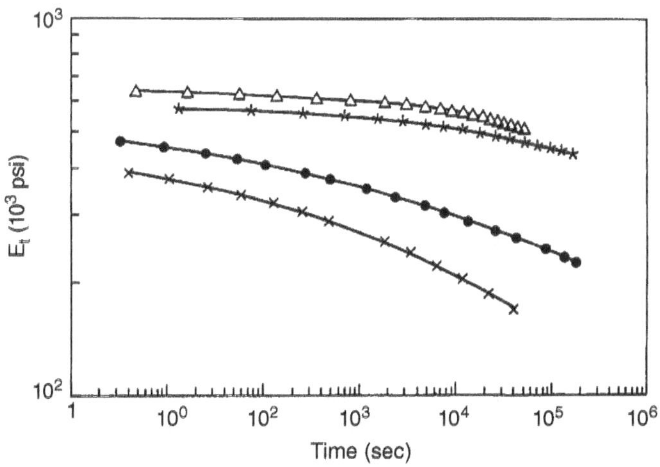


**Figure 1.5** Tensile relaxation moduli at 21 °C for various film types: (•) cellulose acetate, (*) PET, (Δ) MXD nylon, (x) polypropylene. *Source:* Greener et al. [43]/with permission from John Wiley & Sons.

## 1.4 Light Management Films

One key application of polyester films since the late 1990s is as substrates or superstrates for various light management films (LMFs) used in flat panel displays (FPDs) [1–3]. FPDs are now ubiquitous and are all around us. All FPD types, but especially liquid crystal (LC) displays (LCDs), require effective management of light to increase brightness, enhance light efficiency (~increase battery life), and improve overall image quality. A variety of optical films are typically used in different LCDs, with the particular composition and type of films and the LCD structure depending on the end use and market segment served by the display (e.g. TV, tablet, laptop, computer monitor, and smartphone). In portable displays (e.g. smart phones, laptops, tablets, etc.), light efficiency (~battery life) is paramount, while in large stationary displays (e.g. TVs and desktop computer monitors) image quality, brightness, and long-term stability are most critical.

A typical LCD module is shown in Figure 1.6, and a schematic cross-section of a display module is depicted in Figure 1.7. The display usually consists of two

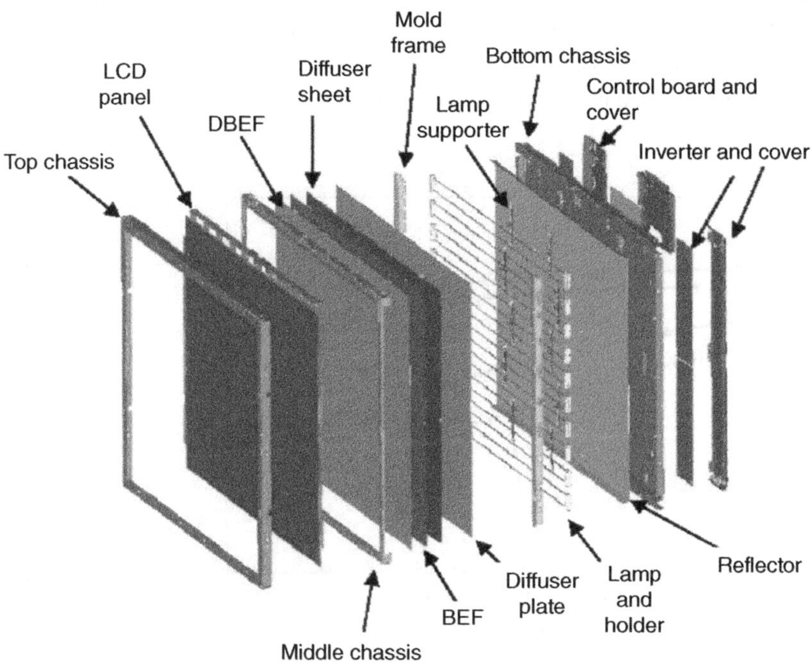


**Figure 1.6** Typical structure of an LCD module. *Source:* Kobayashi et al. [2]/with permission from John Wiley & Sons.

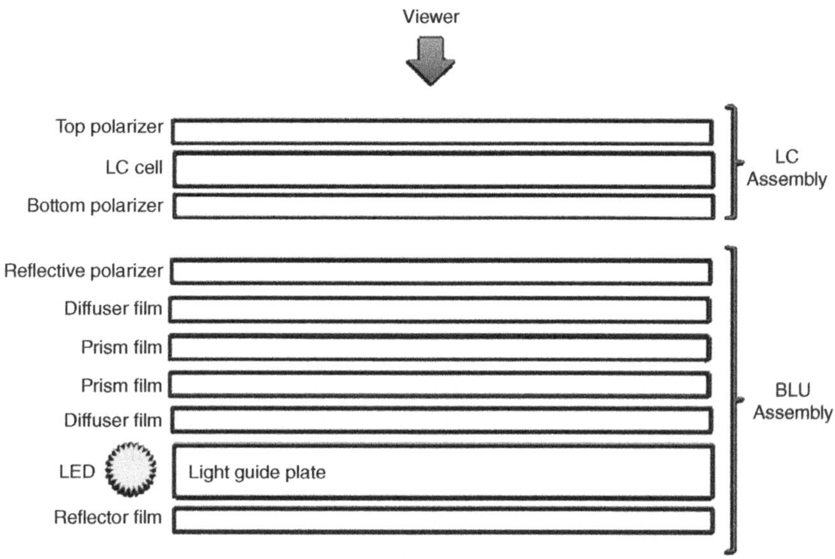


**Figure 1.7** Schematic cross-section of a typical LCD module (not to scale).

main assemblies: the LCD panel (LC assembly) and the backlight unit (BLU assembly). The LC assembly comprises the LC cell, where the actual pixelized image is formed, as well as a number of optical films. The LC cell is bounded by two crossed polarizers, each of which is sandwiched between two CTA (cellulose triacetate)[2] protective films. Additional films usually found in between the polarizers and the LC cell in the LCD panel are compensation films used to extend the viewing angle and improve the color gamut of the display.

The BLU assembly contains the light source (usually fluorescent lights (CCFLs) or light-emitting diodes (LEDs), distributed along the edges of the display) and a variety of LMFs whose main functions are to increase brightness, improve brightness uniformity, and enhance the overall light efficiency of the display. The most common LMFs in the BLU assembly, shown schematically in Figure 1.7, are light guide plates (LGPs), reflector films, brightness films (BEF®), diffuser films, reflective polarizers (DBEF®), and quantum dot (QD) films, although other LMF types have also been used.

The LGP is used to redirect light from the edge light source, usually LEDs placed along the edges of the LGP, distribute it across the area of the display, and direct it as uniformly as possible toward the LC assembly and the viewer. The LGP typically comprises a relatively thick sheet of PMMA or similar acrylic polymers, or it can

2 Sometimes refers to as TAC – tri acetyl cellulose.

have wedge-like configuration [45]. A white reflector film placed at the bottom of the LGP helps redirect incident light by reflection from the edge light source toward the LC assembly and recycle back light reflected from the reflective polarizer.

Diffuser films are used to distribute light as evenly as possible across the display area as well as hide cosmetic defects. Typically light diffusion is achieved by coating scattering beads on a PET substrate (surface diffuser), but other diffuser film types (e.g. volume diffusers) can also be used. Quite often more than one diffuser film is used in a BLU assembly.

The brightness films are sometimes called prism films because they comprise a linear prismatic structure made of photolithographed acrylic layer deposited on a PET substrate (Figures 1.8 and 1.9) [46]. The main function of these films is to collimate the light coming from the LGP toward the viewer. Often two crossed-brightness films are used in order to achieve uniform collimation in both principal directions. Common names for brightness films are prism films or BEF® (brightness enhancement film), but BEF is trademarked by 3M, which pioneered the technology. The effectiveness of the collimation function depends on the prism geometry and refractive index of the photolithographed prism material

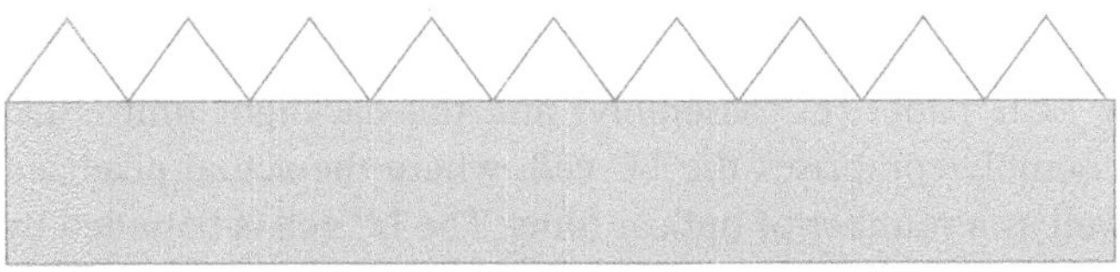

**Figure 1.8** Schematic cross-section of a prism film (not to scale).

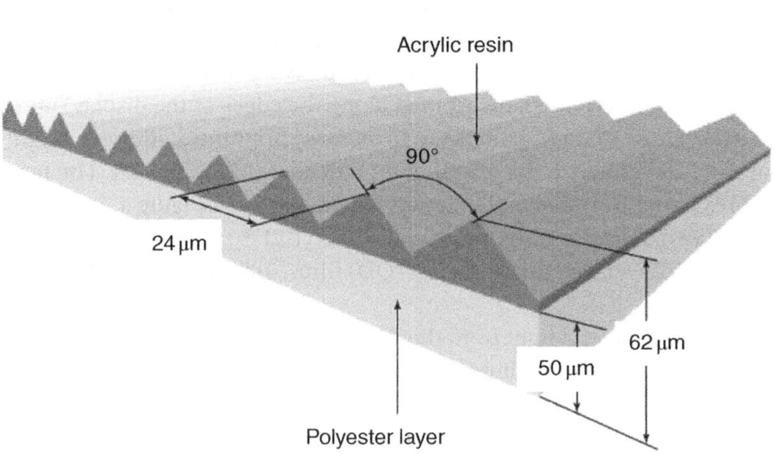


**Figure 1.9** Typical structure of a prism film. *Source:* Hanzawa [46]/with permission from John Wiley & Sons.

(usually a high-index clear acrylic). We note that a similar prismatic structure used in brightness films can also be produced by a suitable hot embossing process of a transparent base film or other methods.

Because of the absorptive nature of the polarizers used in typical LC displays, such displays are inherently light-inefficient, i.e. only light polarized along the polarization plane of the bottom polarizer can pass through, while most of the light directed from the BLU toward the bottom polarizer will be absorbed ("wasted"). One way to mitigate this inherent problem is to add a reflective polarizer just below the bottom polarizer of the LC assembly (see Figure 1.7) with the planes of polarization of the reflective polarizer and bottom polarizer being fully aligned [46]. A well-designed reflective polarizer can substantially enhance the brightness and overall light efficiency of the display by passing light only along one polarization direction, which is aligned with the plane of polarization of the bottom (absorptive) polarizer in the LC assembly. The rest of the light is reflected back to the BLU assembly and recycled. This allows a more efficient use of light by minimizing absorption losses imparted by the absorptive bottom polarizer, thus producing a considerably higher brightness and improving battery life in the case of mobile devices. The most common type of a reflective polarizer is based on the concept of giant birefringent optics in multilayered polymer films [47]. The multilayered film structure in such a component can be manufactured, for example, by the micro-layer coextrusion film technology pioneered by Dow [31] with a typical embodiment comprising alternating layers of uniaxially oriented PMMA and PEN designed to yield the desired reflective polarization effect. The original reflective polarizer film technology was developed by 3M under the trade name DBEF® (Double Brightness Enhancement Film).

Some LC displays with LED backlights (mostly TV's) use QD films placed on top of the LGP to achieve a significantly larger color gamut and hence better image quality [48]. Such films typically comprise a QD film layer sandwiched between two PET-based barrier films.

As is clear from the above, a variety of polyester films play a key role in various components of LC displays as well as other display types. In most embodiments, these films have a passive function serving mostly as substrates or superstrates for various functional layers. However, in some cases, e.g. reflective polarizers, the polyester layer is actively involved in the optical function of the film.

## 1.5 Summary

Polyester film technology is an important component of many established and emerging technologies, including display, microelectronics, photovoltaics, and biomedical. The most prominent examples of film polyesters are PET and PEN, but other synthetic polyester material combinations, e.g. various copolyesters, miscible polyester blends, or polyester composites, can also be used in this space. The main attraction of polyester films is a unique combination of low cost and outstanding physical and mechanical properties, some of which can be attained by a judicious film processing methodology. Indeed, through optimization and control of the tenter-frame process, most commonly used for the manufacturing of biaxially oriented polyester films, it is possible to adjust and optimize the properties of the biaxially stretched film for its intended application. In particular, changes in stretch ratio, stretch temperature, and heat-set temperature can have a profound impact on the microstructure and crystalline morphology of the stretched film and thereby on its overall physical performance. The bulk and surface properties of biaxially stretched polyester films can also be modified and adjusted by copolymerization, miscible and immiscible blending with various value-added components, coextrusion, and coating.

## References


1 MacDonald, B.A., Rollins, K., MacKerron, D. et al. (2005). Engineered films for display technology. In: *Flexible Flat Panel Displays* (ed. G.P. Crawford), 11–33. New York: Wiley.

2 Kobayashi, S., Mikoshiba, S., and Lim, S. (2009). *LCD Backlights*. New York: Wiley.

3 Hildner, M.L. (2005). OLED displays on plastics. In: *Flexible Flat Panel Displays* (ed. G.P. Crawford), 285–312. New York: Wiley.

4 El Chaar, L. and El Zein, N. (2011). *Renew. Sustain. Energy Rev.* 15 (5): 2165–2175.

5 Griffini, G. and Turri, S. (2016). *J. Appl. Polym. Sci.* 133 (11): 43080.

6 Macdonald, W.A. (2015). Latest advances in substrates for flexible electronics. In: *Large Area and Flexible Electronics*, 291–314. New York: Wiley.

7 Dill, E.J., Ajji, A., and Saffar, A. (2023). Chapter 8. Polyester materials in biomedical applications. In: *Polyester Films: Materials, Processes and Applications* (ed. M. Cakmak and J. Greener). New York: Wiley.

8 Yalcin, B. and Cakmak, M. (2018). Chapter 6. Polymer film substrates for roll-to-roll manufacturing. In: *Roll-to-Roll Manufacturing: Process Elements and Recent Advances* (ed. J. Greener, G.H. Pearson and M. Cakmak), 171–224. New York: Wiley.

9 Greener, J., Pearson, G.H., and Cakmak, M. (2018). *Roll-to-Roll Manufacturing: Process Elements and Recent Advances*. New York: Wiley.

10 Massa, D.J. (2023). Chapter 2. Polyester materials and properties. In: *Polyester Films: Materials, Processes and Applications* (ed. M. Cakmak and J. Greener). New York: Wiley.

11 Ji, L.N. (2013). *Appl. Mech. Mat.* 312: 406–410.

**12** Geyer, R., Jambeck, J.R., and Law, K.L. (2017). *Sci. Adv.* 3 (7).

**13** Greener, J., Gillmor, J.R., and Daly, R.C. (1993). *Macromolecules* 26 (24): 6416–6424.

**14** Weiss, R.A. (2023). Chapter 4. Polyester ionomers. In: *Polyester Films: Materials, Processes and Applications* (ed. M. Cakmak and J. Greener). New York: Wiley.

**15** Greener, J., Gillmor, J.R., and Daly, R.C. (1995). *Annual Technical Conference - ANTEC, Conference Proceedings*; Vol. 2, Boston, MA (May 1995).

**16** Kanuga, K. (2023). Chapter 3. Film properties and internal structure development in miscible polyester blends. In: *Polyester Films: Materials, Processes and Applications* (ed. M. Cakmak and J. Greener). New York: Wiley.

**17** Tobita, K., Miki, T., and Takeuchi, N. (1999). Biaxially oriented film. In: *Film Processing* (ed. T. Kanai and G.A. Campbell), 244–280. Munich: Hanser Publishers.

**18** Yamada, T., Nonomura, C., and Kase, S. (1999). Stretching conditions, orientation and physical properties of biaxially oriented films. In: *Film Processing* (ed. T. Kanai and G.A. Campbel), 352–385. Munich: Hanser Publishers.

**19** Tsunashima, K., Toyoda, K., and Yoshii, T. (1999). Theoretical analysis of the tentering process. In: *Film Processing* (ed. T. Kanai and G.A. Campbel), 320–351. Munich: Hanser Publishers.

**20** Cakmak, M. (1999). Influence of processing conditions on structure and physical properties of biaxially stretched engineering thermoplastics. In: *Film Processing*, 281–319. Munich: Hanser Publishers.

**21** Greener, J., Tsou, A.H., and Blanton, T.N. (1999). *Polym. Eng. Sci.* 39 (12): 2403–2418.

**22** MacDonald, B.A., Rollins, K., Eveson, R. et al. (2003). *MRS Online Proc. Libr.* 769.

**23** Cakmak, M., White, J.L., and Spruiell, J.E. (1989). *Polym. Eng. Sci.* 29 (21): 1534–1543.

**24** Ward, I.M. (1985). *Mechanical Properties of Solid Polymers*, 334–337. New York: Wiley.

**25** Greener, J. (2018). Roll-to-roll manufacturing: an overview. In: *Roll-to-Roll Manufacturing: Process Elements and Recent Advances* (ed. J. Greener, G.H. Pearson, and M. Cakmak), 1–17. New York: Wiley.

**26** Rimai, D.S., DeMejo, L.P., and Mittal, K.L. (ed.) (1995). *Fundamentals of Adhesion and Interfaces*. Utrecht: VSP BV.

**27** Jonas, F. and Morrison, J.T. (1997). *Synth. Met.* 85 (1–3): 1397–1398.

**28** Greener, J., Ng, K.C., Vaeth, K.M., and Smith, T.M. (2007). *J. Appl. Polym. Sci.* 106 (5): 3534–3542.

**29** Greener, J. and Pearson, J.M. (1994). A process for making PET photographic film base with improved curl stability. US Patent 5,288,601.

**30** Wang, L., Shogren, R.L., and Carriere, C. (2000). *Polym. Eng. Sci.* 40 (2): 499–506.

**31** Su, T., Al Faisal, A., Zhao, J., and Lee, P. (2023). Chapter 6. Micro-/-nano-layered films and foams. In: *Polyester Films: Materials, Processes and Applications* (ed. M. Cakmak and J. Greener). New York: Wiley.

**32** Lenz, R.W. (1967). *Organic Chemistry of Synthetic High Polymers*. 82–83. New York: Interscience Publishers.

**33** Carothers, W.H. (1929). *J. Am. Chem. Soc.* 51 (2548): 2560.

**34** Swallow, J.C. and Baird, D.K. (1950). Process of producing films. US Patent 2,497,376.

**35** Scarlett, A.C. (1958). Stretching of polyethylene terephthalate film. US Patent 2,823,421.

**36** Heffelfinger, C.J. (1966). Process for stretching polyester film. US Patent 3,256,379.

**37** MacDonald, W.A. (2003). Polyester films. In: *Encyclopedia of Polymer Science and Technology*, 30–41. New York: Wiley.

**38** Cook, J.G., Huggill, H.P.W., and Lowe, A.R. (1948). Manufacture of new highly polymeric linear esters. Br. Pat. 604,073.

**39** MacDonald, W.A. (2004). *J. Mater. Chem.* 14: 4–10.

**40** Graff, G.L., Burrows, P.E., Williford, R.E., and Praino, R.F. (2005). Barrier layer technology for flexible displays. In: *Flexible Flat Panel Displays* (ed. G.P. Crawford), 57–78. New York: Wiley.

**41** Greener, J., Ng, K.C., Vaeth, K.M., and Smith, T.M. (2006). *Digest of Technical Papers - SID International Symposium*; Vol. 37, San Francisco, CA (June 2006).

**42** Greener, J. (2018). Curl effects in roll-to-roll operations. In: *Roll-to-Roll Manufacturing: Process Elements and Recent Advances* (ed. J. Greener, G.H. Pearson and M. Cakmak), 225–250. New York: Wiley.

**43** Greener, J., Tsou, A.H., Ng, K.C., and Chen, W.A. (1991). The bending recovery of polymer films I. a phenomenological model. *J. Polym. Sci. Part B Polym. Phys.* 29 (7): 843–858.

**44** Greener, J. and Gillmor, J.R. (2001). Long term growth of core-set curl in PET film. *J. Polym. Sci. Part B* 39: 1957–1967.

**45** Ishiwatari, Y. (2009). *LCD Backlights* (ed. S. Kobayashi, S. Mikoshiba, and S. Lim), 237–250. New York: Wiley.

**46** Hanzawa, F. (2009). Lens films and reflective polarization films. In: *LCD Backlights* (ed. S. Kobayashi, S. Mikoshiba, and S. Lim), 257–267. New York: Wiley.

**47** Weber, M.F., Stover, C.A., Gilbert, L.R. et al. (2000). Giant birefringent optics in multilayer polymer mirrors. *Science* 287 (5462): 2451–2456.

**48** Moon, H., Lee, C., Kim, J., and Chae, H. (2019). *Adv. Mater.* 31 (34): 1–14.

# 2

# Polyester Materials and Properties

*Dennis J. Massa*

*Massa Science and Technology Consulting, Pittsford, NY, USA*

## 2.1 Introduction

Polyesters are an important class of polymeric materials containing ester groups along their chain backbones. They are synthesized via step-growth polymerization reactions, including those sometimes referred to as polycondensation [1]. Their chain architectures may be linear, branched, hyperbranched, dendrimeric, or crosslinked. Their properties depend upon their chemical composition, chain architecture, microstructure, and macrostructure, as well as their processing and thermal histories. Therefore, an understanding of polyester properties must take these factors into account.

Two broad categories of polyesters are thermoplastics and thermosets. Thermoplastic polyesters can be processed into various forms, among them films, fibers, injection-molded articles, and filled composites. Thermoset polymers are those that are crosslinked in their final form. The variety of polyester structures available allows their use in such diverse fields as films and fibers, packaging and containers, electronics, automotive applications, construction, membranes, and medical uses.

The emphasis of this review is on the relationships between polymer structure and physical properties. Examples illustrating these relationships will be given. Orientation, crystallization, and degradation during processing can also affect the final properties of polyesters and, along with synthesis, are introduced as general background or as they affect specific physical properties. Given the broad scope

*Polyester Films: Materials, Processes and Applications*, First Edition.
Edited by Miko Cakmak and Jehuda Greener.

and utility of polyester materials, there are more topics that can be treated in detail here. The reader is encouraged to seek out extended treatments of these uncovered topics.

## 2.2 Thermoplastic Polyesters: PET, PBT, and PTT

Thermoplastic polyesters are the largest class of polyesters, and of these, terephthalate polyesters are among the most ubiquitous – e.g. poly(ethylene terephthalate), or PET; poly(1,4-butanediol terephthalate), or PBT; and related materials. PET is produced by the polycondensation reaction of terephthalic acid (TPA) or dimethyl terephthalate (DMT) with ethylene glycol (EG) to give the structure shown below. The reaction is assisted by catalysts and, for most common polyesters, is carried out in the melt in the absence of solvent (Figure 2.1) [1].

Analogous polyesters are made, for example, by substituting 1,4-butanediol, 1,3-propanediol, or 1,4-cyclohexanedimethanol (CHDM), instead of EG, to make, respectively, poly(butylene terephthalate), or PBT; poly(trimethylene terephthalate), or PTT; or poly(1,4-cyclohehexanedimethanol terephthalate), or PCT. The example of PBT is given in Figure 2.2. Copolyesters containing various ratios of these glycols can be prepared, as well [1].

A useful comparison of selected physical properties of PET, PTT, and PBT is shown in Table 2.1 [2,3]. Measurements of these properties were done on injection-molded specimens using standard ASTM testing procedures, except for the glass-transition and melting temperatures, $T_g$ and $T_m$, respectively, which were obtained using differential scanning calorimetry [4,8,9].

The thermal transitions of PET, PTT, and PBT are generally found in the order PET > PTT > PBT, as would be expected from the increasing chain length of the aliphatic glycol portion of the polyesters from two to three to four for PET, PTT,

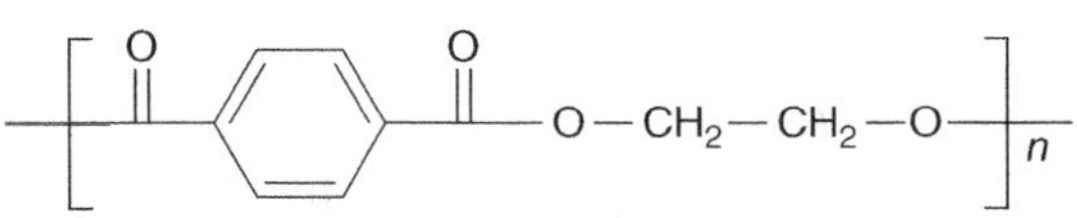


**Figure 2.1** Chemical structure of poly(ethylene terephthalate), or PET.

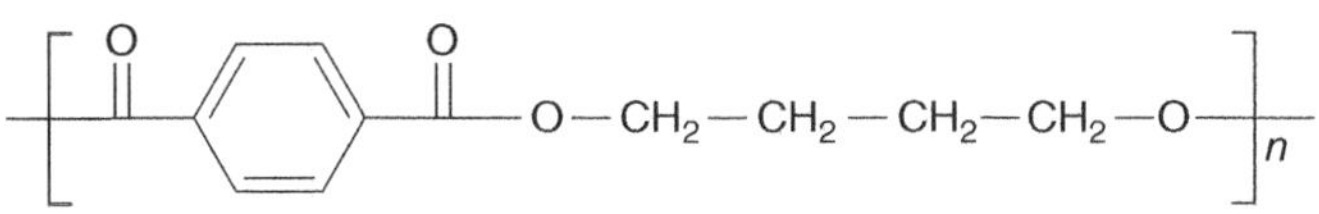


**Figure 2.2** Chemical structure of poly(1,4-butanediol terephthalate), or PBT.

**Table 2.1** Comparison of selected properties of PET, PBT, and PTT.

| Property | PET | PTT | PBT |
|---|---|---|---|
| Glass transition temperature, $T_g$, °C | 80 | 45 | 41 |
| Melting temperature, $T_m$, °C | 252 | 228 | 224 |
| Heat distortion temperature (HDT) at 1.8 MPa, °C | 65 | 59 | 54 |
| Tensile strength, MPa | 73 | 68 | 57 |
| Flexural modulus, GPa | 3.1 | 2.8 | 2.3 |
| Notched Izod impact, J $m^{-1}$ | 37 | 48 | 53 |
| Specific gravity, kg $m^{-3}$ | 1400 | 1350 | 1340 |
| Dielectric constant at 1 MHz | 3.0 | 3.0 | 3.1 |
| Moisture absorption, 14 da, % | 0.49 | 0.15 | 0.12 |

*Source:* Chuah [2], 2003; Dangayach et al. [3], 1997; Petrie [4], 1972; Huang and Chang [5], (2000); Turner [6], (2004); Lotti [7] et al. (2011).

and PBT, respectively. Higher aliphatic content also leads to lower moduli and lower densities, or specific gravities, for PTT and PBT. The values for $T_g$ and $T_m$ reported in Table 2.1 are for initially amorphous polyesters scanned at a heating rate of 10 °C/min through the $T_g$, cold crystallization, and melting regions [4–7]. The $T_g$ does not necessarily correspond to the thermal dimensional stability of the material, which depends upon the material's crystallinity, microstructure, thermal history, and the stresses to which the polyester is subjected. A more practical measurement of thermal dimensional stability is the heat distortion or heat deflection temperature, HDT, which is the temperature at which a standard test bar deflects by 0.25 mm under a standard stress load of 0.46 or 1.8 MPa. As can be seen in Table 2.1, the HDT values follow the trend in glass transition temperatures but differ in magnitude from the $T_g$ values themselves. HDT is higher for semicrystalline and filled polymers, as it is essentially a measure of the bending stiffness, or flexural modulus, of the material under load, which is increased both by the presence of crystallinity and by the addition of rigid filler. For example, adding 30% glass fiber to the polyesters more than triples their flexural moduli and increases their HDT to 224 °C, 216 °C, and 207 °C for PET, PTT, and PBT, respectively, compared to 65 °C, 59 °C, and 54 °C for the unfilled polyesters [2, 3]. Both PBT and PTT have significantly higher crystallization rates during cooling than PET, a property that leads to injection-molded bars of PBT and PTT having higher crystallinity than that of PET [5, 10–12]. Crystallinity increases modulus and decreases moisture absorption, the latter trend shown in Table 2.1. All three polymers have similar notched Izod impact strength values and similar electrical properties.

## 2.3 Thermoplastic Polyesters Containing Cyclic Diols

An interesting class of thermoplastic polyesters comprises those containing cyclic diols such as CHDM. Replacing a portion of or all of the EG in PET with CHDM raises slightly the $T_g$ of PET, increases toughness, and imparts greater stability against hydrolysis. The general structure of polymers of this class is shown below in Figure 2.3.

where $x$ is the glycol mole fraction of EG and $1 - x$ is the glycol mole fraction of CHDM.

Terephthalate copolyesters containing both CHDM and EG provide a variation in materials and properties over the composition range. The glass-transition temperature $T_g$ varies continuously over the composition range according to the Fox equation [13]. The crystalline melting points of the copolymers follow a eutectic-like behavior: the melting point of PET decreases with added CHDM comonomer, and the melting of PCT decreases with added EG comonomer, with a minimum $T_m$ at ca. 30-mol% EG [6]. The reason for this behavior in $T_m$ is that copolymerization introduces disorder into the polymer chain that interferes with chain packing and crystallization. This decreases $T_m$ and increases the crystallization half time, the time at which half the equilibrium crystallinity is attained, by up to four orders of magnitude, to as long as 24 hours or more [6]. At copolymer compositions in the vicinity of 30 mol%, the copolymer is amorphous when extruded or injection-molded. Other copolymer compositions can also be prepared that are amorphous following rapid cooling from the melt state.

That incorporation of comonomers into polyesters decreases $T_m$ and gives $T_g$ values intermediate between those of the homopolyesters according to the Fox equation is a general principle of polyester design. Similarly, incorporating comonomers into crystallizable polyesters decreases the latter's rates of crystallization. For this reason, it may be desirable to incorporate a small amount of comonomer into a polyester solely to control the rate of crystallization, without substantially affecting other physical properties. For example, in the EG-CHDM polymer class discussed above, incorporating isophthalic acid into PCT produces a slower-crystallizing polymer that can be readily quenched into the amorphous

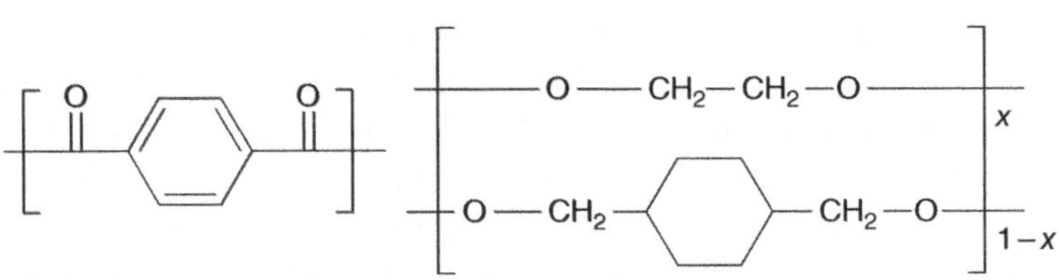


**Figure 2.3** Poly(ethylene glycol-co-1,4-cyclohexanedimethanol terephthalate). *Source:* [6]/John Wiley & Sons.

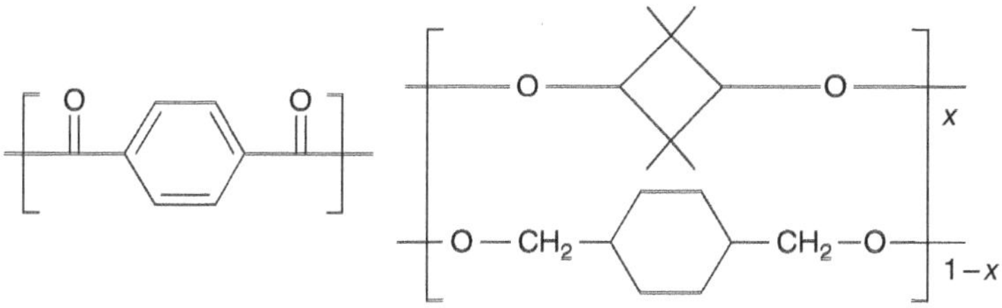

**Figure 2.4** Example of a 2,2,4,4,-tetramethyl-1,3-cyclobutanediol-containing copolyester.

state, while maintaining the polymer's hydrolytic stability. Similarly, incorporating a small amount of CHDM into PET slows its rate of crystallization and improves polymer clarity [6].

Incorporating CHDM into the polymer chain lowers the modulus, relative to that of PET, on account of the motion between conformers of the cyclohexyl ring found at normal measurement temperatures [14–16]. This motion begins at low temperatures (−50 to −100 °C depending upon the deformation rate) and is present throughout normal polyester use temperatures, and is a well-understood property of polyesters containing cyclohexyl rings.

Another cyclic diol that has become commercially important recently is 2,2,4,4-tetramethyl-1,3-cyclobutanediol, or TMCD [17–19]. Incorporation of TMCD into terephthalate polyesters increases their $T_g$ and toughness, providing an alternative to Bisphenol-A, or BPA, polycarbonate for many container and high-performance applications. An example of a TMCD-containing copolyester is given in Figure 2.4.

## 2.4 Thermoplastic Polyester Films: PET and PEN

Thermoplastic polyester films and fibers differ in their physical properties from extrusion-cast or injection-molded polyesters, owing to the orientation process involved in their manufacture. A general film orientation process, which is covered in more detail elsewhere, consists of extrusion and quenching onto a cooling wheel, followed by heating above $T_g$ and orientation in the machine direction (MD) and subsequent widthwise orientation above $T_g$ in the direction transverse to the machine direction (TD) [20]. Other orientation methods are available, including reversing the order of stretching in the two directions, called the inverse process, and simultaneous stretching in both MD and TD [21].

For PET, the orientation ratios are typically in the 200–250% range, at temperatures just above the $T_g$: from 80 to 95 °C in the MD, which is below the temperatures at which thermal crystallization occurs prior to orientation, and usually 5–15 °C higher during TD stretching [21]. Orientation nucleates some crystallinity in the film [21–23]. Following orientation, the film may be heat set and/or heat relaxed at higher temperatures but below $T_m$. Heat-setting increases crystallite size [23], and heat relaxation provides additional dimensional stability by relaxing the oriented amorphous polymer chains between crystallites, along with some ester interchange [24–28].

Although films can be equibiaxially oriented, their properties are never exactly equal in every in-plane direction. The order of stretching affects the final film morphology, as does the stress of conveyance in the MD. The properties can also vary across the width of the film, owing to the stress distribution during manufacturing. This needs to be kept in mind when measuring and specifying the properties of oriented films.

Poly(ethylene naphthalate), whose structure is shown below, can be made into films using the same basic processes as shown in Figure 2.4, using higher temperatures and different draw ratios that reflect both the higher $T_g$of PEN and its different elongational rheology [29, 30]. The chemical structure of PEN is given in Figure 2.5.

Successful polyester stretching above $T_g$ to produce a uniform film requires suitable extensional melt rheology: in particular, stretching should occur in a region of temperatures and rates that satisfy the Considère criterion for stable film orientation, a phenomenon also known as strain hardening [31, 32] or stress hardening [29, 30]. In such a region, nonuniformities are stabilized during stretching, and the fully stretched film is uniform in thickness. For PET and PEN, stable stretching is assisted by orientation-induced crystallization that occurs during stretching [21–23]. In PEN as in PET, orientation-induced crystallinity is increased by heating the film to higher temperatures, or heat-setting, following orientation [24].

Suitable coatings and/or surface treatments may be applied at various stages in the film fabrication process, in order to promote adhesion to subsequent coatings, or to assist in managing conveyance and winding issues [33].

A comparison of typical properties of PET and PEN films is given in Table 2.2 [34–36]. The greater aromaticity and stiffer chain of PEN relative to PET

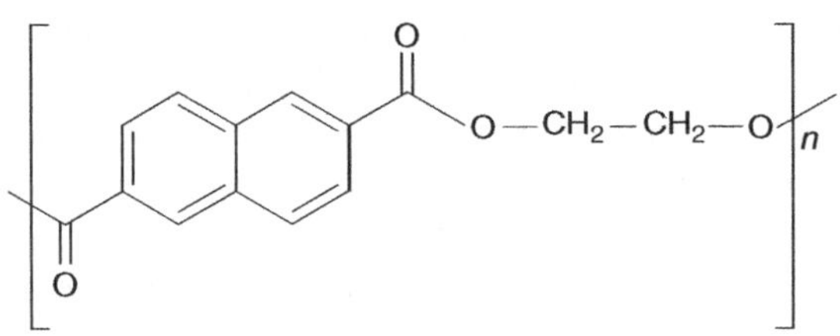


**Figure 2.5** Chemical structure of Poly(ethylene naphthalate), or PEN.

**Table 2.2** Comparison of PET and PEN film properties.

| Property | PET | PEN |
|---|---|---|
| Glass transition temperature $T_g$, °C | 72 | 120 |
| Modulus, GPa | 5.0 | 6.4 |
| Yield strength, MPa | 95 | 130 |
| Break elongation, % | 110 | 70 |
| Break strength, MPa | 173 | 190 |
| Tear strength, g | 74 | 50 |
| Equilibrium water absorption at 100% relative humidity, % | 0.5 | 0.5 |
| Humidity coefficient of linear expansion (in plane), %/%RH | 0.0008 | 0.0008 |
| Density, kg m$^{-3}$ | 1390 | 1360 |

*Source:* Fleischer et al. [34], (1996); Ishigaki and Naoi [35], (1990); Ouchi et al. [36], (1992).

result in higher $T_g$ and higher modulus for PEN. In addition to these properties, PEN has a four- to fivefold lower oxygen permeability than PET, and a 3.7-fold lower water permeability [37]. This low oxygen permeability, together with its suitable melt orientation rheology, make PEN a candidate for high-barrier films and containers. Strictly speaking, the values for modulus and even $T_g$ for PET should be reported for each principal direction, as they are anisotropic – i.e. they have different values in different directions. The modulus and $T_g$ are highest in the direction of highest orientation [26, 38]. The $T_g$ and $T_m$ and copolymers of PET and PEN follow the same trends with polymer composition as was seen for CHDM-EG terephthalate copolymers: $T_g$ vs. composition follows the Fox equation, and $T_m$ vs. composition shows eutectic-like behavior.

The dimensional stability of PET and PEN films can be further enhanced by a process known as physical aging: by annealing below $T_g$ or by slowly cooling films through their $T_g$ to room temperature [39–43]. This process results in relaxation of volume and enthalpy to produce a denser polymer and is a general property of amorphous and semicrystalline materials [4, 39]. Physical aging results in much lower rates of stress relaxation and creep for the annealed or slowly cooled samples [26–28, 40–43]. For example, PET films annealed at 69.2 degrees for five days showed a 33-fold decrease in creep rate compared with that observed after annealing for 45 minutes [26]. This additional dimensional stability due to physical aging manifests itself in superior resistance to residual bending curl, or core set, in wound films, as the core set process has been shown to be governed by stress relaxation [44, 45].

## 2.5 Aromatic Polyesters

Aromatic polyesters are characterized by polymer chains that do not have an aliphatic repeat unit such as EG, 1,4-butane diol, or 1,4-cyclohexane dimethanol. As a result, they can have significantly higher $T_g$ values. An example of this group are polyarylates.

Polyarylates are polyesters that are made from aromatic diacids and bisphenols. A typical chemical structure is shown in Figure 2.6.

Here, the polyester is derived from terephthalic and isophthalic acid derivatives, and Bisphenol-A, or BPA. In polyarylates, it is typical that a portion of the terephthalate group is replaced with isophthalate to give a tougher, more amorphous polymer, as was discussed above, having a slightly lower $T_g$ [46]. When the polymer $T_g$ is not too high, as in the example above ($T_g = 191\,°C$) such polyesters can be made in the melt state and readily extruded without excessive degradation [46].

Higher-$T_g$ polyesters and copolyesters can be made from diacid chlorides and bisphenols, illustrated in Figure 2.7 by the polyester made from phenylindane

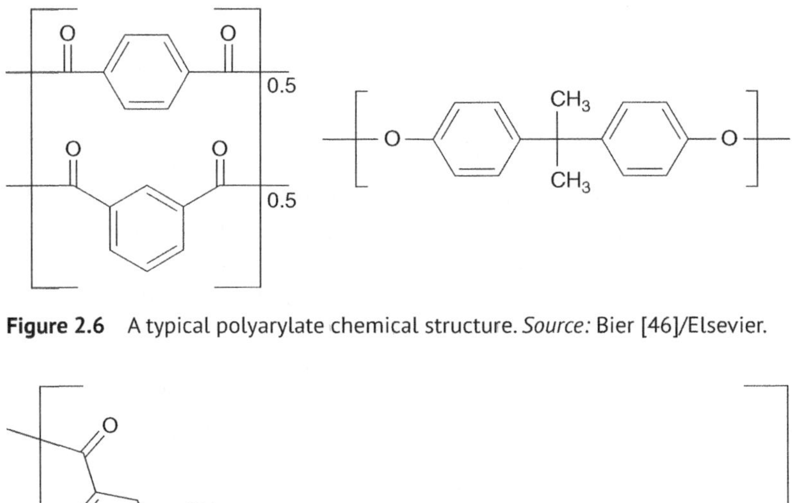


**Figure 2.6** A typical polyarylate chemical structure. *Source:* Bier [46]/Elsevier.

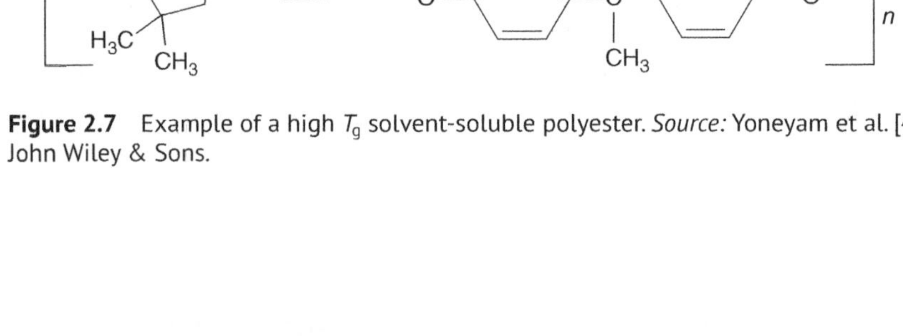


**Figure 2.7** Example of a high $T_g$ solvent-soluble polyester. *Source:* Yoneyam et al. [47]/ John Wiley & Sons.

dicarboxylic acid dichloride and bisphenol-A, using solvent preparative and processing methods [47, 48]. Such polyesters are soluble in common solvents and may have $T_g$ values in excess of 300 °C [48].

## 2.6 Thermotropic Main-Chain Liquid-Crystalline Polyesters

Although monomeric liquid crystals have been known for well over 100 years, thermotropic main-chain liquid-crystalline polymers (LCPs) are a more recently discovered material class. As reviewed by W. J. Jackson, Jr., they were first reported in 1974, when Eastman Chemical Company, then part of Eastman Kodak, reported that copolymers of p-hydroxybenzoic acid with PET were liquid crystalline, and Carborundum company disclosed copolymers of PET, *p*-hydroxybenzoic acid, and 4,4′-biphenol, which were followed by DuPont's substituted hydroquinone LCPs [49]. Shortly thereafter, LCPs based on 2,6-naphthalene dicarboxylic acid were disclosed by Celanese. Since then, a variety of main-chain LCPs and LCP-forming polymers have been synthesized, including bibenzoate-containing polyesters [50–52]. An example of a bibenzoate-containing polyester is the copolymer of PEN with bibenzoate, PENBB, shown in Figure 2.8 [52]. What all these LCPs have in common is the introduction of a mesogenic monomeric unit into the main chain of the polymer (as opposed to side-chain LC polymers) that can lead to the formation of a liquid-crystalline phase in the oriented and/or unoriented state.

Main-chain thermotropic LCPs have unusual physical properties, among them being unexpectedly low viscosity at high shear rates, and high anisotropy in physical properties such as modulus and thermal expansion coefficient when extruded or injection molded [49, 50, 52]. In addition, they are more resistant to chemicals. These attributes have led to their use as specialty polymers for applications requiring the filling and replication of thin-gauge and/or intricate parts, and materials that are exposed to harsh environmental conditions.

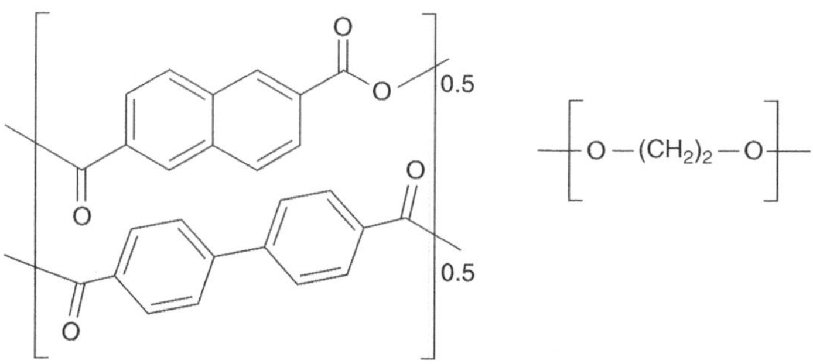


**Figure 2.8** Chemical structure of poly(ethylenenaphthalate-co-bibenzoate) polymer. *Source:* Carr et al. [52]/John Wiley & Sons.

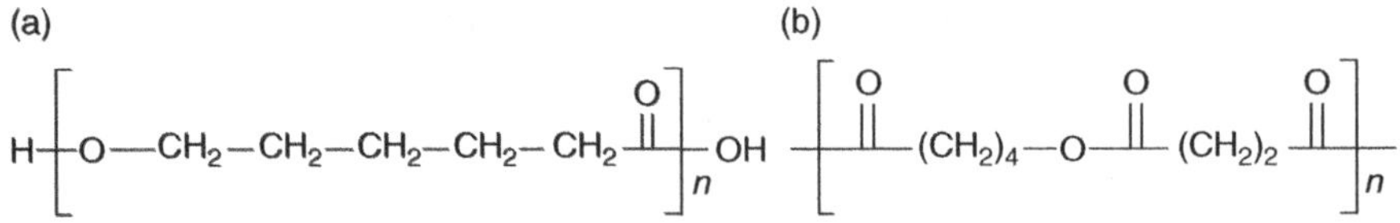

**Figure 2.9** Chemical structures of (a) polycaprolactone and (b) poly(butylene succinate).

## 2.7 Wholly Aliphatic Polyesters

Aliphatic polyesters were among the first polyesters to be synthesized, in the laboratory of Wallace Carothers of DuPont [53]. Owing to the versatility of the ester linkage, there are a large number of wholly aliphatic polyester structures that can be made. Of these, poly(ε-caprolactone), more commonly known as polycaprolactone, or PCL, whose structure is shown in Figure 2.9, is an important example, due to its ease of synthesis [54, 55]; its physical properties [56]; its unusual miscibility with a number of different polymers, including poly(vinyl chloride, acrylonitrile-butadiene-styrene (ABS), and BPA polycarbonate [57–60]; and its biodegradability, both in vivo and in vitro [61]. PCL's advantage for drug delivery and bioresorbable medical uses – for example, as a tissue scaffold – is its long-term degradation timescale [62–66]. PCL can be spun into fibers or electrospun into fiber mats [65, 66]. PCL is also an important component of multiblock polyester urethanes [67–69]. The $T_g$ and $T_m$ of PCL are −65 °C to −60 °C and 56 °C to 65 °C, respectively [55].

Another common aliphatic polyester is poly(butylene succinate), or PBS, shown in Figure 2.9. Owing to fewer methylenes per monomer unit, PBS has a higher $T_g$ and $T_m$ than PCL – i.e. −32 °C and 114 °C, respectively [70]. PBS, PCL, and their copolymers have found use in biomedical and tissue engineering applications [61, 70]. Copolymerization of PBS and PCL with other diacids and diols allows a variety of polyesters having tailored $T_g$, $T_m$, crystallinity, and other physical and chemical properties to be achieved [70].

## 2.8 Branched and Hyperbranched Polyesters

Polyesters made by step-growth reactions typically have molar mass distributions, or polydispersity indices of approximately 2.0 – i.e. the ratio of weight-average to number-average molar mass, $M_w/M_n$, is 2.0 [1]. For many applications, it is desirable to have broader molar mass distributions, in order that the polymers have higher melt elasticity [71, 72]. For polyesters, this can be achieved by introducing trifunctional or tetrafunctional monomer units into the polymerization [71]. Small amounts of trifunctional groups such as glycerol, trimethylolpropane, or trimellitic anhydride lead to long-chain branching; higher amounts lead to short-chain branching and, upon further reaction, crosslinking [71, 73, 74]. Examples of applications of branched polyesters are as varied as thermoplastic elastomers [75], electrophotographic toners [76], and drug delivery binders [77].

Whereas branched polyesters are mostly linear polymers having a small amount of tri- or tetrafunctional branching agent, hyperbranched polyesters are made entirely from multifunctional monomers and are completely branched with no linear portions [78]. Applications of hyperbranched polymers include blends, coatings, rheology modifiers, and thin films [79, 80].

## 2.9 Thermoplastic Polyester Elastomers and Multiblock Copolymers

Thermoplastic polyester elastomers (TPEEs) can be divided into two groups. In the first of these, the hard segment is a thermoplastic polyester as discussed above, and the soft segment is a polymer having a low $T_g$, such as a polyether. Examples of a TPEE having a typical thermoplastic polyester copolymerized with a polyether are poly(butylene terephthalate) with poly(ethylene oxide), or PBT-co-PEO; poly(butylene terephthalate) with poly(tetramethylene oxide), or PBT-co-PTMO; poly(ethylene terephthalate) with poly(tetramethylene oxide), or PET-co-PTMO, and poly(butylene naphthalate with PTMO [75, 81–84]. Multiblock TPEEs can also made from biodegradable polyesters such as copolymers of poly(lactic acid), or PLA, and poly(glycolic acid), or PGA, which will be discussed later [85]. TPEEs are, by the very nature of their synthesis, multiblock copolymers.

In the second group of thermoplastic polyesters, the soft segment itself is the polyester, typically a noncrystalline copolyester whose $T_m$ is below room temperature. Such TPEEs, which can be synthesized from biodegradable and biocompatible polyesters such as PCL, PLA, and PGA, can be tailored to provide soft materials that are useful in soft tissue engineering applications [84, 85].

As with all polymers, both $T_g$ and $T_m$ decrease linearly with $1/M$, where $M$ is the molar mass, so the thermal properties of the soft segments in a multiblock polymer, which may be relatively low M, are often lower than one would expect based on the limiting high-$M$ values for the polymer.

## 2.10 Biodegradable and Compostable Polyesters, and Polyesters from Renewable Feedstocks

Owing in part to concerns about the impact of polymers on the environment, there has been increasing interest in biodegradable and compostable polymers as well as polymers made from renewable feedstocks. These categories are not synonymous. A polymer may be made from monomers of biological origin, and yet not be biodegradable or even compostable, the latter specified by ASTM D6400 [86]. An example is PET, normally made wholly or in large part from fossil fuels, but which can

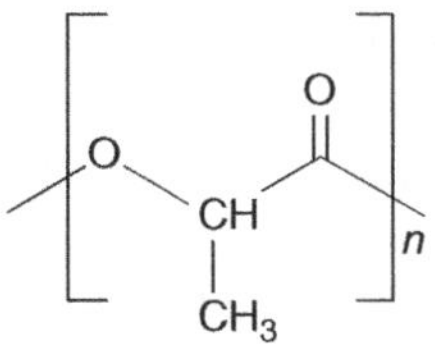


**Figure 2.10** Structure of poly(lactic acid), or PLA.

be made entirely from monomers of biological origin [87]. Regardless of its monomer source, PET has identical physical properties and is neither biodegradable nor compostable. It is, however, recyclable, both physically and chemically [88]. On the other hand, a polymer may be biodegradable and/or compostable and yet be derived entirely from monomers prepared from fossil fuels. Many aliphatic polyesters have this property. Thirdly, a polymer may be both biodegradable and made using monomers of biological origin, poly(L-lactic acid), or PLA, whose structure is shown in Figure 2.10, being the most common example [89].

Polyester bioplastics may be made using well-known synthetic chemical means, such as ring-opening polymerization [89], or they may be obtained by employing microbes, the method used for the polyhydroxyalkanoates [90]. Polyhydroxyalkanoates occur naturally as granular energy storage particles in a number of organisms and can be engineered to comprise up to 90% of their masses [90]. L-lactic acid is the product of a biological pathway from pyruvate, an intermediate in glycolysis, a metabolic process carried out by living organisms [91]. Aliphatic polyesters, including thermosets, targeted for medical use can be made free of catalysts [92].

As the number and variety of polyesters of biological origin are too large to be covered in detail here, the properties of a representative example, PLA, are given in Table 2.3, along with those of PCL, discussed above, and poly(hydroxybutyrate), or PHB, another bioplastic [93]. These, along with the polyhydroxyalkanoates and copolymers with glycolic acid, provide a variety of polyesters that can be designed to give suitable mechanical properties and biodegradability for use in implants, tissue scaffolds, and drug delivery [94–97].

The final category in this section is polyesters derived from renewable feedstocks. These are chemicals derived from present-day biomass, rather than from fossil fuel

**Table 2.3** Selected properties of PLA, PCL, and PHB.

| Property | PLA | PCL | PHB |
|---|---|---|---|
| Glass transition temperature $T_g$, °C | 45–60 | −60 to −65 | 5–15 |
| Melting temperature $T_m$, °C | 150–162 | 58–65 | 168–182 |
| Modulus, GPa | 0.35–3.5 | 0.21–0.44 | 3.5–4.0 |
| Break elongation, % | 2.5–6.0 | 300–1000 | 5.0–8.0 |
| Break strength, MPa | 21–60 | 21–42 | 40 |
| Density, kg m$^{-3}$ | 1210–1250 | 1110–1146 | 1180–1262 |

*Source:* Adapted from [93].

sources, which are biomass containing carbon sequestered and stored during past geological eras. Such polyesters are often said to be sustainable [98]. Except for cellulosics, polymers of biological origin generally do not have high $T_g$ and $T_m$ values. However, research into higher-$T_g$ bioplastics continues to yield polyesters having higher thermal performance [87]. These include 2,5-furandicarboxylate polyesters and polymers derived from lignin, the second-most abundant polymer of biological origin, behind cellulose [98].

Research into polymers made from renewable plant feedstocks is not new. The first plastics were derived from cellulose, and the first commercially important elastomer was found in latex from rubber trees. Polyesters and polyamides made from monomers derived from plant-based oils were reported in the 1960s and 1970s [99–101].

More recent work on renewable polyesters has included aliphatic polyesters derived from vegetable oils [98, 102]; poly(butylene 2,5-furandicarboxlate), whose aromatic structure can be synthesized from C-6 sugars such as fructose via the intermediate 5-hydroxymethylfurfural [103]; polyesters from glycerol, a plentiful byproduct of the hydrolysis of oils and fats to make biodiesel [104, 105], and polyesters from algae [106]. Although the $T_g$ and $T_m$ of poly(butylene 2,5-furandicarboxylate) are 34 °C and 169 °C, respectively, the $T_g$ of 2,5-furandicarboxylates can be increased to 105 °C by substituting cyclic diols instead of butane diol in the polyester [107].

The issue of sustainability is not as straightforward as it might seem. In addition to monomer and polymer properties, one must consider the carbon footprint and net energy balance of the entire process used to make it, in order to establish whether a polymer made from renewable feedstocks is sustainable [108, 109].

## 2.11 Dependence of Polyester Properties on Polymer Entanglement Molar Mass

It is a general property of polymers that their thermal and mechanical properties level off at high molar masses $M$ [110]. For thermal properties, this occurs because the effects of end groups become negligible at higher $M$. Mechanical properties, by comparison, become independent of $M$ at even higher molar masses, when $M$ is three to five times $M_c$, a quantity known as the critical entanglement molecular weight [111]. The critical entanglement molecular weight $M_c$ depends upon polymer structure and chain flexibility. Polyesters are unusual in that their values for $M_c$ are significantly lower than those for vinyl polymers [112].

Another important manifestation of entanglements is that above $M_c$, the melt viscosity increases with the 3.4 power of $M$ [113, 114]. Therefore, one must choose a molar mass $M$ that balances good mechanical and thermal properties at higher

**Table 2.4** Values of $M_c$ and $M_e$ for selected polyesters.

| Polymer | $M_c$ | $M_e$ |
|---|---|---|
| Poly(ethylene terephthalate) | 3270 [112] | 1170 [115] |
| Poly(ethylene naphthalate) | | 1810 [116] |
| Polyarylate (BPA isophthalate/terephthalate) | | 3500 [117] |
| Poly(ethylene 2,5-furandicarboxylate) | | 2400–2700 [118] |
| Poly(ethylene terephthalate-co-hydroxybenzoate) | | 1900 [119] |
| Poly(CHDM 4,4′-bibenzoate/4,3′-bibenzoate) | | 6500–9000 [120] |
| PLA (isotactic) | | 4100 [121] |
| PLA (syndiotactic) | | 11 800 [121] |
| PLA (98/2 L/D) | | 8700 [122] |
| Poly(3-hydroxybutyrate-co-4-hydroxybutyrate) | | 4300 [123] |

*M* with processability from lower melt viscosity. For commercial polyesters, this balance is achieved at significantly lower molar masses than those of vinyl polymers.

Table 2.4 contains values for $M_c$ and $M_e$, the molecular weight between entanglements, for a number of polyesters that have been discussed above. It can be seen that in general the values for $M_c$ and $M_e$ are between 1000 and 5000 g mol$^{-1}$, except for syndiotactic PLA and polyesters containing the highly rigid bisbenzoate group. ($M_c$ is approximately $2M_e$.) By comparison, $M_c$ for polystyrene is approximately 33 000 g mol$^{-1}$ [114].

## 2.12 Polyester Blends

The chemistry and structural variety of polyesters allows them to be blended with a variety of polymers. Such blends may be miscible, having a single phase, or immiscible, having two or more phases [124]. The latter may be reactive in order to produce grafting between the polymer phases and their constituents [125–129]. Polyester blends are new material compositions having unique and useful properties [124].

As mentioned earlier, polycaprolactone, or PCL, is miscible with a variety of polymers, including PVC, ABS, and the polycarbonate of bisphenol-A, or BPA [57–60]. This is attributable to the balance of interactions between the blended polymer and the aliphatic portions of PCL and with the carbonate group [59]. Another amorphous polyester, poly(1,6-hexane succinate), is a miscible polymeric plasticizer for PVC [130]. A similar balance of interactions occurs between CHDM diol polyesters and BPA polycarbonate to form miscible blends [60, 131]. PEN/Polyetherimide and related blends have also been reported to be miscible [132]. Hydrogen bonding promotes blend miscibility with polyesters, as has been shown for blends with poly(vinyl phenol) [133]. The above blends do not require any chemical reaction among the constituents to be miscible. The $T_g$s of the blends vary continuously with blend composition between the limiting $T_g$s of the blend constituents, in accordance with the Fox or Couchman equations [14, 134]. Miscible blends containing crystallizable polyesters crystallize more slowly than their pure constituents [132, 135].

More commonly, blends of polyesters with other polymers are immiscible, having two or more phases present [124]. Such blends are of interest, nevertheless, in part because polyesters can impart chemical resistance and better processability to the blended material. Examples of such polyester blends are PBT with BPA-polycarbonate [136], PET with BPA-polycarbonate [137, 138], polyesters with polyamides [139], and polyesters with other polyesters, including liquid crystalline polyesters [136, 140].

Because polyesters continue to react at high temperature, polyester blend constituents may react with each other during melt processing. When the reaction is unwanted, as in the case of polyester/PC blends, a phosphite catalyst deactivator may be added to minimize reactivity among the constituents, which otherwise may lead to decarboxylation and color [138, 141]. For some blends, interreaction among the blend constituents may be desirable, in order to promote interfacial adhesion. Interfacial reaction among the blend constituents may be promoted by employing reactive polyester groups [125–129].

## 2.13 Ion-containing Polyesters

Polyesters containing ionic groups, sometimes called polyesterionomers, may be synthesized by copolymerizing a nonionic polyester with an ion-containing monomer such as sodium dimethylisophthalate (NaSIP). The base polyester into which the NaSIP is copolymerized may be PET, PBT, PLA, or any suitable polyester [142–145]. Such ion-containing polymers may be used as films, adhesives, and biodegradable polyesters. The counterion in the SIP monomer is typically sodium, but it may include other ions such as lithium [146]. Polyesters having terminal ionic groups at each end are called telechelic polyesters [147].

The incorporation of ions into polyesters changes their physical and chemical properties. Copolymerizing a few percent NaSIP with PET improves fiber dyeability [148]. Incorporating higher amounts of NaSIP, particularly when including a polyether, eventually leads to water solubility [149]. NaSIP-containing polyesters have increased melt strength and enhanced thermal and mechanical properties [150]. Substitution of $Na^+$ with $Li^+$ results in polymers of interest as candidate materials in energy storage devices [146]. Use of divalent cations such as $Ca^{2+}$ instead of $Na^+$ increases melt viscosity by two orders of magnitude and leads to a portfolio of materials of interest in additive manufacturing [149].

## 2.14 Filled Polyesters

The properties of polyesters can be enhanced by the use of fillers [151]. Though a detailed description is beyond the scope of this review, a brief summary is given here. Typical polyester fillers include glass fibers and minerals. Owing to their higher moduli and hardness, fillers increase modulus and higher-temperature dimensional stability, while typically decreasing elongation-at-break. Fillers are usually surface-treated to improve their adhesion to the polyester matrix. Many filled polyesters are unsaturated polyesters reacted in situ to form thermoset composites.

Recent interest has focused on polyester nanocomposites, which are polyesters filled with particles having at least one dimension smaller than 100 nm. Layered silicate clays such as Cloisite™ and Laponite™ are of particular interest as nanofillers. In such materials, it is desirable to intercalate and exfoliate the clay, which is enhanced by ion-containing polyesters [152, 153], telechelic polyesters [154], and star-shaped polymers [155, 156].

Nanocomposites offer the possibility of enhanced physical properties and fire resistance at relatively low levels of filler [157, 158]. If the particles are small enough and dispersed well enough, the composite can remain transparent [157]. Some polyester nanocomposites can be prepared by in-situ polymerization of the polymer in the presence of the nanofiller [153]. Layered silicate nanocomposites have also been shown to have enhanced biodegradability [157].

## 2.15 Polyester Degradation

As mentioned briefly earlier, polyesters are subject to degradation during synthesis, processing, and use which may affect their physical properties. The principle mechanisms of degradation of polyesters are thermal, hydrolytic, and photolytic. The first two can occur during synthesis, processing and use, and the last during exposure to light, typically during long-term use.

It is widely reported in the chemical literature that PET and similar polyesters undergo thermal degradation and can generate byproducts that remain in the polymer after synthesis and processing [159, 160]. For PET, this includes ether linkages such as diethylene glycol, anhydrides, carboxylic and vinyl end groups, and acetaldehyde [161–164]. These are in addition to cyclic oligomers, which are non-degradative byproducts present at equilibrium in polyesters [165, 166]. The presence of degradation byproducts can affect the properties of polyesters in blending and use, particularly in packaging applications, where migration of acetaldehyde and other degradative products is undesirable [167–170]. Cyclic oligomers can also migrate to the surface of polyesters, where they can form crystals or dissolve into surrounding media [167, 168, 171]. Thermal degradation can be ameliorated by adding antioxidants and other thermal stabilizers, but chain scission and other degradative mechanisms that are present at higher temperatures cannot be fully prevented.

Hydrolytic degradation can occur during processing and end use. Hydrolytic degradation during processing is minimized by thoroughly drying the polyester prior to extrusion and processing. Hydrolytic instability limits the use of polyesters in certain applications – for example, exposure to high pH caustic solutions, whereas it is a desirable feature for biodegradability and compostability [96].

Degradation during use is a more complicated phenomenon that cannot be described simply and is therefore beyond the scope of this work. To exemplify, photolysis and hydrolysis combine with thermal and oxidative degradative products present in the polymer as well as oxidative contaminants present in the atmosphere to produce polymer degradation, some auto-oxidative, and others by both radical and nonradical pathways, that may alter the chemical composition, molar mass, solubility, and color [159, 172, 173].

## Acknowledgment

Final chemical structural figures in this work were provided by Dr. Thomas R. Smith.

## References

**1** Odian, G. (2004). *Principles of Polymerization*, 4e. Hoboken: Wiley-Interscience.

**2** Chuah, H. (2003). Poly(trimethylene terephthalate). In: *Encyclopedia of Polymer Science and Technology*, vol. 3 (ed. H.F. Mark and J. Kroschwitz), 544–557. Hoboken: Wiley.

**3** Dangayach, K., Chuah, H., Gergen, W. et al. (1997). *Proc. SPE 55$^{th}$ Ann. Tec. Conf. & Exhibits*, Toronto, Ontario, Canada, p. 2097.

**4** Petrie, S.E.B. (1972). Thermal behavior of annealed organic glasses. *J. Polym. Sci. Polym. Phys. Ed.* 10: 1255–1272.

**5** Huang, J.-M. and Chang, F.-C. (2000). Crystallization kinetics of PTT. *J. Polym. Sci. Part B: Polym. Phys. Ed.* 38: 934–941.

**6** Turner, S.R. (2004). Development of amorphous copolyesters based on 1, 4-cyclohexanedimethanol. *J. Polym. Sci. Part A: Polym. Chem.* 42: 5847–5852.

**7** Lotti, N., Colonna, M., Fiorini, M. et al. (2011). Poly (butylene terephthalate) modified with ethoxylated bisphenol S with increased glass transition temperature and improved thermal stability. *Polymer* 52: 904–911.

**8** Bair, H. (1994). Glass transition measurements by DSC. In: *Assignment of the Glass Transition* (ed. R. Seyler), 50–74. West Conshohocken: ASTM International.

**9** D'Amico, T., Donahue, C.J., and Rais, E.A. (2008). Thermal analysis of plastics. *J. Chem. Educ.* 85: 404–407.

**10** Escala, A. (1978). Studies of the crystallization behavior of the poly(1,4-butylene terephthalate/poly(ethylene terephthalate) blends. PhD Thesis. University of Massachusetts, Amherst.

**11** Escala, A. and Stein, R.S. (1979). Crystallization studies of blends of polyethylene terephthalate and polybutylene terephthalate. *Adv. Chem. Series* 176: 455–487.

**12** Stein, R.S., Khambatta, F.B., Warner, F.P. et al. (1978). X-ray and optical studies of the morphology of polymer blends. *J. Polym. Sci., Polym. Symp.* 63: 313–328.

**13** Fox, T.G. (1956). The composition dependence of glass transition properties. *Bull. Am. Phys. Soc.* 1: 123.

**14** Heijboer, J. (1969). Modulus and damping of polymers in relation to their structure. *Br. Polym. J.* 1: 3–14.

**15** Heijboer, J. (1968). Study of the movements of cycloalkyl side groups in polymethacrylates by dynamic mechanical measurements. *J. Polym. Sci. Part C* 16: 3413–3422.

**16** Liu, Y. and Turner, S.R. (2010). Synthesis and properties of cyclic diester based aliphatic copolyesters. *J. Polym. Sci. Part A* 48: 2162–2169.

**17** Iyer, K.A. (2017). Chain mobility, secondary relaxation, and oxygen transport in terephthalate copolyesters with rigid and flexible cyclic diols. *Polymer* 129: 117–126.

**18** Jackson, W.J. Jr. and Caldwell, J.R. (1967). Antiplasticization. III. Characteristics and properties of antiplasticizable polymers. *J. Appl. Polym. Sci.* 11: 227–244.

**19** Burke, D.J., Kawauchi, T., Kade, M.J. et al. (2012). Ketene-based route to rigid cyclobutanediol monomers for the replacement of BPA in high performance polyesters. *ACS Macro Lett.* 1: 1228–1232.

**20** MacDonald, W.A. (2004). Polyester films. In: *Encyclopedia of Polymer Science and Technology*, vol. 11 (ed. H.F. Mark and J. Kroschwitz), 2–40. Hoboken: Wiley.

**21** Salem, D.R. (1999). Orientation and crystallization in poly(ethylene terephthalate) during drawing at high temperatures and strain rates. *Polym. Eng. Sci.* 39: 2419–2429.

**22** Salem, D.R. (1992). Development of crystalline order during hot-drawing of poly(ethylene terephthalate) film: influence of strain rate. *Polymer* 33: 3182–3188.

**23** Martins, C.I. and Cakmak, M. (2007). Control the strain-induced crystallization of polyethylene terephthalate by temporally varying deformation rates: a mechano-optical study. *Polymer* 48: 2109–2123.
**24** Greener, J., Tsou, A.H., and Blanton, T.N. (1999). Physical and microstructural effects of heat setting in polyester films. *Polymer* 39: 2403–2418.
**25** Dubner, W.S. and Schultz, J.M. (1996). Influence of bond interchange reactions on mechanical behavior of solid poly (ethylene terephthalate). *J. Polym. Sci. Part B: Polym. Phys. Ed.* 34: 2307–2316.
**26** Vallat, M.-F. and Plazek, D.J. (1986). Effects of thermal treatment of biaxially oriented poly (ethylene terephthalate). *J. Polym. Sci. Part B: Polym. Phys. Ed.* 24: 2123–2134.
**27** Vallat, M.-F. and Plazek, D.J. (1988). Effects of thermal treatment on biaxially oriented poly (ethylene terephthalate). II. The anisotropic glass temperature. *J. Polym. Sci. Part B: Polym. Phys. Ed.* 26: 545–554.
**28** Vallat, M.-F., Plazek, D.J., and Bhushan, B. (1988). Effects of thermal treatment on biaxially oriented poly (ethylene terephthalate). III. Creep behavior following various thermal histories. *J. Polym. Sci. Part B: Polym. Phys. Ed.* 26: 555–567.
**29** Cakmak, M. and Lee, S.W. (1995). Molecular mechanism of neck formation in uniaxially stretched poly (ethylene naphthalate) films. *Polymer* 36: 4039–4054.
**30** Cakmak, M., Wang, Y.D., and Simhambhatla, M. (1990). Processing characteristics, structure development, and properties of uni and biaxially stretched poly (ethylene 2, 6 naphthalate) (PEN) films. *Polym. Eng. Sci.* 30: 721–733.
**31** McKinley, G.H. and Hassager, O. (1999). The considere condition and rapid stretching of linear and branched polymer melts. *J. Rheol.* 43: 1195–1212.
**32** Haward, R.N. (1993). Strain hardening of thermoplastics. *Macromolecules* 26: 5860–5869.
**33** de Vasconcelos, V., Vilaça Pinto, G., Moraes de Abreu, C.A. et al. (1999). Effect of the corona treatment and of the 1, 4-cyclohexanedimethanol on the surface characteristics of the poly (ethylene terephthalate) film. *Macromol. Symp.* 148: 333–343.
**34** Fleischer, C.A., Bauer, C.L., Massa, D.J., and Taylor, J.F. (1996). Film as a composite material. *MRS Bull.* 21: 14–19.
**35** Ishigaki, K. and Naoi, T. (1990). Method of making a rolled silver halide element. U. S. Patent No. 4,933,267.
**36** Ouchi, I., Hosoi, M., and Takashi, T. (1992). Hygroscopic expansion of polyester films. *Jpn. J. Appl. Phys.* 31: 2502–2507.
**37** Kit, K.M., Schultz, J.M., and Gohil, R.M. (1995). Morphology and barrier properties of oriented blends of poly (ethylene terephthalate) and poly (ethylene 2, 6-naphthalate) with poly (ethylene-co-vinyl alcohol). *Polym. Eng. Sci.* 35: 680–692.
**38** Lewis, E.L.V. and Ward, I.M. (1980). The measurement of shear compliances for oriented polyethylene terephthalate sheet. *J. Mater. Sci.* 15: 2354–2364.

**39** Struik, L.C.E. (1978). *Physical Aging in Polymers and Other Amorphous Materials*. Amsterdam: Elsevier.

**40** Struik, L.C.E. (1977). Physical aging in plastics and other glassy materials. *Polym. Eng. Sci.* 17: 165–173.

**41** Cerrada, M. and McKenna, G. (2000). Creep behavior in amorphous and semicrystalline PEN. In: *Time Dependent and Nonlinear Effects in Polymers and Composites* (ed. R. Schapery and C. Sun), 47–69. West Conshohocken: ASTM International.

**42** Gilmor, J. and Greener, J. (1999). Stress relaxation in semi-crystalline PEN films: physical aging effects. *J. Plastic Film & Sheeting* 15: 37–46.

**43** Schrader, R.W. and Carroll, J.F. Jr. (1979). Process for reducing core-set curling tendency and core-set curl of polymeric film elements. US Patent 4,141,735.

**44** Greener, J., Tsou, A.H., Ng, K.C., and Chen, W.A. (1991). The bending recovery of polymer films. I. A phenomenological model. *J. Polym. Sci. Part B: Polym. Phys. Ed.* 29: 843–858.

**45** Tsou, A.H., Greener, J., and Smith, G.D. (1995). Stress relaxation of polymer films in bending. *Polymer* 36: 949–954.

**46** Bier, G. (1974). Polyarylates (polyesters from aromatic dicarboxylic acids and bisphenols). *Polymer* 15: 527–535.

**47** Yoneyama, M., Kuruppu, K.D.A., Kakimoto, M.-A., and Imai, Y. (1989). Preparation and properties of polyarylates and copolyarylates from phenylindanedicarbonyl chloride and bisphenols. *J. Polym. Sci. Part A: Polym. Chem.* 27: 979–988.

**48** Elman, J. F. and Massa, D.J. (2005). Liquid crystal cell with compensator layer and process. US Patent 6,853,424.

**49** Jackson, W.J. Jr. (1989). Liquid crystal polymers. XI. liquid crystal aromatic polyesters: early history and future trends. *Mol. Cryst. Liq. Cryst.* 169: 23–49.

**50** Lenz, R.W. (1985). Structure-order relationships in liquid crystalline polyesters. *Pure Appl. Chem.* 57: 1537–1544.

**51** Heifferon, K.V., Mondschein, R.J., Talley, S.J. et al. (2019). Tailoring the glassy mesophase range of thermotropic polyesters through copolymerization of 4, 4'-bibenzoate and kinked isomer. *Polymer* 163: 125–133.

**52** Carr, P.L., Nicholson, T.M., and Ward, I.M. (1997). Mesophase structures in poly (ethylene terephthalate), poly (ethylene naphthalate) and poly (ethylene naphthalate bibenzoate). *Polym. Advan. Technol.* 8: 592–600.

**53** Carothers, W.H. (1931). Polymerization. *Chem. Rev.* 8: 353–426.

**54** Sisson, A.L., Ekinci, D., and Lendlein, A. (2013). The contemporary role of ε-caprolactone chemistry to create advanced polymer architectures. *Polymer* 54: 4333–4350.

**55** Labet, M. and Thielemans, W. (2009). Synthesis of polycaprolactone: a review. *Chem. Soc. Rev.* 38: 3484–3504.

56 Woodruff, M.A. and Hutmacher, D.W. (2010). The return of a forgotten polymer – polycaprolactone in the 21st century. *Prog. Polym. Sci.* 35: 1217–1256.

57 Koleske, J.V. (1978). Blends containing poly(ε-caprolactone) and related polymers. In: *Polymer Blends*, vol. 1 (ed. D.R. Paul and S. Newman). New York: Academic Press.

58 Eastmond, G.C. (1999). Poly (ε-caprolactone) blends. *Advan. Polym. Sci.* 149: 59–223.

59 Woo, E.M., Barlow, J.W., and Paul, D.R. (1985). Thermodynamics of the phase behaviour of poly (vinyl chloride)/aliphatic polyester blends. *Polymer* 26: 763–773.

60 Cruz, C.A., Barlow, J.W., and Paul, D.R. (1979). The basis for miscibility in polyester-polycarbonate blends. *Macromolecules* 12: 726–731.

61 Siddiqui, N., Asawa, S., Bhaskar, B. et al. (2018). PCL-based composite scaffold matrices for tissue engineering applications. *Mol. Biotechnol.* 60: 506–532.

62 Dash, T.K. and Badireenath, K.V. (2012). Poly-є-caprolactone based formulations for drug delivery and tissue engineering: a review. *J. Control. Release* 158: 15–33.

63 Vuyiswa, M.J. and Sinha, R.S. (2014). Poly (ε-caprolactone) nanocomposite scaffolds for tissue engineering: a brief overview. *J. Nanosci. Nanotechnol.* 14: 535–545.

64 Monique, M. and Hutmacher, D.W. (2007). Biodegradable polymers applied in tissue engineering research: a review. *Polym. Int.* 56: 145–157.

65 Suwantong, O. (2016). Biomedical applications of electrospun polycaprolactone fiber mats. *Polym. Advan. Technol.* 27: 1264–1273.

66 Janmohammadi, M. and Nourbakhsh, M.S. (2018). Electrospun polycaprolactone scaffolds for tissue engineering: a review. *Int. J. Polym. Mater.* 68 (9): 527–539.

67 Pereira, I.M. and Orefice, R.L. (2010). In situ evaluation of structural changes in poly (ester-urethanes) during shape-memory cycles. *Polymer* 51: 1744–1751.

68 Bogdanov, B., Toncheva, V., Schacht, E. et al. (1999). Physical properties of poly(ester-urethanes) prepared from different molar mass polycaprolactone-diols. *Polymer* 40: 3171–3182.

69 Critchfield, F.E., Koleske, J.V., Magnus, G., and Dodd, J.L. (1973). Effect of short chain diol on properties of polycaprolactone based polyurethanes. *Adv. Urethane Sci. Technol.* 2: 141–152.

70 Gigli, M., Fabbri, M., Lotti, N. et al. (2016). Poly (butylene succinate)-based polyesters for biomedical applications: a review. *Eur. Polym. J.* 75: 431–460.

71 McKee, M.G., Unal, S., Wilkes, G.L., and Long, T.E. (2005). Branched polyesters: recent advances in synthesis and performance. *Prog. Polym. Sci.* 30: 507–539.

72 Münstedt, H. (2010). Rheological properties and molecular structure of polymer melts. *Soft Matter* 7: 2273–2283.

73 Patton, E.V., Wesson, J.A., Rubinstein, M. et al. (1989). Scaling properties of branched polyesters. *Macromolecules* 22: 1946–1959.

**74** Lusignan, C.P., Mourey, T.H., Wilson, J.C., and Colby, R.H. (1995). Viscoelasticity of randomly branched polymers in the critical percolation class. *Phys. Rev. E* 52: 6271–6280.

**75** Davis, B., Gray, T.F., and Masser, H.R. (1982). Copolyesterethers. US Patent 4,349,469.

**76** DeMejo, L.P., Massa, D.J., Wesson, J.A., and Wilson, J.C. (1990). Influence of rheology and molecular architecture on the fusing behavior of toners. *Proc. SPIE* 1253: 85–95.

**77** Dailey, L.A., Wittmar, M., and Kissel, T. (2005). The role of branched polyesters and their modifications in the development of modern drug delivery vehicles. *J. Control. Release* 101: 137–149.

**78** Brenner, A.R., Brigitte Voit, D.J., Massa, D.J., and Turner, S.R. (1996). Hyperbranched polyesters: end group modification and properties. *Macromol. Symp* 102: 47–54.

**79** Massa, D.J., Shriner, K.A., Turner, S.R., and Voit, B.I. (1995). Novel blends of hyperbranched polyesters and linear polymers. *Macromolecules* 28: 3214–3220.

**80** Voit, B., Beyerlein, D., Eichhorn, K.-J. et al. (2001). Functional hyperbranched polyesters with applications in blends, coatings, and thin films. *Chem. Ing. Tech.* 73: 1492–1596.

**81** Souffie, R.D. (1982). Segmented polyether-ester copolymers – thermoplastic elastomers. *J. Elastom. Plast.* 14: 166–186.

**82** Fakirov, S., Fakirov, C., Fischer, E.W. et al. (1993). Reversible morphological changes in poly (ether ester) thermorplastic elastomers during deformation as revealed by small-angle X-ray scattering. *Coll. Polym. Sci.* 271: 811–823.

**83** Muramatsu, S. and Lando, J.B. (1998). Reversible crystal deformation of poly (tetramethylene terephthalate) segments in semicrystalline segmented poly (ether ester) thermoplastic elastomers. *Macromolecules* 31: 1866–1870.

**84** Ye, H., Zhang, K., Kai, D. et al. (2018). Polyester elastomers for soft tissue engineering. *Chem. Soc. Rev.* 47: 4545–4580.

**85** Zeng, X., Wu, B., Wu, L. et al. (2014). Poly(l-lactic acid)-block-poly(butylene succinate-co-butylene adipate) multiblock copolymers: from synthesis to thermo-mechanical properties. *Ind. Eng. Chem. Res.* 53: 3550–3558.

**86** American Society for Testing and Materials (2013). *Annual Book of ASTM Standards 2013*, Sect. 8, 08.03. West Conshohocken: ASTM International.

**87** Nguyen, H.T.H., Qi, P., Rostagno, M. et al. (2018). The quest for high glass transition temperature bioplastics. *J. Mater. Chem. A* 6: 9298–9331.

**88** Ragaert, K., Delva, L., and Van Geem, K. (2017). Mechanical and chemical recycling of solid plastic waste. *Waste Manage.* 69: 24–58.

**89** Södergard, A. and Stolt, M. (2002). Properties of lactic acid based polymers and their correlation with composition. *Prog. Polym. Sci.* 27: 1123–1163.

**90** Laycock, B., Halley, P., Pratt, S. et al. (2013). The chemomechanical properties of microbial polyhydroxyalkanoates. *Prog. Polym. Sci.* 38: 536–583.

**91** Verhees, C.H., Kengen, S.W.M., Tuininga, J.E. et al. (2003). The unique features of glycolytic pathways in Archaea. *Biochem. J.* 375: 231–246.

**92** Tham, W.H., Wahit, M.U., Kadir, M.R.A., and Wong, T.W. (2016). Polyol-based biodegradable polyesters: a short review. *Rev. Chem. Eng.* 32: 201–221.

**93** Farah, S., Anderson, D.G., and Langer, R. (2016). Physical and mechanical properties of PLA, and their functions in widespread applications – a comprehensive review. *Adv. Drug Deliv. Rev.* 107: 367–392.

**94** Sanhueza, C., Acevedo, F., Rocha, S. et al. (2019). Polyhydroxyalkanoates as biomaterial for electrospun scaffolds. *Int. J. Biol. Macromol.* 124: 102–110.

**95** Sun, X., Xy, C., Wu, G. et al. (2017). Poly(lactic-co-glycolic acid): applications and future prospects for periodontal tissue regeneration. *Polymers* 9: 189–208.

**96** Woodard, L.N. and Grunlan, M.A. (2018). Hydrolytic degradation and erosion of polyester biomaterials. *ACS Macro Lett.* 7: 976–982.

**97** Barouti, G., Jaffredo, C.G., and Guillaume, S.M. (2017). Advances in drug delivery systems based on synthetic poly (hydroxybutyrate)(co) polymers. *Prog. Polym. Sci.* 73: 1–31.

**98** Vilela, C., Sousa, A.F., Fonseca, A.C. et al. (2014). The quest for sustainable polyesters–insights into the future. *Polym. Chem.* 5: 3119–3141.

**99** Awl, R.A., Neff, W.E., Weisleder, D., and Pryde, E.H. (1976). Poly (amide-acetals) and poly (ester-acetals) from polyol acetals of methyl 9 (10)-formylstearate: Preparation and physical characterization. *J. Am. Oil Chem. Soc.* 53: 20–26.

**100** Neff, W.E., Awl, R.A., Pryde, E.H., and Cowan, J.C. (1973). Poly (amide-acetals) and poly (ester-acetals) from 9 (10)-formylstearic acid as stationary phases for gas chromatography. *J. Am. Oil Chem. Soc.* 50: 235–239.

**101** Miller, W.R., Pryde, E.H., Awl, R.A. et al. (1971). Nylon-9 via 9-aminononanoic acid from soybean oil. *Ind. Eng. Chem. Res.* 10: 442–447.

**102** Jesmy, J., Pourfallah, G., Merkley, D. et al. (2014). Thermoplastic polyesters and Co-polyesters derived from vegetable oil: synthesis and optimization of melt polycondensation for medium and long chain poly(ω-hydroxyfatty acid)s and their ester derivatives. *Polym. Chem.* 5: 3203–3213.

**103** Papagiorgiou, G.Z., Tsanaktsis, V., Papagiorgiou, D.G. et al. (2014). Evaluation of polyesters from renewable resources as alternatives to the current fossil-based polymers. Phase transitions of poly (butylene 2, 5-furan-dicarboxylate). *Polymer* 55: 3846–3858.

**104** Valerio, O., Misra, M., and Mohanty, A.K. (2018). Poly(glycerol-co-diacids) polyesters: from glycerol biorefinery to sustainable engineering applications, a review. *ACS Sustain. Chem. Eng.* 6: 5681–5693.

**105** Miri, M.J., Nori, K.E., Andrew, R.M., and Ge, G. (2011). Copolyester synthesis using glycerol from biodiesel production. *ACS Symp. Ser.* 1063: 11–35.

**106** Noreen, A., Zia, K.M., Zuber, M. et al. (2016). A critical review of algal biomass: a versatile platform of bio-based polyesters from renewable resources. *Int. J. Biol. Macromol.* 86: 937–949.

**107** Wang, J., Mahmud, S., Zhang, X. et al. (2019). Biobased amorphous polyesters with high Tg: trade-off between rigid and flexible cyclic diols. *ACS Sustain. Chem. Eng.* 7: 6401–6411.

**108** Patzek, T.W. and Pimentel, D. (2005). Thermodynamics of energy production from biomass. *Crit. Rev. Plant Sci.* 24: 327–364.

**109** Pimentel, D. and Patzek, T.W. (2005). Ethanol production using corn, switchgrass, and wood; biodiesel production using soybean and sunflower. *Nat. Resour. Res.* 14: 65–76.

**110** Teegarden, D. (2004). *Polymer Chemistry: Introduction to an Indispensable Science*. Arlington: NSTA Press.

**111** Martin, J.R., Johnson, J.F., and Cooper, A.R. (1972). Mechanical properties of polymers: the influence of molecular weight and molecular weight distribution. *J. Macromol. Sci., Rev. Macromol. Chem.* 8: 57–199.

**112** Aharoni, S.M. (1986). Correlations between chain parameters and the plateau modulus of polymers. *Macromolecules* 19: 426–434.

**113** Ferry, J.D. (1980). *Viscoelastic Properties of Polymers*. New York: Wiley.

**114** Graessley, W.W. and Segal, L. (1969). Flow behavior of polystyrene systems in steady shearing flow. *Macromolecules* 2: 49–57.

**115** Fetters, L.J., Lohse, D.J., Richter, D. et al. (1994). Connection between polymer molecular weight, density, chain dimensions, and melt viscoelastic properties. *Macromolecules* 27: 4639–4647.

**116** Youk, J.H., Ho, W.J., and Yoo, D.I. (1997). Molecular dimensions of poly (ethylene isophthalate) and poly (ethylene naphthalene-2, 6-dicarboxylate). *Polym. Bull. (Berl.)* 39: 257–263.

**117** Zabala, M.J., Santamaria, A., Pena, J.J., and Munoz, M.E. (1987). Flow properties of concentrated solutions of polyarylate in m-cresol. *Macromol. Chem.* 188: 1129–1136.

**118** van Berkel, J.G., Guigo, N., Visser, H.A., and Sbirrazzuoli, N. (2018). Chain structure and molecular weight dependent mechanics of poly (ethylene 2, 5-furandicarboxylate) compared to poly (ethylene terephthalate). *Macromolecules* 51: 8539–8549.

**119** Izu, P., Munoz, M.E., Ena, J.J., and Santamaria, A. (1994). Viscoelastic measurements on main-chain thermotropics. *Polymer* 35: 2422–2427.

**120** Mondschein, R.J., Dennis, J.M., Liu, H. et al. (2019). Influence of bibenzoate regioisomers on cyclohexanedimethanol-based (co)polyester structure–property relationships. *Macromolecules* 52 (3): 835–843.

**121** Chile, L.-E., Mehrkhodavandi, P., and Hatzikiriakos, S.G. (2016). A comparison of the rheological and mechanical properties of isotactic, syndiotactic, and heterotactic poly (lactide). *Macromolecules* 49: 909–916.

**122** Dorgan, J.R., Williams, J.S., and Lewis, D.N. (1999). Melt rheology of poly (lactic acid): Entanglement and chain architecture effects. *J. Rheol.* 43: 1141–1155.

**123** Zhu, Z., Dakwa, P., Tapadia, P. et al. (2003). Rheological characterization of flow and crystallization behavior of microbial synthesized poly(3-hydroxybutyrate-co-4-hydroxybutyrate). *Macromolecules* 36: 4891–4897.

**124** Paul, D.R. and Newman, S. (1978). *Polymer Blends*, vol. 1 and 2. New York: Academic Press.

**125** Legros, A., Carreau, P.J., Favis, B.D., and Michel, A. (1997). Morphology modification by interfacial chemical reaction in a polyester/ethylene vinyl acetate/polyethylene blend. *Polymer* 38: 5085–5089.

**126** Oyama, T., Kobayashi, S., Okura, T. et al. (2019). Biodegradable compatibilizers for poly(hydroxyalkanoate)/poly(ε-caprolactone) blends through click reactions with end-functionalized microbial poly(hydroxyalkanoate)s. *ACS Sustain. Chem. Eng.* 7: 7969–7978.

**127** Chen, Q., Shan, P., Tong, C. et al. (2019). Influence of reactive blending temperature on impact toughness and phase morphologies of PLA ternary blend system containing magnesium ionomer. *J. Appl. Polym. Sci.* 136: 47682.

**128** Sui, G., Want, K., Xu, S. et al. (2019). The combined effect of reactive and high-shear extrusion on the phase morphologies and properties of PLA/OBC/EGMA ternary blends. *Polymer* 169: 66–73.

**129** Calderon, B.A., Soule, J., and Sobkowicz, M.J. (2019). Synthesis and characterization of compatibilizers for blends of polypropylene carbonate and polybutylene succinate via free-radical grafting of maleic anhydride. *J. Appl. Polym. Sci.* 136: 47553.

**130** Gao, C., Wang, B., Hu, Z. et al. (2019). Effect of the molecular weight on the plasticization properties of poly(hexane succinate) in poly(vinyl chloride). *J. Appl. Polym. Sci.* 136: 47081.

**131** Mohn, R.N., Paul, D.R., Barlow, J.W., and Cruz, C.A. (1979). Polyester–polycarbonate blends. III. Polyesters based on 1, 4-cyclohexanedimethanol/terephthalic acid/isophthalic acid. *J. Appl. Polym. Sci.* 23: 575–587.

**132** Bicakci, S. and Cakmak, M. (1998). Uniaxial deformation behaviour of ternary blends of poly (ethylene naphthalate), poly (ether imide) and poly (ether ether ketone) films. *Polymer* 39: 5405–5420.

**133** Landry, M.R., Massa, D.J., Landry, C.J.T. et al. (1994). A survey of polyvinylphenol blend miscibility. *J. Appl. Polym. Sci.* 54: 991–1011.

**134** Couchman, P.R. and Karasz, F.E. (1978). A classical thermodynamic discussion of the effect of composition on glass-transition temperatures. *Macromolecules* 11: 117–119.

**135** Barnum, R.S., Barlow, J.W., and Paul, D.R. (1982). Melting and crystallization behavior of miscible copolyester–polycarbonate blends. *J. Appl. Polym. Sci.* 27: 4065–4080.

**136** Porter, R.S. and Wang, L.H. (1992). Compatibility and transesterification in binary polymer blends. *Polymer* 33: 2019–2030.

**137** Linder, M., Henrichs, P.M., Hewitt, J.M., and Massa, D.J. (1985). Use of carbon–carbon nuclear spin diffusion for the study of the miscibility of polymer blends. *J. Chem. Phys.* 82: 1585–1598.

**138** Henrichs, P.M., Tribone, J., Massa, D.J., and Hewitt, J.M. (1988). Blend miscibility of bisphenol A polycarbonate and poly (ethylene terephthalate) as studied by solid-state high-resolution carbon-13 NMR spectroscopy. *Macromolecules* 21: 1282–1291.

**139** Ellis, T.S. (1999). Interactions and phase behavior of polyester blends. *Plast. Eng.* 52: 209–233.

**140** Kim, S.H. (2004). Thermotropic liquid crystal polymer reinforced polyesters. In: *Modern Polyesters: Chemistry and Technology of Polyesters and Copolyesters* (ed. J. Scheirs and T.E. Long), 565–596. Chichester: Wiley.

**141** Wooten, W. C. and Zannucci, J.S. (1983). Deactivation of polyester catalyst residues. US Patent 4,401,804.

**142** Chisholm, B.J., Sisti, L., Soloveichik, S., and Gillette, G. (2003). Hydrolytic stability of sulfonated poly (butylene terephthalate). *Polymer* 44: 1903–1910.

**143** Han, S.-I., Yoo, Y., Kim, D.K., and Im, S.S. (2004). Biodegradable aliphatic polyester ionomers. *Macromol. Biosci.* 4: 199–207.

**144** Lu, M., Runt, J., and Painter, P. (2009). An infrared spectrocopic study of a polyester copolymer ionomer based on poly (ethylene oxide). *Macromolecules* 42: 6581–6587.

**145** Zhang, M., Zhang, M., Moore, R.B., and Long, T.E. (2013). Influence of charge placement on the thermal and morphological properties of sulfonated segmented copolyesters. *Polymer* 54: 3521–3528.

**146** Fragiadakis, D., Dou, S., Colby, R.H., and Runt, J. (2009). Molecular mobility and conduction in polyester copolymer ionomers based on poly(ethylene oxide). *J. Chem. Phys.* 130: 064907–064917.

**147** Lin, Q., Unal, S., Fomof, A.R. et al. (2006). Synthesis and characterization of telechelic phosphine oxide polyesters and cobalt (II) chloride complexes. *Polymer* 47: 4085–4093.

**148** Pal, S.K., Gandhi, R.S., and Kothari, V.K. (1996). Effect of comonomer on structure and properties of textured cationic dyeable polyester. *J. Appl. Polym. Sci.* 61: 401–406.

**149** Pekkanan, A.M., Zawaski, C., Stevenson, A.T. et al. (2017). Poly (ether ester) ionomers as water-soluble polymers for material extrusion additive manufacturing processes. *ACS Appl. Mater. Inter.* 9: 12324–12331.

**150** Bautista, M., Martinez de Ilarduna, A., Alla, A., and Munoz-Guerra, S. (2013). Sulfonated poly (hexamethylene terephthalate) copolyesters: enhanced thermal and mechanical properties. *J. Appl. Polym. Sci.* 129: 3527–3535.

**151** Nielsen, L.E. and Landel, R.F., *Mechanical Properties of Polymers and Composites*, (2, rev. and expanded), M. Dekker, New York, 1994.

**152** Barber, G.D., Calhoon, B.H., and Moore, R.B. (2005). Poly(ethylene terephthalate) Ionomer Based Clay Nanocomposites Produced via Melt Extrusion. *Polymer* 48: 6706–6714.

**153** Hwang, S.-Y. and Im, S.-S. (2011). PET-ionomers and PETi/MMT nanocomposites: a comparison of the effect of ionic groups on their crystallinities and physical properties. *Polym. Compos.* 32: 259–267.

**154** Zhulina, E., Singh, C., and Balazs, A.C. (1999). Attraction between Surfaces in a polymer melt containing telechelic chains: guidelines for controlling the surface separation in intercalated polymer–clay composites. *Langmuir* 15: 3935–3943.

**155** Singh, C. and Balazs, A.C. (2000). Effect of polymer architecture on the miscibility of polymer/clay mixtures. *Polym. Int.* 49: 469–471.

**156** Robello, D.R., Yamaguchi, N., Blanton, T., and Barnes, C. (2004). Spontaneous formation of an exfoliated polystyrene–clay nanocomposite using a star-shaped polymer. *J. Am. Chem. Soc.* 126: 6118–6119.

**157** Maiti, P., Batt, C.A., and Giannelis, E.P. (2007). New biodegradable polyhydroxybutyrate/layered silicate nanocomposites. *Biomacromolecules* 8: 3393–3400.

**158** Gilman, J.W., Jackson, C.L., Morgan, A.B. et al. (2000). Flammability properties of polymer– layered-silicate nanocomposites. Polypropylene and polystyrene nanocomposite. *Chem. Mater.* 12: 1866–1873.

**159** Reich, L. (1968). Polymer degradation by differential thermal analysis techniques. *J. Polym. Sci. Macromol. Rev.* 3: 49–112.

**160** Shukla, S., Lofgren, E.A., and Jabarin, S.A. (2005). Effects of injection-molding processing parameters on acetaldehyde generation and degradation of poly (ethylene terephthalate). *Polym. Int.* 54: 946–955.

**161** Hornof, C. and Zeman, L. (1978). La formation des liaisons éthériques lors de la production du polytéréphthalate d'éthylène glycol. *Can. J. Chem.* 56: 2703–2706.

**162** Khemani, K.C. (1999). A novel approach for studying the thermal degradation, and for estimating the rate of acetaldehyde generation by the chain scission mechanism of ethylene glycol based polyesters and copolyesters. *Polym. Degrad. Stab.* 67: 91–99.

**163** Kim, T.Y. and Jabarin, S.A. (2003). Solid-state polymerization of poly (ethylene terephthalate). III. Thermal stabilities in terms of the vinyl ester end group and acetaldehyde. *J. Appl. Polym. Sci.* 89: 228–237.

**164** Halek, G.W. (1986). The zero-order kinetics of acetaldehyde thermal generation from polyethylene terephthalate. *J. Polym. Sci., Polym. Symp.* 74: 83–92.

**165** Semlyen, J.A. (2000). *Cyclic Polymers*, 2e. Dordrecht: Kluwer Academic.

**166** Holland, B.J. and Hay, J.N. (2002). The thermal degradation of PET and analogous polyesters measured by thermal analysis–Fourier transform infrared spectroscopy. *Polymer* 43: 1797–1804.

**167** Abboudi, M., Odeh, A., and Aljoumaa, K. (2016). Carbonyl compound leaching from polyethylene terephthalate into bottled water under sunlight exposure. *Toxicol. Environ. Chem.* 98: 167–178.

**168** Matsuga, M., Tojima, T., Kawamura, Y., and Tanamoto, K. (2005). Survey of formaldehyde, acetaldehyde and oligomers in polyethylene terephthalate food-packaging materials. *Food Addit. Contam.* 22: 783–789.

**169** Blanco, I., Cicala, G., Restuccia, C.L. et al. (2012). Role of 2-hydroxyethyl end group on the thermal degradation of poly (ethylene terephthalate) and reactive melt mixing of poly (ethylene terephthalate)/poly (ethylene naphthalate) blends. *Polym. Eng. Sci.* 52: 2498–2505.

**170** Bandi, S., Mehta, S., and Schiraldi, D.A. (2005). The mechanism of color generation in poly (ethylene terephthalate)/polyamide blends. *Polym. Degrad. Stab.* 88: 341–348.

**171** Edge, M., Mohammadian, M., Hayes, M. et al. (1992). Aspects of polyester degradation: motion picture film and videotape materials: conservation of images. *J. Imaging Sci. Technol.* 36: 13–20.

**172** Sylvester, P.D., Ryan, H.E., Smith, C.D. et al. (2013). Perylene-based profluorescent nitroxides for the rapid monitoring of polyester degradation upon weathering: an assessment. *Polym. Degrad. Stab.* 98: 2054–2062.

**173** Goeschen, C. and Wille, U. (2013). Damage of polyesters by the atmospheric free radical oxidant NO3: a product study involving model systems. *Beilstein J. Org. Chem.* 9: 1907–1916.

# 3

# Film Properties and Internal Structure Development of Miscible Polyester Blends

*Karnav Kanuga*

*Medtronic Inc., Dublin, Ireland*

## 3.1 Introduction

Blends of thermoplastic polyesters for film applications are based on two major engineering polymers, i.e. poly(ethylene terephthalate) (PET) and poly(ethylene 2,6-naphthalate) (PEN). The chemical structure of PEN and PEN is shown in Figure 3.1. The substitution of the phenyl ring of PET by the naphthalene double ring of PEN has some effect on the melting point ($T_m$) but has significant effect on the glass transition temperature ($T_g$), which is about 40 °C higher for PEN.

The challenges in blending thermoplastic polyesters with other polymers for generating performance benefits are similar to those encountered in blending other thermoplastics. The basic requirement for the development of blends with synergistic properties is miscibility which is quite rare most of the times. The thermodynamic criterion of polymer-polymer miscibility (on a molecular level) is that the free energy of mixing, $\Delta G_{mix}$, should be negative, $\Delta G_{mix} < 0$ [1]. It is known that in majority of cases, the above criterion of polymer-polymer miscibility is not satisfied, and the blends phase separate under such situation with the formation of undesirable product morphology. A thermodynamically miscible polymer pair is normally characterized by single $T_g$ and any deviation from a single $T_g$ or broadening of the $T_g$ indicates partial miscibility, or in cases with two distinct $T_g$'s it indicates immiscibility. Often two immiscible materials are compatibilized to get performance benefits. By compatibilization, the interfacial properties of an immiscible polymer blend are suitably modified to obtain desired properties by addition of a compatibilizer either physically or non reactively compatibilized or in situ or reactively compatibilized [2]. The focus of this chapter is on miscible

*Polyester Films: Materials, Processes and Applications*, First Edition.
Edited by Miko Cakmak and Jehuda Greener.

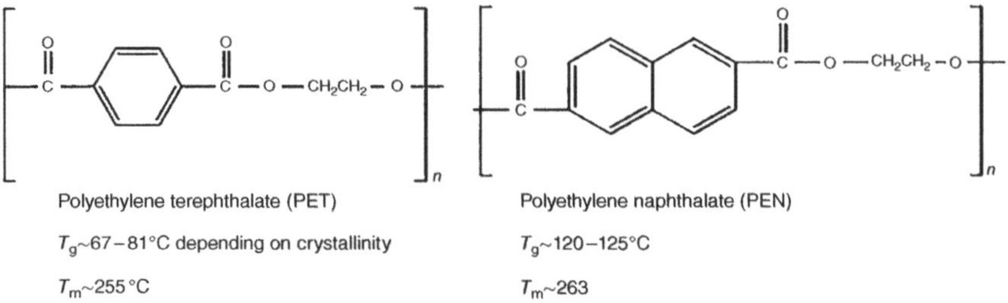


**Figure 3.1** Chemical structure of PET and PEN.

or partially miscible blend films of polyesters. Most common industrial method of blend preparation involves melt blending in a twin-screw extruder. Optimum processing conditions, and rheological properties of the blend components are needed to ensure adequate mixing and development of useful properties in the blends. Polymers that are blended with polyesters to develop synergistic miscible or partially miscible blend films are polyetherimide (PEI), polyethylene naphthalate (PEN) and polyether ether ketone (PEEK).

### 3.1.1 Thermal Transitions in Polyester

Polyesters such as PET and PEN fall in the category of semi-crystalline thermoplastic polymers. As-cast films of PET and PEN exhibit three thermodynamic transitions as shown in Figure 3.2, $T_g$ at ~70 °C for PET and 120 °C for PEN, respectively. This is followed by a cold crystallization temperature ($T_{cc}$) at ~150 °C for PET and 200 °C for PEN, and lastly, a melting temperature ($T_m$) of 245 °C for PET and 275 °C for PEN, respectively. Initially, amorphous polyester films are typically stretched between $T_g$ and $T_{cc}$, which is the rubbery state for the material and ideal for large deformation. Material at this stage undergo strain hardening and orientation-induced crystallization. Polymer chain orientation and subsequent crystallization imparts the mechanical strength to polyester films but

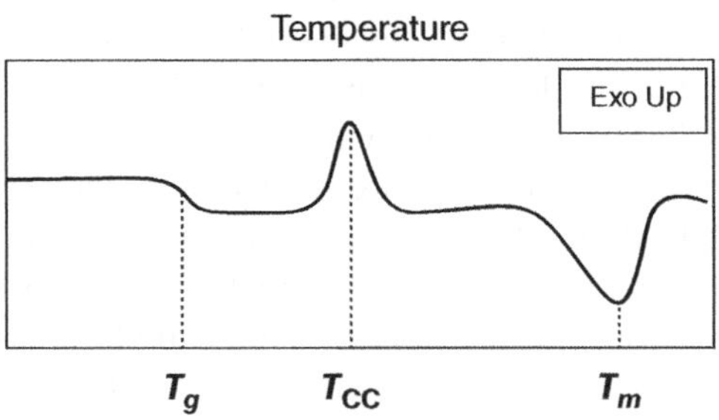


**Figure 3.2** Schematic diagram of a DSC thermogram showing three thermodynamic transitions in as-cast amorphous polyester films.

since the orientation-induced crystals are typically very small, they do not affect transparency of the films. Crystallinity at this stage does not typically exceed 25%, and films contain a great deal of oriented amorphous chains. It is important to note that orientation-induced crystallization is different from thermal crystallization, and the later occurs around $T_{cc}$. Stretched films are annealed further to thermally stabilize and prevent high shrinkage, and the process is also called heat-setting. Annealing of the film is undertaken between $T_{cc}$ and $T_m$ to achieve the maximum crystallization rate to boost the crystallinity induced by thermal crystallization resulting in dimensionally stable films nearly up to their crystalline melting range.

### 3.1.2 PET/PEN Blends

Blends of PET and PEN have been attracting increasing interest because they combine superior properties of PEN with the economy of PET. Melt processing of PET and PEN leads to a transesterification reaction resulting in the formation of PET and PEN copolymers (first block copolymers and then random copolymers), which has been extensively studied [3, 4]. The copolymers formed during the melt mixing act as compatibilizers for the starting materials [4]; thus, miscible blends of PET and PEN can be obtained through the formation of copolymers of PET and PEN. The blend performance will be governed by the degree of transesterification and the blend composition. PEN possesses higher gas barrier properties compared to PET. Biaxially oriented films of PET/PEN blends have shown to significantly improve barrier properties over PET films. Wu et al. reported a four to five times reduction in air permeation for partially transesterified biaxially oriented PET/PEN films compared to PET films of the same thickness and was strongly influenced by the PEN content in the blends [5]. Figure 3.3 shows a comparison of the air permeability of PET/PEN films with different PEN concentrations. The air permeation of PET/PEN films decreased with increasing PEN concentration as shown in Figure 3.4.

PEN also has a significantly higher $T_g$ and melting temperature compared to PET, which gives it superior temperature resistance. However, PEN is an expensive polyester so blending of PEN with PET can prove synergistic and prove economically viable for various applications. PET and PEN blends have been shown to be miscible as noted earlier and transparent films can be obtained. Various researchers have found that the $T_g$ of PET/PEN blends changed almost linearly with NDC content. In the case of miscible blends of polymers or random copolymers, the $T_g$ of the blends or random copolymers can be expressed by the empirical Fox equation, which was calculated, and results are shown in Figure 3.5.

Because of the formation of copolymer during the transesterification reaction, blend properties resemble those of copolymers of PET and PEN. A study by Jabarin [3] has found that blend transesterification level is much less than that of

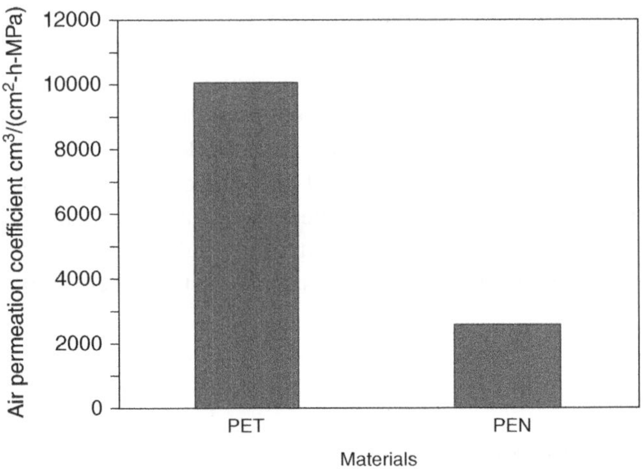


**Figure 3.3** Comparison of the air permeability of BOPET and BOPEN films.

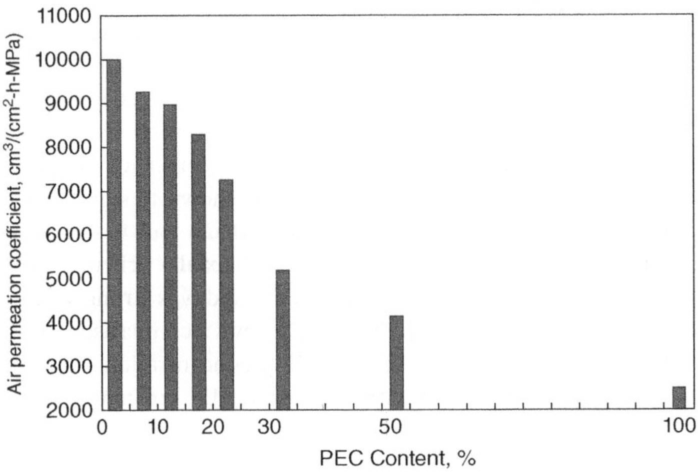


**Figure 3.4** Influence of the PEN concentration on the air permeability of BOPET/ BOPEN films.

totally random (100%) materials and, thus, the sequence length of the naphthalate or terephthalate units in blends is much longer than that in random copolymers. However, it seems that $T_g$ values are not very different for random copolymers and short-block copolymers. Once a certain degree of randomness has been achieved, these values seem insensitive to further changes of the sequence length (block length). The glass-transition temperatures of PET/PEN blends that have achieved this critical level of randomness follow both the Fox equation and the

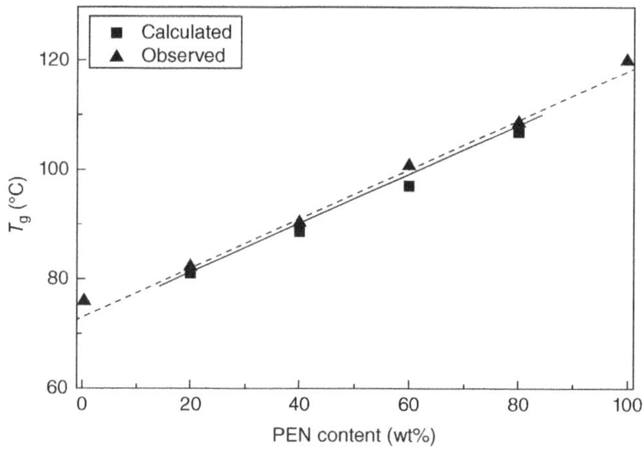


**Figure 3.5** $T_g$ of blends as a function of PEN content.

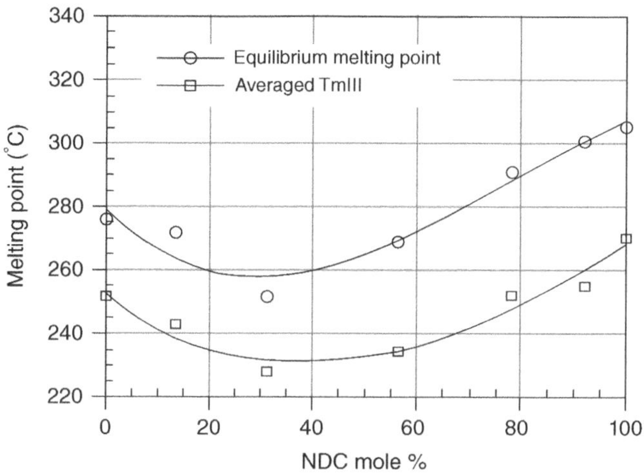


**Figure 3.6** Melting points vs. NDC content for PET/PEI blends.

Gibbs–DiMarzio equation very well, indicating that they behave in a manner similar to that of random copolymers. In terms of melting temperature, the melting points of both PET and PEN were depressed because of the formation of miscible blends and copolymers (Figure 3.6). The depression of the melting point increases with the increased content of the other component. For PET-rich blends, PET is considered to be the crystalline component, whereas PEN is considered to be the amorphous component rejected from the crystalline phase. In PEN-rich blends, PET is thought to be excluded from the crystalline phase as defects or diluents.

Interestingly the type of crystallization in transesterified PET/PEN blends depends on the method of crystallization. It has been shown that quiescent isothermal crystallization led to independent crystallization of the major component, while strain-induced crystallization under uniaxial drawing led to co-crystallization of both components [6]. The differences created by the two modes of crystallization have been attributed to their crystallization kinetics, since strain-induced crystallization is a much faster process compared to quiescent crystallization.

### 3.1.3 PET/PEI Blends

A synergistic miscible blend is formed by compounding polyetherimide (PEI) into PET. PEI is a linear amorphous high-performance polymer with a high $T_g$ of approximately 215 °C. Ever since Martínez et al. [7] reported that PET/PEI blends were miscible in the melt-state over the whole composition range, the blends have been studied by several research groups within a relatively short period of time due to its potential advantages on blending. The bulky aromatic imide group provides PEI with high rigidity, creep resistance, and high thermal stability, while the flexible ether linkage allows for good melt processability. Choi et al. reported the several benefits of PEI in PET-rich blends not only for the final utility of the films but also during film stretching and heat-setting processes during their manufacture [8, 9]. The increase in $T_g$ of the films due to the presence of PEI is desirable since it widens the application window of the films by increasing their use temperature (Figure 3.7). Besides the increase in $T_g$, $T_{cc}$ also increases as shown. Although the increased $T_{cc}$ is accompanied by the increased $T_g$ with increased PEI concentration, the magnitude of $T_{cc}$ is even larger at a given PEI concentration.

For PET/PEI blends the single $T_g$ shifts to higher temperature values with the increase of PEI weight fraction as shown in Figure 3.8. The $T_{cc}$ also shifts to a higher temperature as increasing fraction of non-crystallizable PEI chains suppresses the crystallizability of PET by decreased mobility and dilution effect. At the same time the increase in $T_{cc}$ is quite significant compared to $T_g$, broadening the rubbery state processing temperature window between $T_g$ and $T_{cc}$. $\Delta T_g$ ($T_g$ width), which is full width at half height of the first derivative of the DSC curve at $T_g$, indicates that there is a broadening of $T_g$ with the increase of PEI concentration illustrating the development of micro-heterogeneities in the material with concentration of PEI greater than 20 wt%.

Addition of PEI significantly influences PET film structure and properties. PEI blending results in a wider processing window, defined by the temperature difference between $T_g$ and $T_{cc}$, in which the process (i.e. deformation, followed by heat-setting) operates in a more favorable environment. The crystallization rate of PET in the blend films was dominated by two independently counteractive factors: PEI concentration and preferred orientation. PEI lowers the crystallization rate due mainly to the dilution effect that spatially prevents the crystallizable polymer

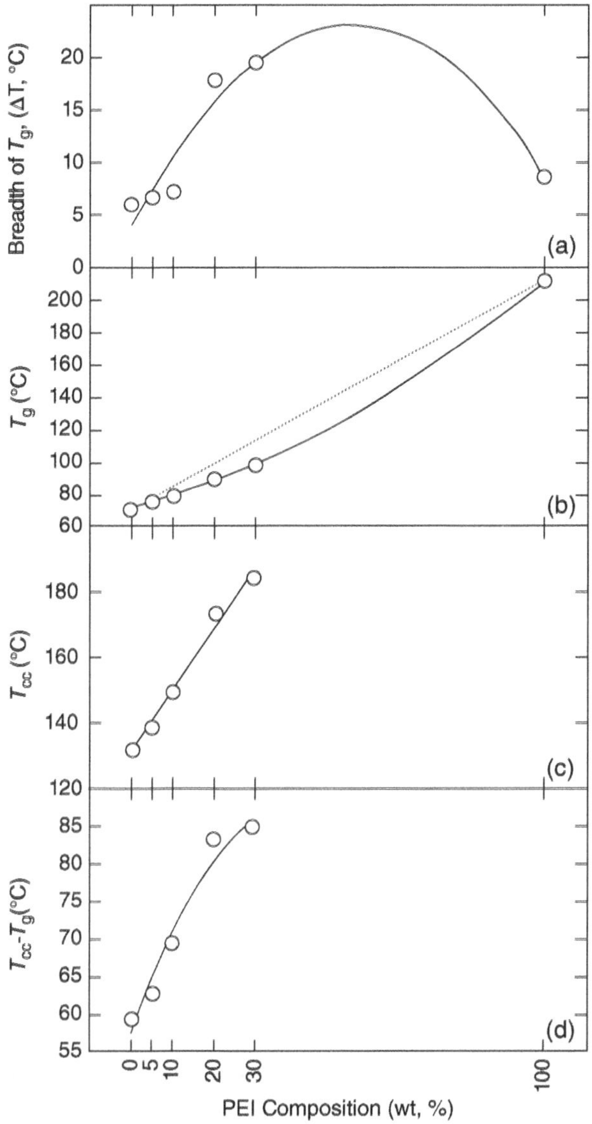


**Figure 3.7** Breadth of $T_g$ range, $T_g$, $T_{cc}$, and $T_{cc}$-$T_g$ for PET/PEI films as a function of PEI composition determined by DSC measurements at the heating rate of 10 C/min. *Source:* Choi et al. [9] / with permission from Elsevier.

chains from coming together, while preferred orientation boosts it by reducing the activation energy for transport for crystallization. Although the order of crystalline structure decreases on stretching as the PEI concentration increases, the stress level becomes higher, giving rise to strain hardening at smaller stretch ratios.

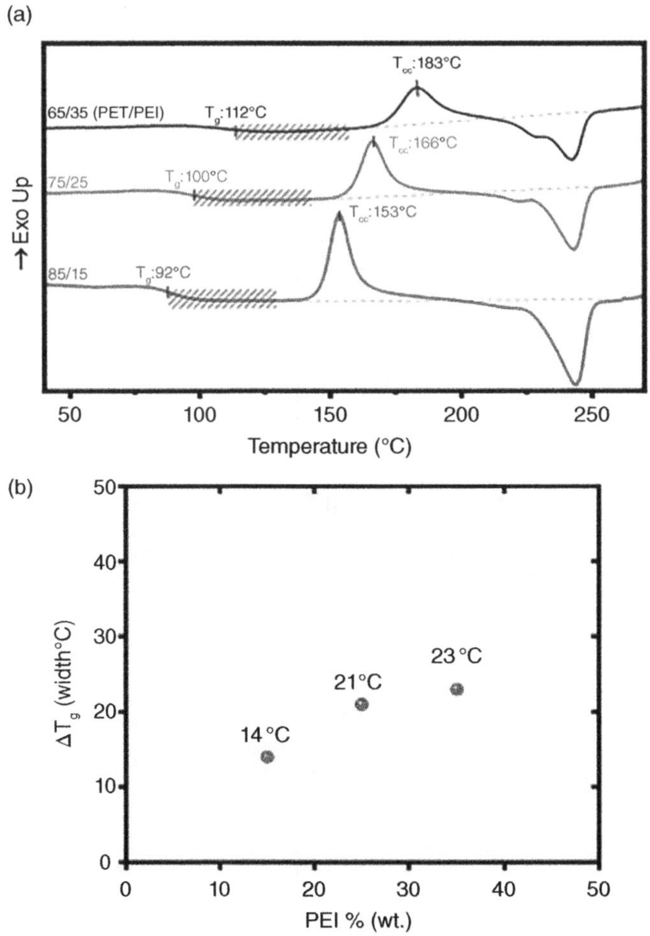


**Figure 3.8** (a) DSC thermogram of compression-molded and water-quenched PET/PEI blends at heating rate of 10 C/min (b) $\Delta T_g$ (width C) of the blends.

This deformation behavior has a benefit of being able to strain harden the films with the higher degree of amorphous orientation without locking in the structure by crystallization, which may impart more flexibility to the films during deformation stage, that is tenter-frame processing. The authors found that the increase in PEI concentration led to increase in highly defective crystallites where bulky PEI chains intrude into the PET crystalline regions (see WAXD patterns in the Figure 3.9). Above 30% PEI, the blend crystallizability disappears. Three categories of distinct structure evolution were observed during stretching of PET/PEI blends with 100/0, 90/10, and 70/30. In 100/0, amorphous orientation takes place

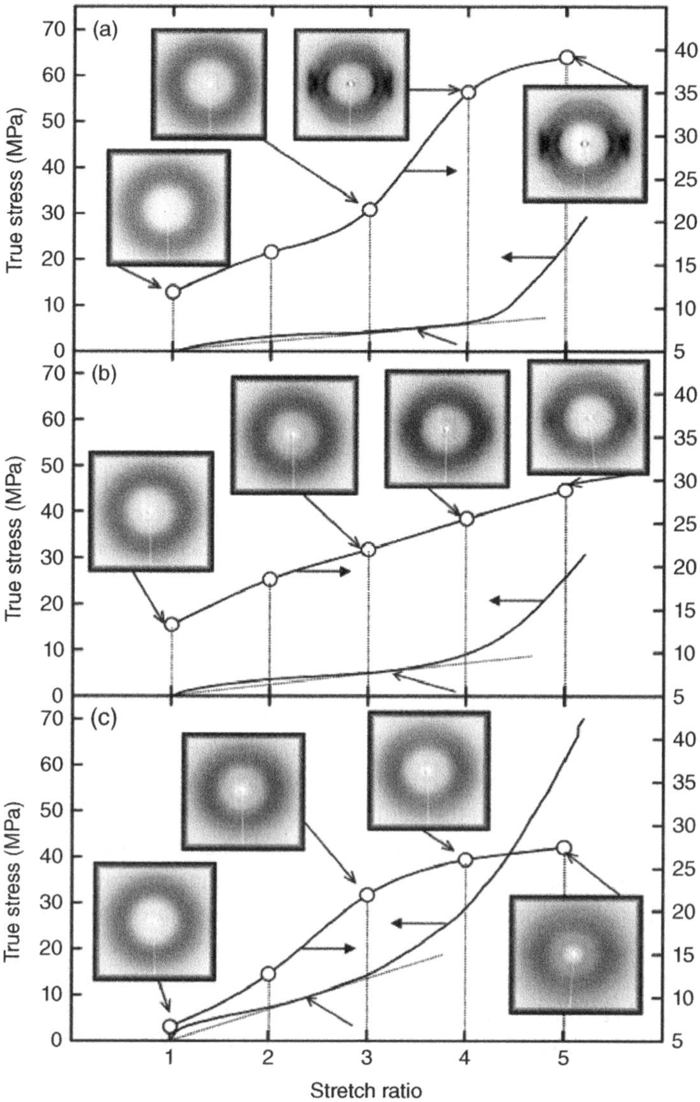


**Figure 3.9** True stress, degree of crystallinity, and WAXS of PET/PEI films as a function of stretch ratio: (a) 100/0, (b) 90/10, and (c) 70/30.

rapidly, followed by crystallization. In 90/10, amorphous PET and PEI intermingled molecules steadily orient with suppressed crystallization of PET due to the dilution effect and insufficient stress level to overcome it. At 70/30 PET/PEI composition, amorphous molecules orient significantly even at small stretch ratios due to the high-stress level; however, the crystallization is greatly restricted because of the high dilution effect, forming a defective crystalline structure. The slowing down of the crystallization kinetics by the addition of PEI, also shown by the increase in $T_{cc}$ (Figure 3.4), allows a prolonged deformation process that is conducive to increased molecular order without locking in the structure and hence improved physical and gas barrier properties on which an enhanced utility of the films may be achieved [10].

### 3.1.4 PEN/PEI Blends

PEN (poly(ethylene naphthalate)) is a slow crystallizing polymer similar to PET with a moderately higher $T_g$ of around 120 °C. The chemical structure of PEN as described earlier in the chapter is similar to PET with the exception of the naphthalene rings, which makes PEN chains stiffer than PET. This polymer, like PET, can be quenched into an amorphous state and it can also be crystallized either by slow cooling from the melt or by stretching in rubbery state in between $T_g$ and $T_{cc}$. One of the unusual characteristics of PEN is that it shows necking upon stretching from the amorphous state above the $T_g$ [11] caused by a highly localized cooperative orientation of the naphthalene planes parallel to the surface of the film resulting in uneven thickness distribution that is detrimental to the appearance and the final properties of the product.

It has been shown in the literature that PEN also forms a miscible blend with PEI and has synergistic effects such as necking elimination in PEN [11]. Authors found a rapid drop in the measured stretching force accompanying the initial neck formation suggesting that when the naphthalene planes are aligned parallel to the broad surface of the sample, the interchain friction is greatly reduced as a result of this "graphite-like" orientation behavior of the naphthalene planes parallel to the broad surfaces of the film. To counteract this behavior, it has been shown that PEN can be blended with PEI in order to increase the friction forces between the PEN chains by introducing stiff and bulky PEI chains into their environment. Kanuga and Cakmak [12] further studied the influence of PEI concentration on the thermal behavior of as-cast PEN/PEI films as shown in Figure 3.10a. The $T_g$ increases slightly with the addition of PEI. The melting temperature was observed to increase first and then decrease as the PEI fraction in the blend increased. The authors attributed this behavior to the relatively low melt temperature (310 °C) used for the film-casting process as compared with its equilibrium melting point of $T_m = 337$ °C. PEN/PEI blends do not follow the normal melting point depression phenomenon until they are melted over 340 °C. The presence of PEI increases the viscosity of the melt, which in turn helps the PEN crystals to survive in the form of small crystallites or frozen-ordered state above the normal melting temperature. These remaining crystallites or the frozen-ordered states in the quenched

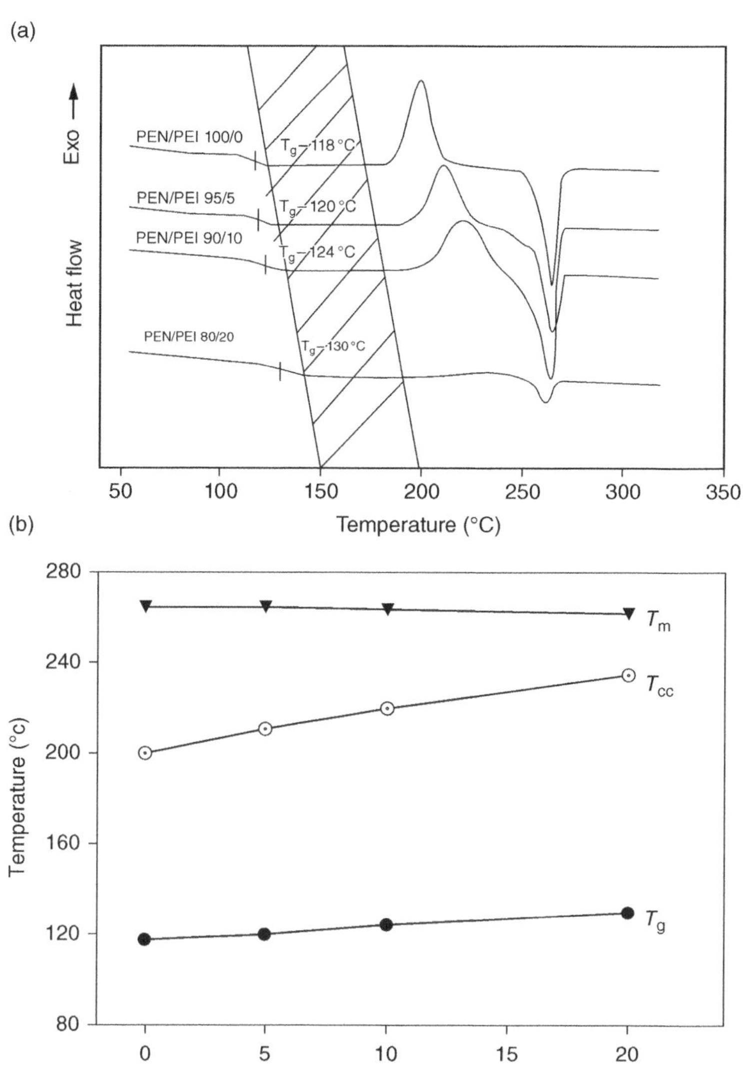


**Figure 3.10** (a) DSC thermograms of melt-cast films. The dotted area marks the stretching window. (b) Change in $T_g$, $T_{cc}$, and $T_m$ as a function of blend composition.

as cast films act as nuclei during the subsequent crystallization stage at lower temperatures. Because of this earlier start of crystallization, crystals grow bigger and more perfect, which in turn increases the melting point of the blend. Above 5% PEI concentration, another mechanism, dilution effect, starts to decrease the crystallizability of PEN similar to what is seen with PET/PEI blends as discussed earlier in this chapter. Above 20% PEI, PEN significantly loses its ability to crystallize. Crystallization from the amorphous state during heating is delayed with increasing PEI content as the mobility of PEN chains is suppressed with the addition of PEI, consequently the $T_{cc}$ moves from 200 °C for 100/0 PEN/PEI to 235 °C for 80/20 PEN/PEI blend (Figure 3.10b). Interestingly it was shown that the increase in $T_{cc}$ is larger than the increase in $T_g$ with increasing PEI concentration, leading to a widening of the $T_g$-$T_{cc}$ window over which optimum stretching can be accomplished. The melting temperature $T_m$ decreases as shown in Figure 3.10b with increasing PEI content following the normal melting point depression [12].

For PEN/PEI blends the degree of crystallinity developed during stretching is studied. For 20% PEI the crystallinity developed is 18%, which is lower by about 40% when compared to 5% PEI sample, confirming that PEI suppresses crystallization of PEN due to the dilution effect (Figure 3.11). At low PEI (5 wt%) concentration the stiff PEI molecules are well dispersed in PEN, and hence they suppress the formation of a physical network formed mostly of PEN crystallites as the PEI chains act as disruptors slightly delaying the onset of strain hardening. However, further addition of PEI causes a decrease in the onset of strain hardening as the mobility of the PEN is hindered due to the "rigidification effect."

Kanuga and Cakmak [12] have further studied the stress-optical behavior of PEN-PEI blends during stretching using online birefringence and true mechanical measurements to unravel the complex structural changes in PEN as influenced

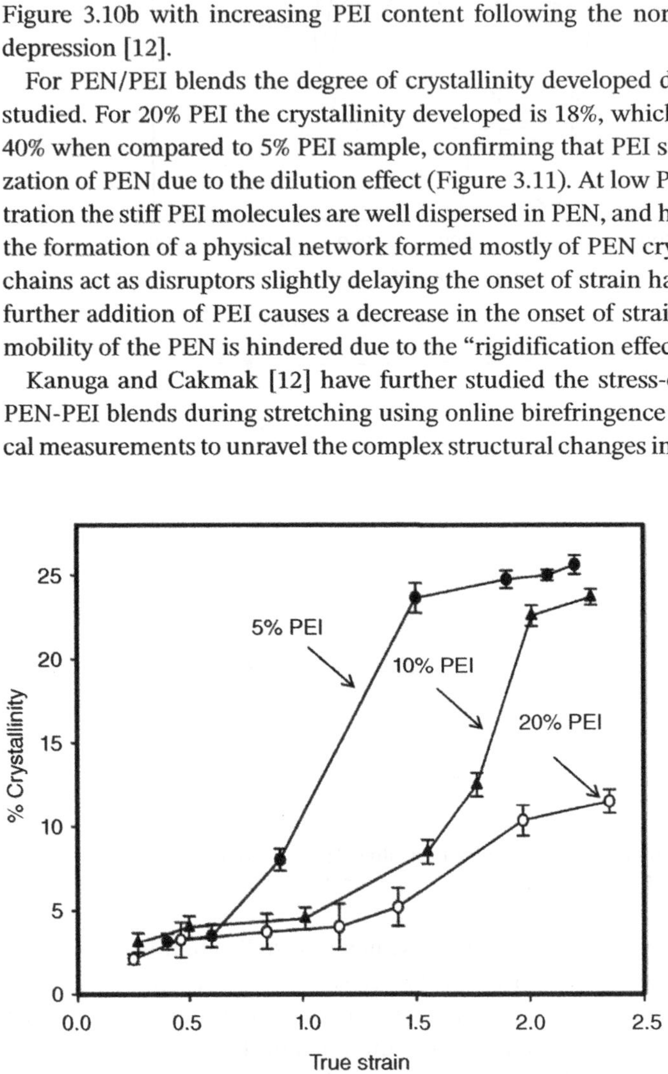


**Figure 3.11** % Crystallinity vs. true strain for PEN/PEI blends.

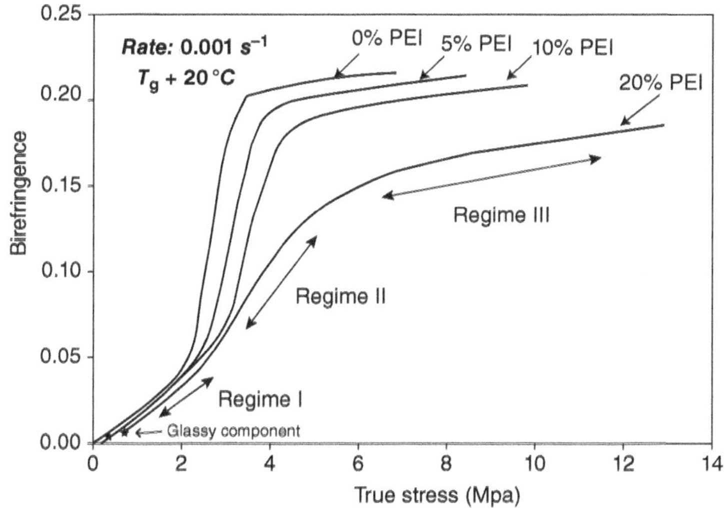


**Figure 3.12** Effect of blend composition on the stress-optical behavior at engineering strain rate of 0.001 s−1 and stretch ratio 4×.

by the addition of PEI as shown in Figure 3.12. PEN-PEI blend compositions stretched at $T_g + 20\,°C$ show three distinct stress-optical regimes. Regime I is typically amorphous orientation and allows development of critical network formation. On further stretching the material follows Regime II, marked by strain-induced crystallization and accompanied by a steep increase in birefringence. Finally, there is Regime III behavior, which represents strain hardening of the material with negligible increase in birefringence.

As the PEI content increases from 5 to 20%, an additional glassy component with a much lower slope appears initially but stretching beyond this glassy regime, the material follows traditional three-regime behavior as shown in Figure 3.13. The large glassy component is observed closer to the $T_g$, and gets smaller as temperature is increased. Such behavior is also observed for PEN (Figure 3.14) at stretching temperatures closer to $T_g$ ($T_g \sim 120\,°C$) but is non-existent at and beyond $T_g + 20\,°C$ (140 °C) stretching temperatures [13]. Additionally, the authors have shown that the overall birefringence decreases with increasing PEI content, and this was attributed to PEI having lower intrinsic birefringence, decrease in the degree of crystallinity of PEN, and lower crystalline as well as amorphous orientation of PEN molecules with the addition of PEI. The authors have gone further and mapped out the stretching behavior in three distinct dynamic phases (Figure 3.15). The first phase is the nematic phase where the material stays in a "nematic-like" order and shows no strain-induced crystallization after stretching. This is exhibited at temperatures close to the $T_g$, and the stress-optical

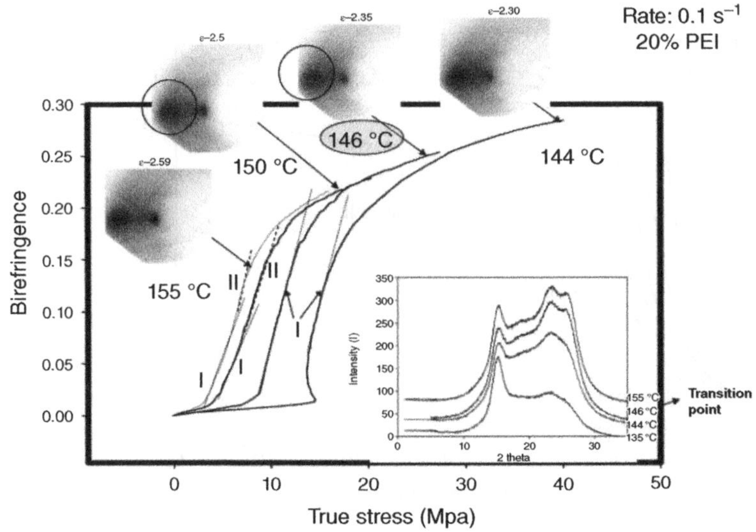


**Figure 3.13** Stress-optical behavior of 20% PEI blend stretched at engineering strain rate of 0.1 s$^{-1}$ and final stretch ratio 4× with corresponding WAXD patterns indicating nematic-to-crystalline transition.

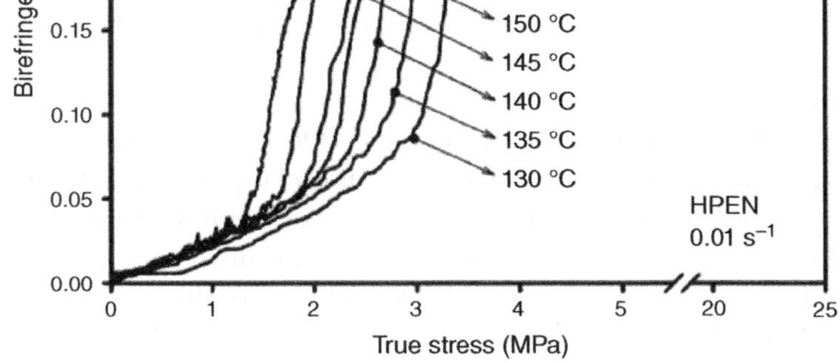


**Figure 3.14** Temperature effect on the mechano-optical behavior of HPEN stretched at rate of 0.01 s$^{-1}$ to a final draw ratio of 5x.

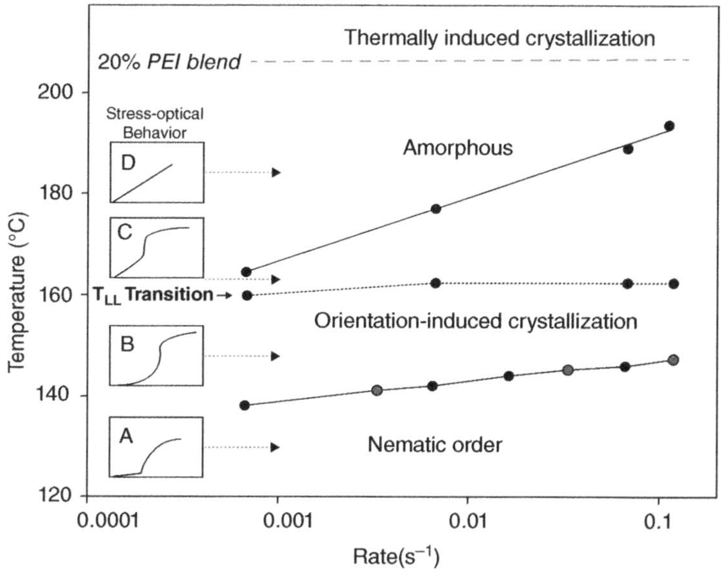


**Figure 3.15** Dynamic Phase Diagram for 20% PEI blend as a function of temperature and engineering strain rate.

behavior has a significant initial glassy component as shown in Figure 3.13. PEN and 5% PEI blend films do not show this nematic behavior and it is only observed at 10 and 20% PEI concentration. The appearance of nematic order is attributed to suppression of orientation in the presence of stiff PEI chains in PEN. With increasing stretch temperatures all blend films exhibit strain-induced crystallization during stretching in the second phase. At high-stretch temperatures relaxation dominates and the material goes through a crystalline-amorphous transition beyond which the films remain amorphous during stretching marking the third phase. A new transition is also observed below which the stress-optical behavior includes an initial glassy component attributed to stiff segmental correlations. Increasing the temperature beyond this transition, these segmental correlations break down, or in other words "segmental melting" occurs and the initial glassy behavior disappears, and the material shows rubbery deformation with the stress-optical behavior in Regime I follow the SOR. This transition is discussed in detail in the next section.

#### 3.1.4.1 Liquid-Liquid Transition Temperature ($T_{ll}$)

It is observed by researchers that for PET, PEN, as well as their blends with PEI, a kinetic transition is observed. The material does not abruptly go from glassy amorphous to true fluid when stretching temperature is increased past $T_g$. Instead, a gradual structural transition region is observed where there is decreasing fraction of glassy behavior as the temperature is increased and beyond the critical temperature it disappears (Figure 3.13 and 3.14). This temperature corresponds to approximately 1.2 times of the materials $T_g$ (K). These observations point to and correlate precisely with the 'liquid-liquid transition temperature ($T_{ll}$)' occuring at 1.2 x $T_g$ as described by Boyer and later on by Frenkel [14]. According to Boyer, $T_{ll}$ represented a temperature above which the entire polymer chain is free to move as a unit and below which the material has some structure and is represented as a 'hindered liquid' or a 'fixed liquid state'. Frenkel finally associated the concept of 'segmental melting' at $T_{ll}$, above which the entropy of the polymer chains gives the driving force to free itself from the segmental restraints. Similar behavior is observed here for PEN and PEN-PEI blends. At low temperatures closer to $T_g$, there is glassy component. Increasing the temperature further, the photoelastic constant increases as well as the regime I slope decreases and they merge to a single value at the $T_{ll}$ and it remain constant beyond this temperature (Figure 3.16). This happens to occur at 1.25 times $T_g$ (C). Stretching below the $T_{ll}$ in PEN/PEI blends, the prevailing segmental interactions do not allow the material to exhibit rubbery deformation because the energy applied to deform the material at this point is used to break these segmental correlations, leading to high stresses in the system with little increase in birefringence. As soon as these preexisting

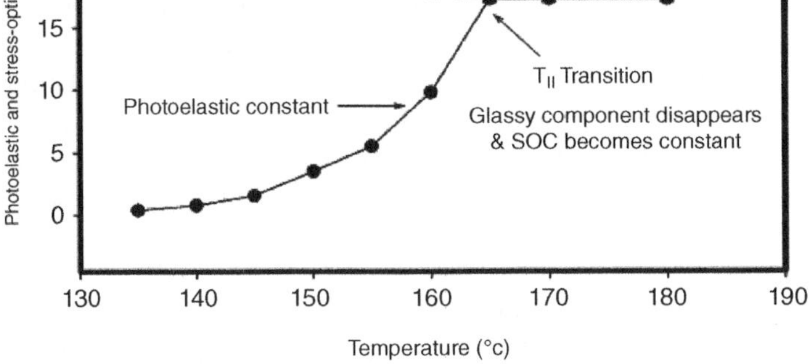


**Figure 3.16** Photoelastic and stress optical constant as a function of temperature for 20% PEI blend stretched at engineering strain rate of 0.01 $s^{-1}$ and final stretch ratio 4×.

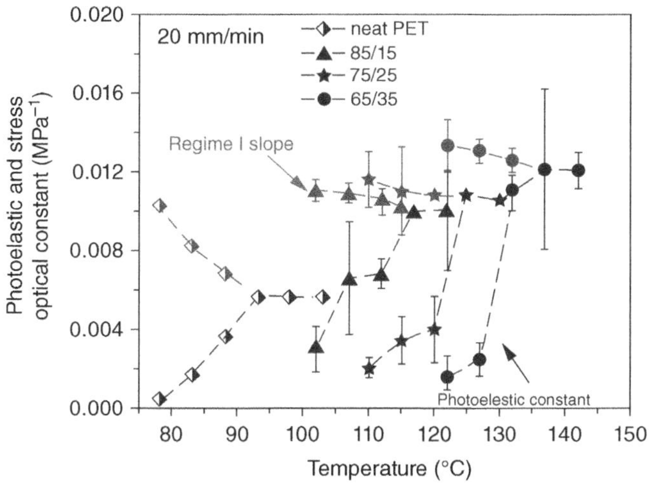


**Figure 3.17** Photoelastic and stress optical constant for PET/PEI blends.

"segmental crystals" disintegrate beyond the $T_{ll}$, the material starts showing rubbery deformation with the regime I, following the stress-optical rule, and the strain optical behavior after this transition becomes completely linear, indicating no inherent network holding the material anymore which makes it behave as a true fluid. Similar behavior is also shown in Figure 3.17 for PET/PEI blends [14].

### 3.1.5 PEN/PEI/PEEK Ternary Blends

Another synergistic blend is formed by the addition of PEEK to PEN/PEI blend creating a miscible ternary blend depending on the blend composition. The addition of PEI into PEN eliminates necking; however, it has some disadvantages. PEI being an amorphous material decreases the crystallizability of PEN, rendering the material completely amorphous beyond 20% PEI content in the blend due to the dilution effect of PEI and increased viscosity. This absence of crystallinity in the material is detrimental to the processability and final product properties, specifically tensile strength and dimensionality stability of films and materials. Bicakci et al. [15] blended PEEK which is a crytallizable polymer with PEN/PEI mixture and investigated the uniaxial deformation of tertiary blends with compositions shown in Figure 3.11. The influence of stretching temperature on the stress–strain behavior of selected PEN-rich ternary blends is shown in Figure 3.18. At $T_g + 10\,°C$, the stress–strain curve exhibits distinct yielding followed by stress drop, which accompanies the neck formation. At the later stages of deformation, the rapid rise in stress signifies rapid stress hardening, which occurs as a result of

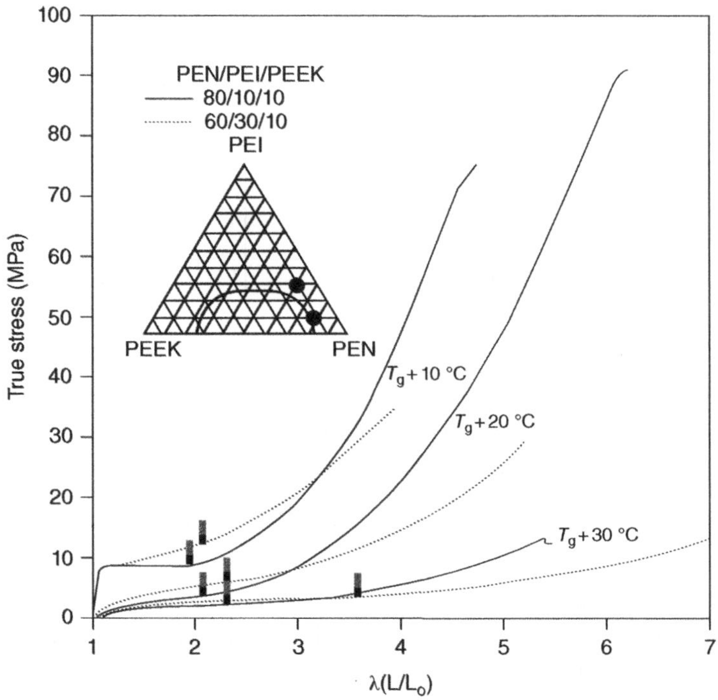


**Figure 3.18** Effect of temperature on the uniaxial deformation behavior of PEN/PEI/PEEK (80/0/10), (60/30/10) films stretched at 250 mm $min^{-1}$. The arrows indicated the onset of stress hardening.

stress-induced crystallization. When the temperature is raised to $T_g$ + 20 °C, yielding followed by stress drop is essentially eliminated, and the entire stress–strain curve resembles affinely deforming cross-linked rubber. Normally, PEN and PEN/PEI blends exhibit necking under these conditions. The critical strain at which the substantial stress hardening occurs is indicated by arrows within the figure.

At $T_g$ + 30 °C, this increase occurs around 3.5–4x stretch ratio. When the proportion of PEI is increased at the expense of PEN while keeping the proportion of PEEK the same, there are distinct changes in the deformation behavior. One change is that the stress drop after the yield point that was observed in PEN and 80/10/10 PEN/PEI/PEEK films at $T_g$ + 10 °C is eliminated, but the elongation at break is also reduced. However, at $T_g$ + 20 °C and $T_g$ + 30 °C, ideal stress–strain curves are observed. The increased proportion of PEI leads to high stresses at low deformations as a result of increased friction between the polymer chains by the presence of stiffer and bulkier PEI chains. But this comes at the expense of reduced stresses at high strains since the stress-induced crystallization process is detrimentally influenced by the increased proportion of non-crystallizable PEI chains. This is clearly shown in figure comparing stress–strain curves of 80/10/10 and 60/30/10 blends for $T_g + 20\,°C$ processing temperature. In order to counteract this reduction of stress at high strains, the authors increased the proportion of PEEK to 30%. The overall goal of increasing the $T_g$ while reducing the critical strain beyond which substantial stress hardening occurs was thus accomplished. The authors have shown that the addition of PEEK recovered the strain hardening and associated self-leveling behavior while boosting the $T_g$ of the final films. Additionally, the addition of PEEK helps recover some of the stress hardenability in the ternary blends containing a large amount of polyetherimide. This allows one to produce films with higher $T_g$ and uniform thickness as the critical strain for strain hardening is reduced which is essential for self-leveling of the films.

## References


1 Nadkarni, V.M. and Rath, A.K. Blends of thermoplastic polyesters. In: *Handbook of Thermoplastic Polymers: Homopolymers, Copolymers, Blends and Composites* (ed. S. Fakirov), 835. Wiley-VCH.

2 Nadkarni, V.M. and Rath, A.K. Blends of thermoplastic polyesters. In: *Handbook of Thermoplastic Polymers: Homopolymers, Copolymers, Blends and Composites* (ed. S. Fakirov), 837. Wiley-VCH.

3 Jabarin, A. and Yu, S. (2001). Transesterification reaction kinetics of poly(ethylene terephthalate)/poly(ethylene 2,6-naphthalate) blends. *Journal of Applied Polymer Science* 81: 11–22.

4 Stewart, M.E., Cox, A.J., and Taylor, D.M. (1993). Reactive processing of poly(ethylene 2,6-naphthalene dicarboxylate)/poly(ethylene terephthalate) blends. *Polymer* 34: 4060.

5 Wu, W., Wagner, M.H., Qian, Q. et al. (2006). Morphology and barrier mechanism of biaxially oriented poly(ethylene terephthalate)/poly(ethylene 2,6-naphthalate) blends. *Journal of Applied Polymer Science* 101: 1309–1316.

6 Patcheak, T.D. and Jabarin, S.A. (2001). Structure and morphology of PET/PEN blends. *Polymer* 42: 975–8985.

7 Martinez, J.M., Eguiazabal, J., and Nazabal, J. (1993). Miscibility of poly(ether imide) and poly(ethylene terephthalate). *Journal of Applied Polymer Science* 48: 935–937.

8 Choi, J., Serhatkulu, T., and Cakmak, M. (2016). Temporal evolution of structure in uniaxially stretched PET/PEI blends during constrained annealing: a real time birefringence study. *Polymer* 97: 569–579.

**9** Choi, J. and Cakmak, M. (2016). Morphological evolution during thermal and strain induced crystallization in poly (ethylene terephthalate)/poly (ether imide) blend films. *Polymer* 84: 10–20.

**10** Greener, J., Pearson, G., and Cakmak, M. Roll-to-roll manufacturing: process elements and recent advances. In: *Polymer Film Substrates for Roll-to-Roll Manufacturing: Process-Structure-Property Relationships*, 1e; Chapter 6 (ed. B. Yalcin and M. Cakmak). Wiley.

**11** Cakmak, M. and Kim, J.C. (1997). Necking mechanism and its elimination in uniaxially drawn films of poly(ethylene naphthalate) (PEN)/polyetherimide (PEI) blends. *Journal of Applied Polymer Science* 65 (11): 2059–2074.

**12** Kanuga, K. and Cakmak, M. (2005). Nonlinear mechanooptical behavior of poly(ethylene naphthalate)/poly(ether imide) blends. Dynamic phase behavior. *Macromolecules* 38 (23): 9698–9710. also in Kanuga, K.D. (2006). A fundamental study on the non-linear mechano-optical behavior of polyethylene naphthalate, its blends with polyetherimide and its nanocomposites. Ph.D. Dissertation. University of Akron, Akron, OH.

**13** Martins, C.I. and Cakmak, M. (2005). Large deformation mechano-optical and dynamical phase behavior in uniaxially stretched poly(ethylene naphthalate). *Macromolecules* 38: 4260–4273.

**14** Boyer, R.B. (1998). Polymer yearbook. 234–337.

**15** Bicakci, S. and Cakmak, M. (1998). Uniaxial deformation behavior of ternary blends of poly(ethylene naphthalate), poly(ether imide) and poly(ether ether ketone) films. *Polymer* 39: 5405–5420.

# 4

# Polyester Ionomers

*Robert A. Weiss*

*Department of Chemical Engineering, University of Connecticut, Storrs, CT, USA*

## 4.1 Introduction

Ionomers are amphiphilic polymers wherein a modest concentration of acid or ionic groups bonded to a low dielectric constant polymer produce substantial changes in the physical and mechanical properties [1]. Intermolecular supramolecular bonds between acid or salt groups (e.g., hydrogen bonds, ionic bonds, or dipolar interactions) produce physical, reversible crosslinks that affect the material properties. Because of the thermal reversibility of the crosslinks, most ionomers are thermoplastics – that is, they exhibit melt flow at elevated temperatures and can be processed by conventional polymer melt processing techniques, such as extrusion or injection molding. The first commercial ionomers were statistical copolymers of ethylene and methacrylic acid that were partially neutralized with sodium or zinc ions [2]. Those ionomers were first introduced commercially in 1964 by DuPont using the tradename *Surlyn*. The term ionomer is now a generic designation for numerous ionic polymers that have been developed since the advent of Surlyn, though the number of commercial ionomers is actually small.

Most ionomers are nanostructured materials, similar to block copolymers and nanocomposites. In this case, microphase separation of an ion-rich phase occurs as a consequence of the immiscibility of the polar ionic species and the relatively hydrophobic polymer chains. The evidence for ion-rich nanodomains, often referred to as *ionic aggregates* or *clusters*, is usually a peak corresponding to a characteristic distance of ~1–5 nm in a small angle X-ray scattering (SAXS) pattern. In general, ionomers have higher melt and solution viscosities, higher modulus and better toughness compared to an analogous non-ionic polymer with the same

*Polyester Films: Materials, Processes and Applications*, First Edition.
Edited by Miko Cakmak and Jehuda Greener.

backbone chemistry. The ionic groups can also provide enhanced chemical properties and functionality to the polymer, e.g., ionic conductivity and novel transport, optical, and surface properties.

Other classes of synthetic polymers with bonded ionic groups are ion-exchange resins and polyelectrolytes. Those materials, however, are usually more highly functionalized polymers than ionomers and/or are crosslinked thermoset polymers. Early definitions of ionomers restricted the concentration of the ionic species to less than 15 mol% of the polymer repeat units [3]. That definition, however, expanded as research in the field increased and new backbones and fixed ionic functionalities (anionic and cationic) were used to develop new materials. Another definition that has been used to distinguish ionomers from polyelectrolytes is the molar ratio of the ionic species to the polymer repeat units, which for an ionomer is generally much less than unity and for a polyelectrolyte is near unity. That definition has been loosely adopted in this review. In most instances, ionomers are not water-soluble, though some of the ionic polyesters discussed herein are water-dispersible or even water-soluble, even though the mole fraction of ionic groups is considerably less than unity.

This chapter is primarily focused on polyester main-chain polymers. It does not include polymers that have polyester side-chains, nor does it include segmented polyurethanes with polyester soft segments. In general, the chapter considers two major types of main-chain polyester ionomers: (i) thermoplastic polyesters and (ii) liquid crystalline polymers.

LCPs with polyester backbones. Research on thermoset main-chain polyesters represents a very small fraction of the publications on polyester ionomers, and where appropriate, those materials are also included in this review. The thermoplastic polyesters include statistical ionomers where the ionic groups are randomly or statistically placed along the chain, segmented polymers (usually poly(ether ester) ionomers, where polyalkylene glycol chains are attached to the monomer or chain segments containing the ionic species), and telechelic ionomers where the ionic groups are located only at one or more chain ends. The LCP ionomers include materials where the ionic groups are pendent to the chain or contained in a short side chain.

This chapter on polyester ionomers is divided into two parts: (i) the peer-reviewed literature and (ii) U. S. patents. The reason for the division is that the peer-reviewed literature is generally more specific as to the synthesis, structure, and properties of the ionomers, and the material has been evaluated by peer review. The patent literature is extremely valuable with regard to general classes of polyester structures and applications of polyester ionomers. However, except for the examples provided in the patents, it is not evident what structures described or claimed have actually been characterized or even synthesized. Also, the properties and structures of the ionomers described in patents are much less rigorously described, if even available, and the chemistry and properties provided have not been peer-reviewed, so are, in general, less reliable as to their accuracy or reproducibility.

The literature on polyesters, mostly sulfonate-containing polyesters (sulfopolyesters) is quite large, so only English language papers or patents (original or translation) were used in preparing this review. However, when sufficient details of a synthesis and/or properties of a polyester ionomer were available in abstracts for work in other languages, that work was included herein. Several strategies were used to search the literature with ACS SciFinder Scholar. Given the immense literature on the subject, especially the patent literature, and the fact that not all papers/patents actually identify their polymers as ionomers, this chapter is probably not exhaustive, though it is probably the most complete review of the subject that has heretofore been published.

In this chapter, the ionic group concentrations are reported two different ways, mol% ionic groups (based on the total monomer feed or measured copolymer composition) and ion-exchange capacity (IEC). More often, authors use mol% ionic groups to describe their ionomers, but that measure can be ambiguous for copolymers with a variety of repeat units and/or large differences in the formula weight of the different repeat units. In such cases, ionomers may have small differences in mol% values, but large difference in their IEC values. In this review, IEC is preferred as the measure of the ionic group concentration, since it is more useful for comparing the influence of ionic groups on properties of different materials. When the authors reported mol% in their paper, both quantities are included in the review.

## 4.2 Random and Segmented Polyester Ionomers

The most well-known commercial polyester is poly(ethylene terephthalate), PET, which is forecast to have a global market of over $42 billion in 2023 [4], so it is not surprising that ionomers based on terephthalate and similar aromatic-based polyesters (e.g., isophthalate) have been studied. For example, sulfonate-containing ionomers based on PET (SPET) can be synthesized [5, 6] by the copolymerization of terephthalic acid, TA (or dimethyl terephthalate, DMT), dimethyl 5-sulfoisophthatlate, DMSI) and ethylene glycol (EG), Scheme 4.1.

Timm and Hsieh [7] studied the effects of melt-spinning and hot-drawing of a commercial SPET fiber (Dacron 64 from DuPont[1]) that had 2–3 mol% of the

---

1 This produce is no longer available from DuPont. DuPont divested their Dacron polyesters to Invisita (Koch Industries) in 2004, and Invista sold their PET business to Inforama Corporation. It is not clear whether Inforama produces sulfopolyesters.

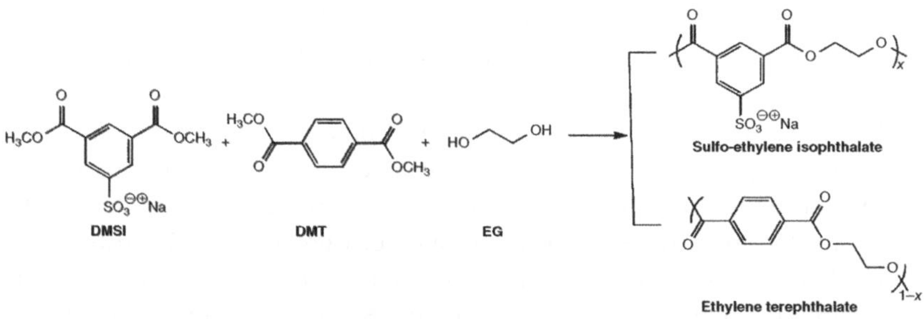


**Scheme 4.1** Synthesis of SPET – a random copolymer of (ethylene terephthalate) and (ethylene sulfoisophthalate). The polymerization byproducts are not shown [5, 6].

aromatic rings sulfonated (IEC = 0.10–0.15 meq $g^{-1}$). The properties of fibers spun from SPET and a PET with about twice the viscosity-average molecular weight ($M_v$) were very different, though some of the differences were attributed to the differences in the molecular weights. The authors discussed the differences in the thermal properties of spun, drawn, crimped and relaxed fibers of PET and SPET, but there was no mention of what specific effects the physical associations of the ionic groups had on the melt viscosity of the SPET or properties of the fibers. Those authors also studied the effects of solvents (dimethyl sulfoxide, pyridine, and methylene chloride) on the crystal structure in SPET fibers [8].

Ingamells et al. [9, 10] studied the rate of hydrolysis in water and the effect of plasticization on the exchange of sodium ions with regard to dyeing a commercial polyester fiber, Dacron 65T (DuPont) (see footnote 1). The papers, however, provided no information on the composition of the fibers, though Figure 4.1 of

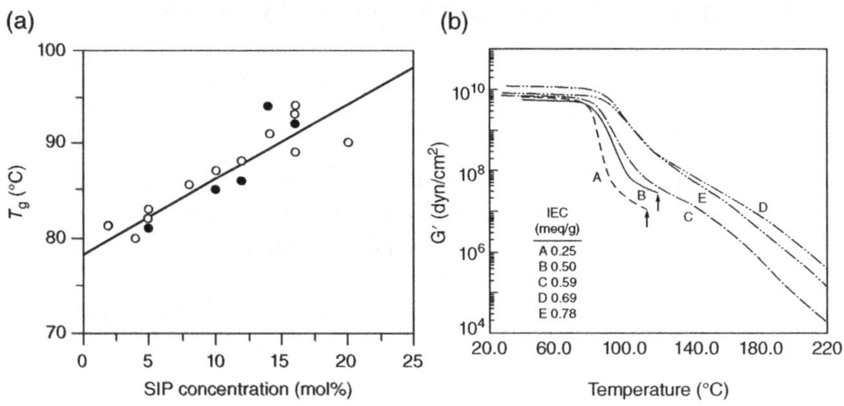


**Figure 4.1** (a) Effect of SIP concentration on the $T_g$ of SPET ionomers. (b) Effect of IEC on the dynamic shear modulus of SPET ionomers. *Source:* Greener et al. [5]/With permission from American Chemical Society.

Ref. [9] indicated that the fibers were SPET. The first step of the hydrolysis was the exchange of sodium ions (presumably from sodium-5-sulfoisophthalate residues) by hydrogen ions, and the authors concluded that as long as the dyeing process was carried out in the presence of at least 0.1M electrolyte, hydrolysis was suppressed [9]. Plasticization of the fibers with phenol during the dyeing process greatly improved the accessibility of the sulfonate groups and facilitated ion-exchange [10], e.g., $K^+$ for $Na^+$.

Ostrowska-Gumkowska and Ostrowska-Czubenko [11] studied the molecular interactions in the sulfonic acid derivative of SPET by Fourier transform infrared (FTIR) spectroscopy. The free-acid derivative was prepared by ion-exchange of K-SPET ionomers containing 1.9–10.1 mol% potassium sulfonate groups (IEC = 0.098–0.50 meq $g^{-1}$). They reported that the sulfonic acid groups associated through hydrogen bonds, but the interpretation of the infrared spectra was complicated by the fact that the sulfonic acid groups also interacted with other proton acceptor and donor groups on the polyester chains. Associated dimers, higher order aggregates of sulfonic acid groups and hydrogen-bonded complexes of $SO_3H$ with the carbonyl groups along the chain and carboxyl and hydroxyl groups at the chain ends were identified.

Greener et al. [5] characterized the oscillatory and steady-state shear rheological properties of SPET ionomer melts with sulfoisophthaloyl (SIP) concentrations of 2–20 mol% (IEC = 0.10–0.96 meq $g^{-1}$). Sulfonation increased the glass transition temperature ($T_g$), shear viscosity, flow activation energy and the plateau modulus, which is consistent with the formation of temporal crosslinks by intermolecular physical associations of the sulfonate groups [12]. See Figure 4.1. For compositions with greater than 10 mol% SIP (IEC > 0.50 meq $g^{-1}$), time-temperature superposition failed, which the authors attributed to the formation of ion-rich aggregates, i.e., *ionic clusters*. Note that it appears from this paper and other papers reviewed herein that the onset ion concentration at which polyester ionomers exhibit ionic clustering is higher than for other well-studied ionomers, e.g., those based on polystyrene, polyethylene, or poly(tetrafluoroethylene), which may be a result of the higher dielectric constant of a polyester than for polyolefins.

Guo et al. [6] studied the thermal properties of SPET ionomers with SIP concentrations of 2–9 mol% (IEC = 0.10–0.45 meq). $T_g$ of the ionomers decreased for compositions less than 4 mol% SIP and then increased at higher SIP concentrations, which contrasts with the results for most ionomers, where $T_g$ shows a monotonic increase with increasing ionic group concentration [13]. No reasonable explanation was provided for the observed $T_g$ behavior, but the authors did use a high boiling solvent, trifluoroacetic acid, to prepare their samples, and it is possible that their washing and drying procedure did not completely remove the trifluoroacetic acid. As a result, residual solvent may have suppressed the ionic interactions in the ionomer, which led to an underestimation of the $T_g$ values for their ionomers. The melting temperature, $T_m$, defined as the maximum melting rate determined by differential scanning calorimetry (DSC), and the crystallinity decreased with increasing sulfonation. Lee et al. [14] reported that ionic aggregates in SPET increased the crystallization rate and produced finer crystallites without spherulite formation.

Zhang et al. [15] compared the melt viscosity and fiber properties of PET, SPET and a sulfonated PET ionomer (SPETT) prepared from DMT, EG, and dimethyl 2-sulfoterephthalate (DMST). The SPET (1.5–4.5 mol% SIP; IEC = 0.078–0.23 meq $g^{-1}$) and the SPETT (3.0 mol% DMST; IEC = 0.15 meq $g^{-1}$) were obtained from the Tianjing Institute of Chemical Fiber and Suzhou Factory of Chemical Fiber, respectively. The melt viscosity and flow activation energy of the ionomers increased with increasing IEC. The shrinkage ratio of the fibers increased, and the residual extension ratio decreased as the annealing temperature increased.

Hsiao et al. [16] studied the properties of blends of nylon 6 with commercial "cationic dyeable PETs" (LeaLea Enterprise Co., Ltd, Taiwan). Although the chemical structures of the modified PETs were not disclosed in that paper, the authors indicated that it contained SIP residues, so it would appear that the ionomers were SPET. The SIP concentrations were 2.0–10 mol% (IEC = 0.10–0.50 meq $g^{-1}$). The polyamide and the SPET ionomer were immiscible and exhibited phase-separated morphologies with poor interfacial adhesion. The melt viscosity and the tensile strength of the blends exhibited minimum values at a 50/50 blend composition. Shu and Hsiao [17] characterized the thermal and flow properties of blends of SPET and a "cationic dyeable nylon 6" that also contained SIP. Despite the fact that both polymer components contained sulfonate groups that could form intermolecular supramolecular bonds, the modified nylon 6/SPET blends were also immiscible and the melt viscosity and tenacity of extruded fibers exhibited minimum values at the 50/50 blend composition.

Hsiao et al. [18] characterized the alkaline hydrolysis of hollow fibers prepared from PET, SPET, and SPET/PET blends. The hydrolysis rate of the hollow fibers increased with SPET concentration. Similarly, the alkaline hydrolysis of SPET and a SPET modified with polyethylene glycol (PEG-x), where $x$ = molecular weight of the PEG, 400 or 2000 g $mol^{-1}$ [19]. The chemistry of the modification was not described, but it would appear that those modified polymers were poly(ester ether) ionomers, which are discussed later in this review. The rate of alkaline hydrolysis increased with increasing SIP concentration and increasing PEG molecular weight.

A commercial SPET from Invista (Spartansburg, SC)[2] containing 2.29 mol% SIP (IEC = 0.11 meq $g^{-1}$) was used as a compatibilizer for blends of PET [20] and

2 The author could not find such a product on the Invista website, so it may no longer be available from Invista.

nylon 6 and PET with an aromatic polyamide [21]. Biaxial orientation of those compatibilized blends produced high aspect ratio, two-dimensional polyamide platelets that reduced oxygen permeability of the blend by a factor three. Barber et al. [22] used SPETs from Eastman Chemical Co. with sulfonate concentrations of 1.8–5.8 mol% (IEC = 0.093–0.29) to prepare nanocomposites with two different organically-modified montmorillonite clays (OMC) by melt compounding a mixture of the ionomer and 5 wt.% OMC in a twin-screw extruder. The SPET greatly improved the intercalation of the OMC compared to what was achieved with unmodified PET and significant exfoliation of the OMC was also reported, as evidenced by X-ray diffraction, transmission electron microscopy and tensile property measurements.

A commercial SPET, Gerol PS50 (Kemira Chemical, Belgium) was also used to compatibilize nanocomposites of an organoclay (Cloisite 20A, Southern Clay Products) and PET [23]. The addition of the SPET compatibilizer improved intercalation of the clay and produced partial exfoliation.

Paranhos et al. [24] studied the microstructure of hydrogels formed from a blend of polyvinyl alcohol (PVA) and SPET and nanocomposites prepared from mixtures of the polymer blend with a natural montmorillonite clay. The objective in that research was to control the crystallinity of the PVA. The addition of SPET to PVA reduced $T_g$, increased the chain mobility of the PVA, decreased the size of the PVA crystallites, and increased the crystallinity. The addition of the montmorillonite clay decreased the polymer chain mobility, increased the size of the PVA crystallites, and increased free volume.

Another commercially important phthalate-based polyester is poly(butylene terephthalate), PBT, and sulfonated PBT ionomers (SPBT) have been synthesized by the reaction of 1,4-butanediol (BD) with mixed monomers of DMT and DMSI [25–28], Scheme 4.2. Gorda and Peiffer [25] characterized the crystalline morphology and tensile properties of SPBTs with 1.8–13.5 mol% SIP (IEC = 0.081–0.54 meq $g^{-1}$). As with other semi-crystalline ionomers, the ionic species was excluded from the crystalline phase and resided in the amorphous phase. Increasing the sulfonate concentration lowered $T_m$ and the degree

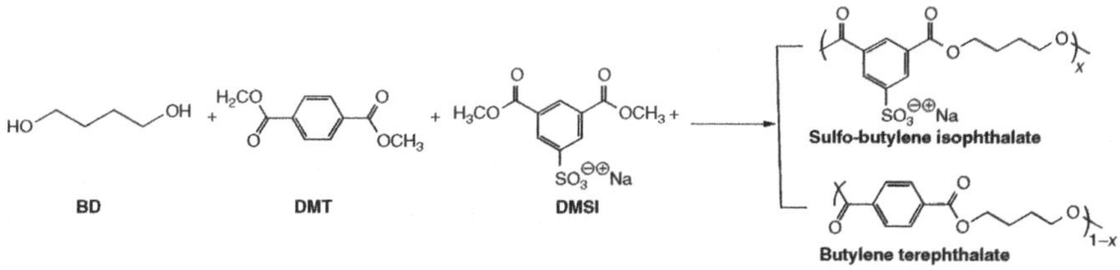


**Scheme 4.2** Synthesis of sulfonated poly(butylene terephthalate) ionomer (SPBT). The polymerization byproducts are not shown [25–28].

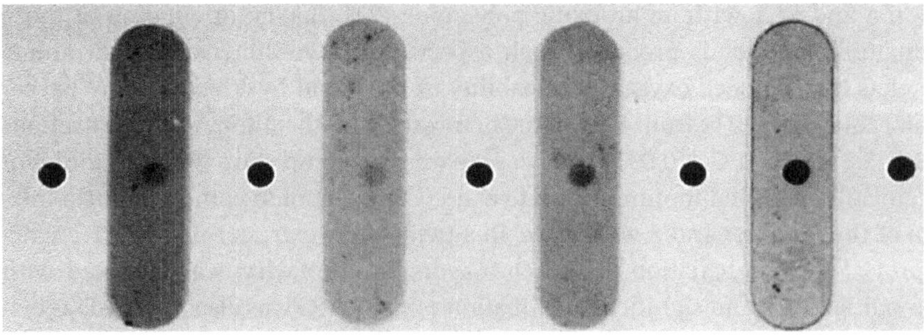


**Figure 4.2** Increasing clarity of SPT ionomer with increasing concentration of sulfonate groups (numbers represent the IEC in meq/g. All the samples were compression molded and had comparable thickness, 36 μm. *Source:* Reproduced with permission from Ref. [25], John Wiley & Sons.

of crystallinity. The SPBT ionomers were optically clearer than PBT or copolymers of butylene terephthalate and butylene isophthalate, due to the lower crystallinity in the ionomers, Figure 4.2. The semi-crystalline ionomers had a spherulitic morphology, and although the spherulite radius did not change with sulfonate concentration, the sulfonate groups did perturb the internal structure of the spherulites.

For low SIP concentrations, <5 mol% (IEC < 0.22 meq $g^{-1}$), an *unusual* [29] spherulite structure was observed, where the optic axis was oriented at 45° to the spherulite radius. At higher sulfonate concentrations, a *normal* spherulite structure was observed, where the optic axis was oriented parallel to the spherulite radius. The SPBT containing 13.5 mol% SIP (IEC = 0.54 meq $g^{-1}$) was completely amorphous and optically clear. The direct effect of the sulfonate group associations on the mechanical properties was complicated by the fact that the sulfonate concentration and the degree of crystallinity were coupled – i.e., crystallinity decreased with increasing sulfonation. However, the authors concluded that the sulfonate groups tended to improve the tensile properties of the material.

Nicholson et al. [26] reported that blends of SPBT and bisphenol A polycarbonate (PC) were partially miscible. Using melting point depression data and the theory of Nishi and Wang [30], those authors calculated a polymer-polymer interaction parameter of $\chi_{1,2} = -0.04$. A negative value of $\chi_{1,2}$ generally promotes miscibility. Light microscopy studies indicated that the maximum miscibility between PC and the amorphous phase of the ionomer occurred when the sulfonation level was between 7.9 and 11.1 mol% (IEC = 0.35 and 0.48, respectively). In a subsequent neutron scattering study of SPBT and PC blends using deuterated PC, this

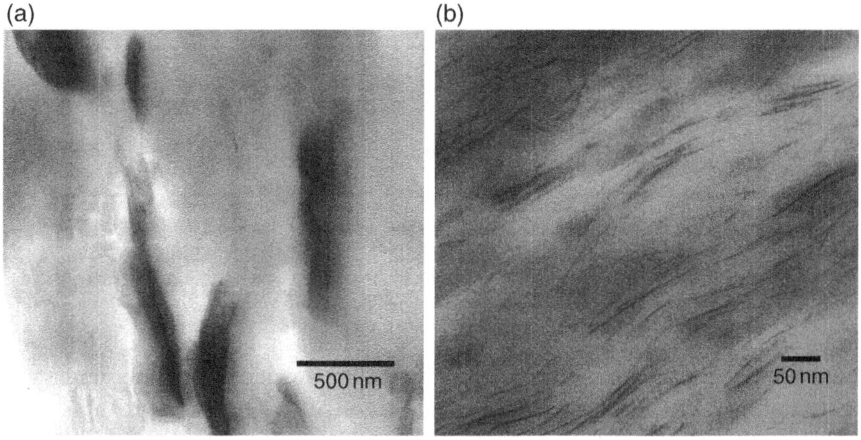


**Figure 4.3** Transmission electron micrograph of SPBT-0.22 nanocomposites with (a) 5 wt.% sodium montmorillonite clay and (b) 5 wt.% OMC compounded with a twin-screw extruder. The nanocomposite in (b) is highly exfoliated. *Source:* Reproduced with permission from Ref. [31], American Chemical Society.

same research group [27] determined that $\chi_{1,2}$ was positive (~ 0.01–0.05, depending on the sulfonation level of the ionomer and the blend composition) and that the optimum ionomer and blend composition for miscibility was a 50/50 (w/w) blend of PC with SPBT ionomer with 7.9 mol% SIP (IEC = 0.35 meq $g^{-1}$).

Chisholm et al. [31] studied the effects of sulfonate functionality on the properties of PBT/sodium montmorillonite and OMC nanocomposites. The use of OMC and SPBT with 1.0–5.0 mol% sulfonate groups (IEC = 0.045–0.22 meq $g^{-1}$) produced highly exfoliated nanocomposites, Figure 4.3. Although the nanocomposite morphology did not change significantly with sulfonate concentration, increasing the sulfonate concentration increased the modulus at room temperature and at high temperature.

Sulfonated poly(butylene isophthalate) ionomers, SPBI, with ~1.96–16.6 mol% SPI, based on the feed composition, (IEC = 0.088–0.71 meq $g^{-1}$) were synthesized from the polycondensation of dimethyl isophthalate (DMI), DMSI and BD [32], Scheme 4.3. Increasing the feed concentration of DMSI led to a decrease in molecular weight and an increase in the ratio of carboxylic acid to hydroxyl end-groups. The number-average molecular weight, $M_n$, of their ionomers ranged from 7000–20 000 g $mol^{-1}$. The ionomers became more soluble in polar solvents as the ionic group concentration increased, and the SPBI ionomer with 16.6 mol% SPI (IEC = 0.71 meq $g^{-1}$), was completely amorphous and water-dispersible. Like the SPBT ionomers, $T_g$ increased, and $T_m$ and crystallinity decreased with increasing SIP concentration for the SPBI ionomers. Thermogravimetric analysis (TGA)

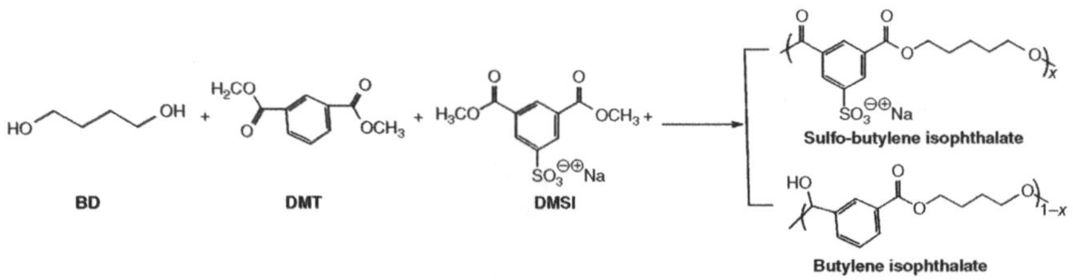


**Scheme 4.3** Synthesis of sulfonated poly(butylene isophthalate) ionomer (SPBI). The polymerization byproducts are not shown. [32, 33].

indicated that the sulfonic acid form of the SPBI was unstable at high temperatures, but the thermal stability of the sodium sulfonate derivative was comparable to that of neat PBI. The crystallization rate of the SPBI ionomers was suppressed by the sulfonate groups [33], because of a decrease in the molecular mobility due to the physical crosslinks from the intermolecular associations of the sulfonate groups. Associations of the ionic groups increased the melt viscosity of SPBI ionomers and the flow activation energy increased linearly with sodium sulfonate concentration [34]. Also, time-temperature superposition failed for the rheological properties for these ionomers.

Chen et al. [35] synthesized a modified sulfonated PET ionomer, MCDP, using an esterification reaction of TA with two diols, EG and 2-methyl-1,3-propanediol (MPD), followed by a polycondensation reaction with sodium-5-sulfo-bis(hydroxyethyl)-isophthalate (SBHEI), Scheme 4.4. The use of the MPD reduced the melt viscosity and the crystallinity of the ionomer compared to SPET. Blends of MCDP and cellulose acetate butyrate (CAB) were melt-spun into fibers, and the dyeability and mechanical properties of the fibers were evaluated. With regard to the nomenclature used to describe the ionomer synthesis, note that those authors refer to the SBHEI monomer as "sodium-5-sulfo-isophthalate" and use the

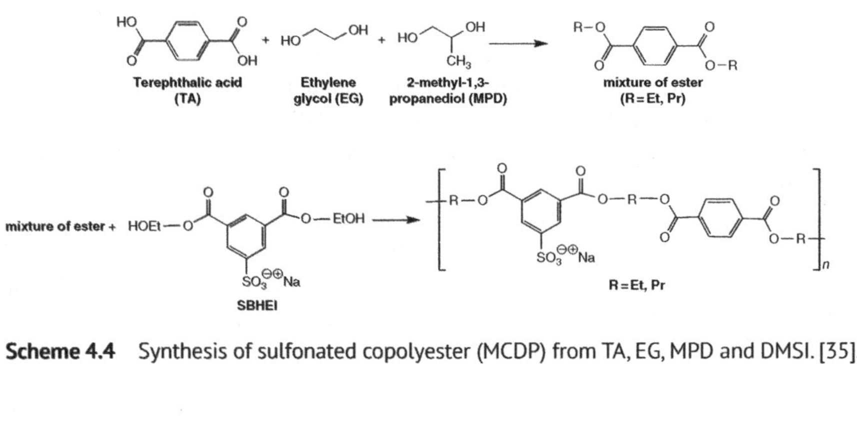


**Scheme 4.4** Synthesis of sulfonated copolyester (MCDP) from TA, EG, MPD and DMSI. [35].

nomenclature “SIP” for that monomer. In this review, SIP was used to denote the sodium-5-sulfoisophthaloyl residue in the sulfopolyesters.

Wang et al. [36] studied the effects of pH and temperature on the adsorption of a cationic dye, C. I. Basic Red, by the MCDP ionomer. The rate, diffusion constant and equilibrium adsorption increased with increasing temperature. The relatively high activation energy led the authors to conclude that this was a chemisorption process, and a Langmuir adsorption model adequately described the adsorption kinetics.

Fu and Gu [37] synthesized sulfonated similar polyester ionomers as shown in Scheme 4.4, except that they used a diol mixture of EG and one other diol (either 1,3-propanediol (PD), 2-methyl-1,3-propanediol (MPD), or 2,2-dimethyl-1,3-propanediol (NPG)) in the first reaction. For all the different ionomers synthesized using combinations of EG with PD, MPD or NPG, the composition of the ionomer was about 47 mol% EG, 3.1 mol% diol #2, 0.73 mol% SIP and 49 mol% TA, which corresponds to an IEC of ~0.077 meq $g^{-1}$. Only the PET segments crystallized and as the second diol was changed from PD, MPD to NPG, the ionomer produced successively more irregular chain structures that depressed crystallization. The ionomers that contained PD or MPD exhibited good spinning behavior, while the ionomer with NPG had poor spinnability. The tensile modulus of the ionomers decreased by about 30% from that of PET and the tensile breaking strength decreased about 15%, but the tensile elongation improved by 40-50%. The mechanical properties were considered to be sufficient for textile applications, and the ability to dye the fibers with cationic dyes improved with the incorporation of the PD, MPD and NCDP diols.

Wang et al. [38] also prepared sulfopolyesters from TA, EG, SBHEI and a polyol(s). When the SIP concentration was 3 mol% and the polyol concentration was 0.8 mol%, the surfaces of copolyester films transformed from hydrophobic to hydrophilic.

Subramani et al. [39] synthesized a polyester ionomer from a combination of a non-ionic phthalate (DMT) and an ionic phthalate (DMSI) with three different glycols, diethylene glycol (DEG), propylene glycol (1,2-propanediol, PG) and dipropylene glycol (DPG), to form polyester microparticles (1–8 μm), Scheme 4.5. The concentrations shown in Scheme 4.5 were calculated using the feed composition reported in the paper. However, a large excess of glycols was used in the synthesis (mole ratio of hydroxyl/carboxylate in the feed was >2), so it is not clear from the paper what was the actual copolymer composition. Nikolić et al. [40] synthesized a similar copolymer of DMT and DMSI, but with only two glycols, EG (85%) and PG (15%), Scheme 4.6. Their work mainly concerned simulations of the self-assembly of the ionomer into micellar particles in aqueous solution.

Baik and Kim [41] synthesized sulfonate-containing copolyester ionomers from DMT, DMI, DMSI, EG and DEG for application as low temperature sizing agents

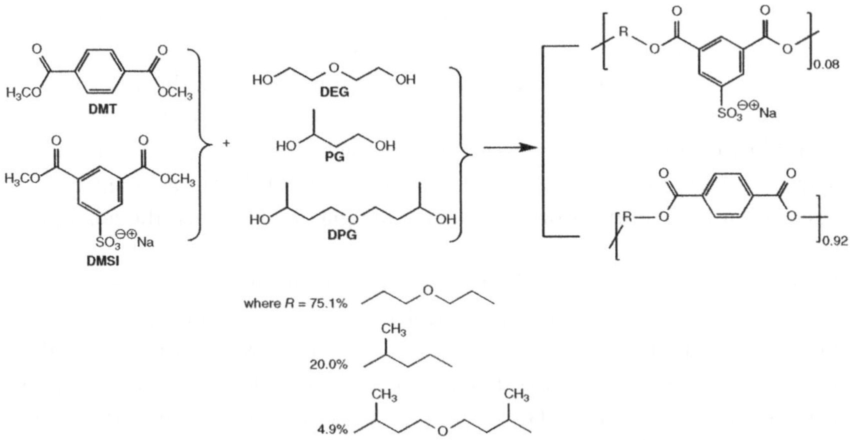

**Scheme 4.5** Synthesis of sulfonated polyester copolymer from the reaction of DMT, DMSI, DEG, PG and DPG. The copolymer composition was calculated from the feed composition, but a 130% excess of glycol was used in the polymerization (see text). [39].

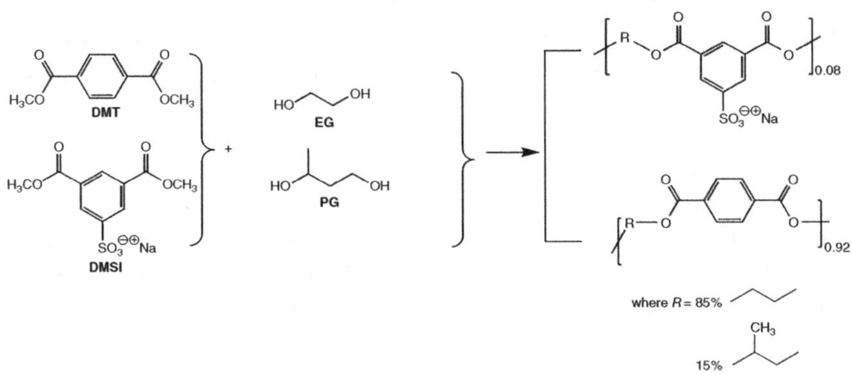

**Scheme 4.6** Synthesis of sulfonated polyester copolymer from the reaction of DMT, DMSI, EG and PG. The copolymer composition was calculated from the feed composition. [40].

for weaving. Those ionomers were dyeable with cationic dyes, water-dispersible, had low viscosity and good adhesion properties.

Phosphonated PET ionomers prepared from melt copolymerization of sodium-dimethyl-5-phosphoisophthalate (DMPI) with DMT and EG, and ethylene glycol, Scheme 4.7, were reported by Ju et al. [42]. The introduction of 0.5–4.0 mol% disodium phosphonate (IEC = 0.052–0.41 meq $g^{-1}$ – assuming that each phosphonate groups represents two equivalents) decreased $T_m$, the crystallinity and the crystallization rate. The tensile modulus, yield stress and melt viscosity also increased relative to pure PET. The phosphonated PET ionomers worked well as compatibilizers

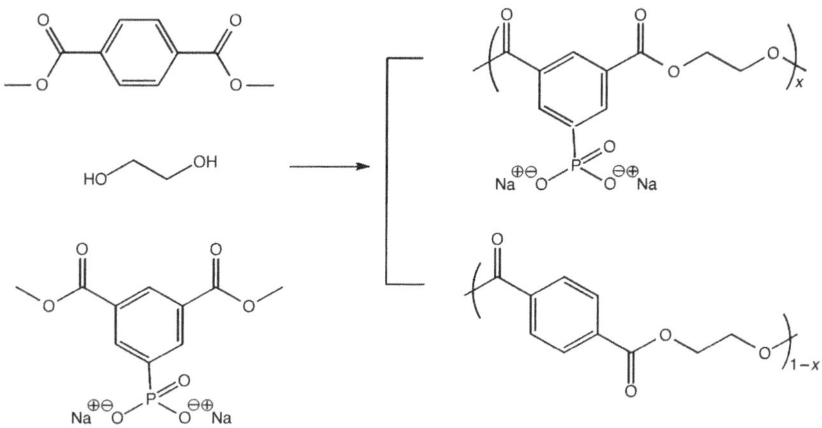

**Scheme 4.7** Synthesis of random copolymer of ethylene terephthalate and Na-phosphonated ethylene isophthalate, where $x = 0.005-0.04$). Condensation by-products are not shown. [42].

for blends of an amorphous polyester and a semicrystalline polyamide [43]. Compatibilization was a consequence of the metal phosphonate groups physically bonding to the amide groups of the polyamide and the PET segments of the ionomer mixing with the polyester. A comparison of the compatibilization ability of the phosphonated PET ionomer with that of a sulfonated PET ionomer indicated that the disodium-phosphonate groups were three times more effective at compatibilization than Na-sulfonate groups.

Eastman Kodak (now Eastman Chemical Co.) developed slow crystallizing variants of PET by introducing a non-crystallizable monomer, 1,4-cyclohexanedimethanol (CHDM), which is made by first reducing dimethyl terephthalate (DMT) to dimethyl-1,4-cyclohexanedicarboxylate (DMCD) and then reducing DMCD to obtain a 70/30 trans/cis mixture of CHDM [44]. Commercial PETG amorphous polyesters are produced by Eastman Chemical Co. using polycondensation reactions of DMT with mixed glycols of EG and CHDM, with less than 50% CHDM) [45], Scheme 4.8.

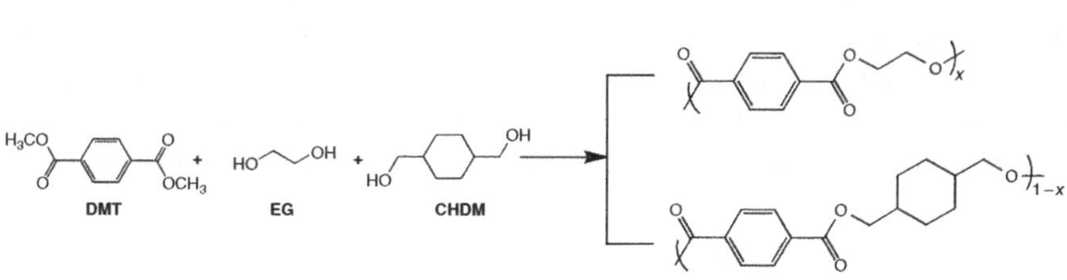


**Scheme 4.8** Eastman PETG amorphous copolyester, where $x > 0.5$. [45].

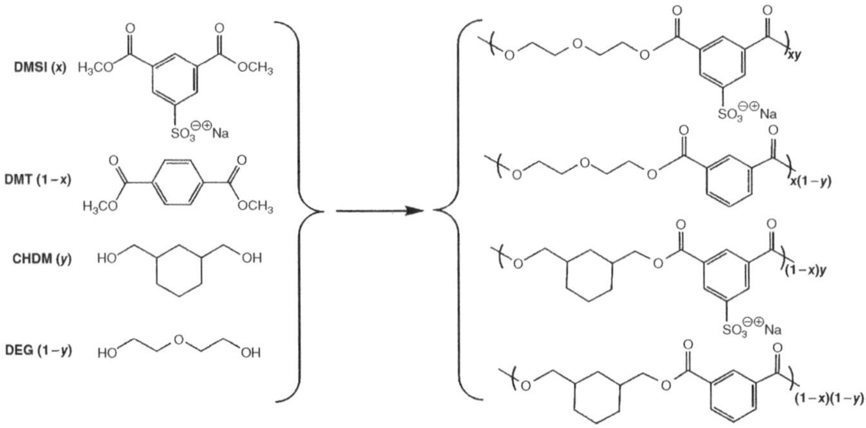


**Scheme 4.9** General structure of AQ55 a commercial amorphous polyester from Eastman Chemical, as reported in Refs. [46, 48] showing all the possible repeat units. x and y denote the mole fraction of DMSI (based on total diacids) and mole fraction of DEG (based on total diols), respectively.

Commercial AQ sulfopolyester ionomers, sold by Eastman Chemical are reported [46, 47] to be copolymers of DMI, DMSI, DEG and CHDM, Scheme 4.9, where $x$ is the mole fraction of DMI (based on the total diacid feed) and y is the mole fraction of DEG (based on the total diol feed). Unfortunately, neither the actual compositions of these ionomers (i.e., the values of $x$ and $y$), nor the equivalent weights or ion-exchange capacity are published. Moore and coworkers [48], however, reported that the product AQ55 has an equivalent molecular weight (EW) of 1500 g $mol^{-1}$, which corresponds to an IEC = 0.67 meq $g^{-1}$.[3] The AQ ionomers are water-dispersible and form colloidal dispersions in water [47] that are used commercially in a variety of applications, including inks, adhesives, paints, coatings, and personal care products [47].

The AQ55 ionomer (Eastman Chemical Co.) has been shown to be an effective compatibilizer for nylon 66 blends with PET [48, 49], poly(styrene-co-4-vinyl pyridine) [50] and PBT [51]. Grady et al. [46] compared the sulfonate microstructure in cadmium and nickel neutralized AQ55 ionomers with sulfonated polystyrene and ethylene-methacrylic acid copolymer ionomers using extended X-ray absorption fine-structure (EXAFS) measurements. The first shell around the sulfur groups was similar in all the Cd-ionomers, but there were differences in the microstructures of the different ionomers neutralized to the Ni-salt. Boykin and Moore [52] studied the effect of transesterification reactions on the properties of blends of PET

3 Equivalent molecular weight (EW) is defined as the mass of one equivalent. Therefore, IEC = 1/(EW).

and AQ55. In that case, the transesterification reactions essentially create functionalized copolyester compatibilizers *in situ*. The focus of that work was the effect of different cations used for the AQ55 ionomer on the transesterification kinetics.

Ammala et al. [53] used the AQ55 ionomer to disperse an OMC or a synthetic fluoromica clay in an aqueous dispersion. They then coated the dispersion onto PET, dried it and prepared a PET/clay nanocomposites by melt extrusion. The use of the sulfonate polyester ionomers improved exfoliation of the clay particles, which was demonstrated by the clarity of the composites, as well as transmission electron microscopy and X-ray diffraction data. The compatibilized nanocomposite also had a higher melt viscosity than similar PET/clay compositions prepared without the sulfopolyester.

Tudryn et al. [54] studied the counterion dynamics and linear viscoelastic behavior of a polyester ionomer with the structure shown in Scheme 4.10 using dielectric spectroscopy and oscillatory shear rheology. The authors reported that the ionomer was obtained from Sigma-Aldrich, but no details of the synthesis were reported, and the material is not currently listed on the Sigma Aldrich website. Based on the monomers that would produce the repeat units shown in Scheme 4.10, the ionomer was probably AQ55, though the copolymer composition was not clear. The mole fractions shown for the terephthaloyl (0.67) and SIP (0.33) most likely represent the mole fractions of DMT and DMSI normalized for the total concentration of diacid ester monomers in the feed or copolymer, but the origin of those values was not provided in the paper. Similarly, the corresponding values for the CHDM and DEG residues – that is, mole fraction of diols in the feed or copolymer, were also not provided. The sodium cations in this ionomer were exchanged for various ammonium-based ionic liquid counterions, which produced as much as a $10^4$ increase in conductivity compared to the ionomer with sodium counterions, which was attributed to the weaker interactions between the ionic liquid sulfonate groups. Large, sterically hindered ionic liquid counterions suppressed the formation of physical crosslinks, which depressed $T_g$ and significantly decreased the terminal relaxation time of the ionomer.

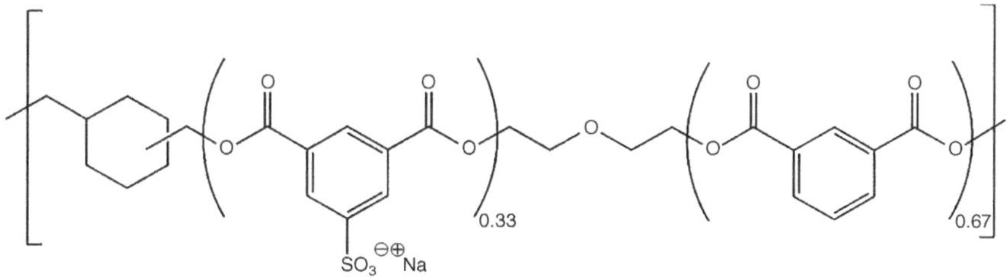


**Scheme 4.10** Polyester ionomer structure reported in Ref. [54].

Yeom and Pourdeyhimi [55] developed unique shaped, fibers and nonwovens using a bicomponent spunbond method consisting of a sacrificial sulfopolyester sheath and a shaped core. The fibers were touted as having potential for filtration media and scaffolds for tissue engineering. The authors described the sulfopolyester as an "EastOne" product of Eastman Chemical Co, but no such product was found by an internet search or on the Eastman Chemical website. However, as noted in the discussion above, Eastman does sell water-dispersible sulfopolyesters products under the AQ-ionomers tradename.

Noonan and McConkey [56, 57] developed water-soluble polyester ionomer adhesives, Scheme 4.11. The P1 ionomer was prepared by the melt condensation of 1,4-bis(2-hydroxyethoxy) cyclohexane diol (BHEC) (50 mol% of total monomers) with dimethyl succinate (35 mol%) and the sodium salt of dimethyl-3,3′-bis(imino)-disulfonyl dibenzoate, DMISDBNa, (15 mol%). P1 produced an excellent adhesive, but when the BHEC was replaced with either CHDM or EG, P2 and P3, respectively, the ionomers had poorer water solubility and adhesive

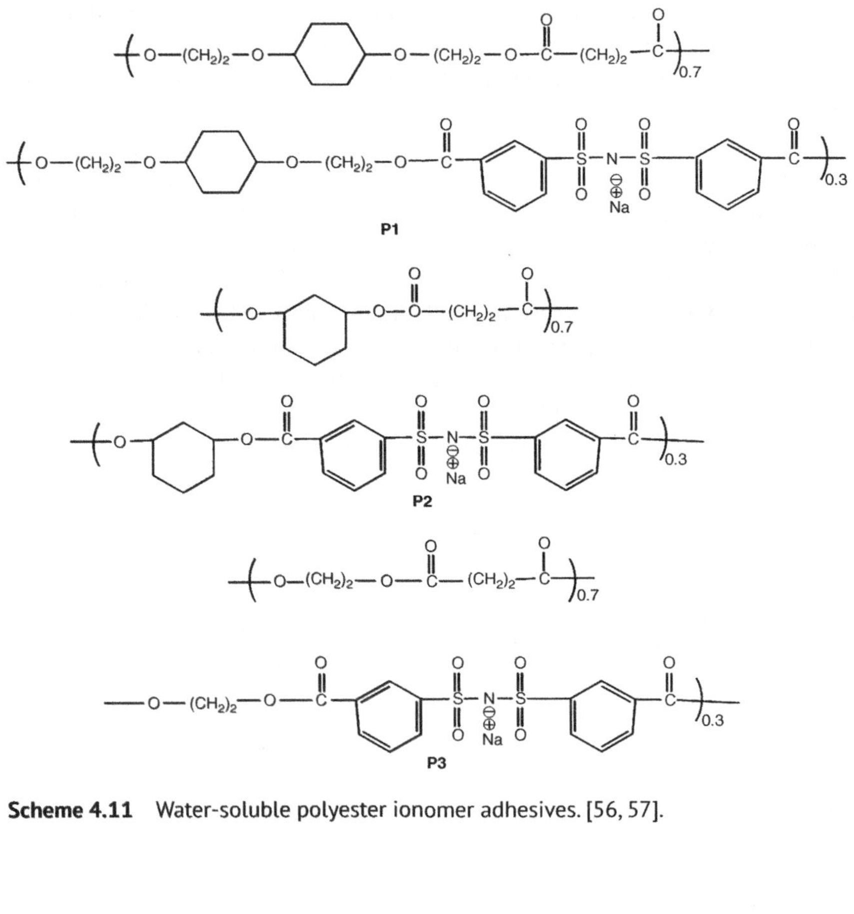


**Scheme 4.11** Water-soluble polyester ionomer adhesives. [56, 57].

strength. When the BHEC was replaced with either DEG or tetraethylene glycol, the resulting polymers were readily soluble in water and the adhesive strength was similar to that of P1. When hexamethylene glycol was used, the polyester ionomer was not water soluble, but if 25 mol% of the BHEC was replaced with any of the other three diols, e.g., with hexamethylene glycol, Scheme 4.12, the adhesive strength of the resulting polyester ionomer, P4, was very high.

The adhesive strengths of the polymers shown in Schemes 4.11 and 4.12 were relatively low at temperatures greater than 100 °C, but that problem was resolved by crosslinking the ionomer. That required the introduction of a photo-crosslinkable monomer, diester diethyl *p*-phenylenebis(acrylate), Scheme 4.13.

Connelly et al. [58] studied the linear viscoelastic behavior and the finite strain, steady shear and uniaxial extensional rheology of polyester ionomers prepared by

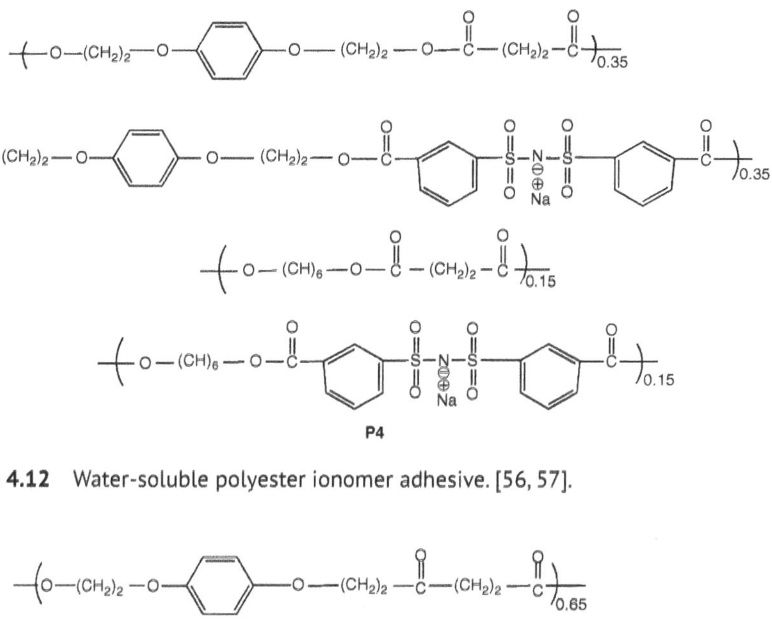

P4

**Scheme 4.12** Water-soluble polyester ionomer adhesive. [56, 57].

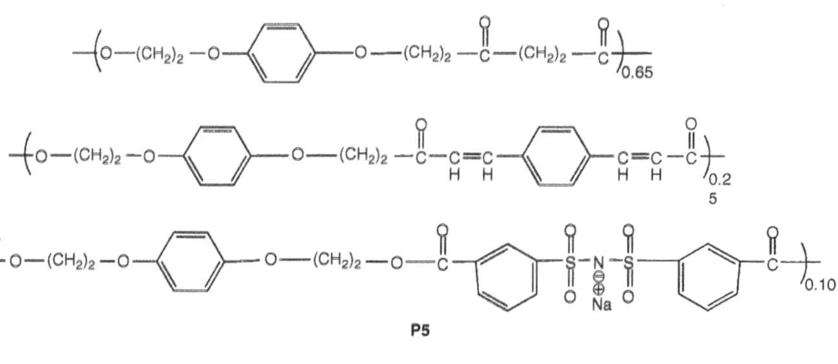

P5

**Scheme 4.13** UV-crosslinkable polyester ionomer adhesive. [56, 57].

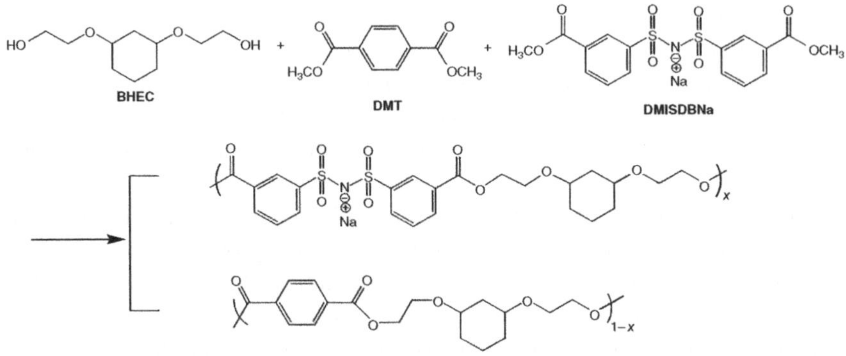


**Scheme 4.14** Polyester ionomer prepared from polycondensation of DMT, BHEC and DMISDBNa. $x$ = 0,0.01, 0.03, 0.05 and 0.07. [58].

the polycondensation of DMT, DMISDBNa) and BHEC containing 1–7 mol% DMISDBNa (IEC = 0–0.13 meq g$^{-1}$, respectively), Scheme 4.14. The four ionomers and the neutral polyester (i.e., when $x = 0$ in Scheme 4.14) had nearly the same intrinsic viscosities ($[\eta] \approx 0.5\,\text{dL g}^{-1}$) measured in a 50:50 mixture of phenol and chlorobenzene. The $T_g$ of the ionomer increased with increasing IEC. Similarly, As IEC increased, the zero-shear viscosity ($\eta_o$) and the terminal relaxation time ($\tau$) of the melt increased when the viscoelastic measurements were made at temperatures corresponding to the same difference between the experimental temperature and $T_{g,}$ i.e., ($\Delta T = T_{exp}{-}T_g$). A form of the BKZ single integral constitutive model [59] modified by Kaye and Kennett, [60] was fit to the experimental shear and extensional rheology data. Although the model produced reasonable predictions for the shear rate dependence of viscosity ($\eta\,(\dot{\gamma})$) and the first normal stress difference ($N_1(\dot{\gamma})$) response, the fits to the extensional flow experiments were less satisfactory. The modified BKZ model fit the data for the neutral polyester, but the predictions of the model for the ionomers became increasingly poorer with increasing IEC. For large strain, steady flow experiments, the ionomers became less strain dependent with increasing IEC.

Ferrar et al. [61] incorporated 1,4,5,8-naphthalene-tetracarbonyl-bis(5-hydroxy pentyl)-imide (NB5) groups into a polyester ionomer, Scheme 4.15, for use as hole-blocking layers in negative-charging multilayer photoreceptors. The very high concentration of SIP used in the synthesis shown in Scheme 4.15 produced a very high melt viscosity that complicated the synthesis. The high melt viscosity was resolved by adding sebacic acid to the feed, Scheme 4.15, but that ionomer was not as stable in aqueous solution as the alicyclic-containing ionomer (Scheme 4.15).

Poly(butylene succinate) ionomers were synthesized using sulfonated dimethyl fumarate, SDMF as the ionic monomer [62, 63], Scheme 4.16. SDMF was prepared

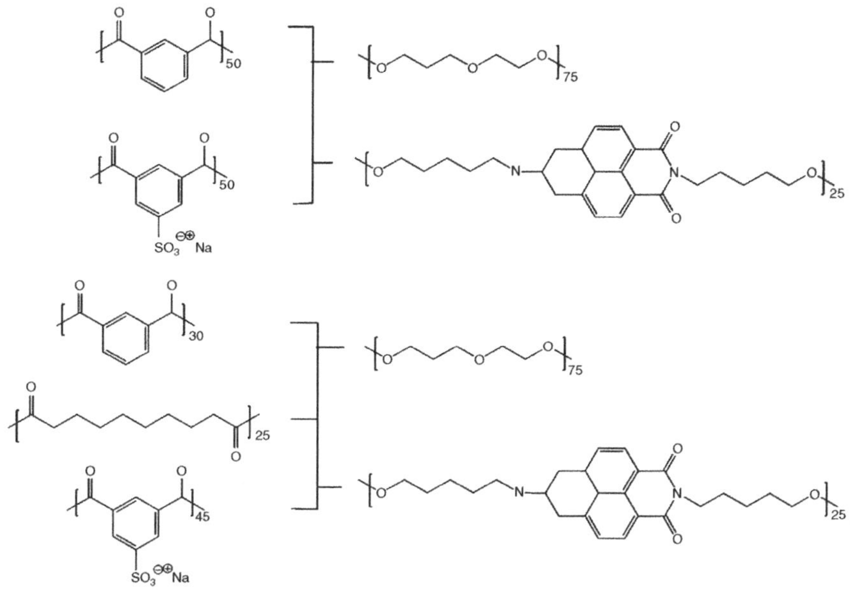

**Scheme 4.15** Two examples of polyester ionomers containing NB5 groups. The subscripts denote the molar percentage of each species normalized for the total dicacid(ester) monomer on the left and diol and the right. [61].

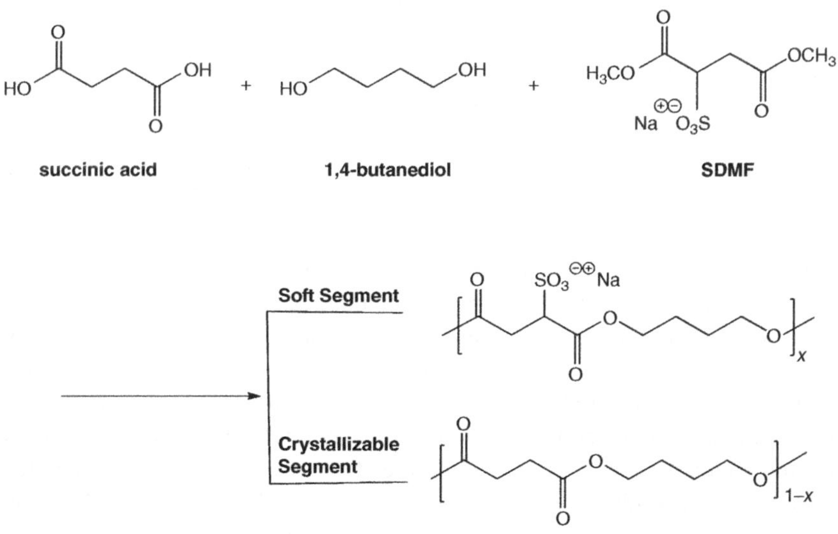


**Scheme 4.16** Poly(butylene succinate) sulfonate ionomer using SDMF. [62, 63].

by the reaction of sodium bisulfite and dimethyl fumerate. The ionomers covered a range of sulfonate concentration from 0.30 to 3.0 mol% (IEC = 0.017–0.17 meq $g^{-1}$, respectively). The 0.30 mol% ionomer (IEC = 0.017 meq $g^{-1}$) had a dynamic melt viscosity, measured at low frequency and 150 °C, of nearly an order of magnitude greater than the 1.2 mol% (IEC = 0.069 meq $g^{-1}$) ionomer, which had a viscosity that was greater by half an order of magnitude than a non-ionic poly(butylene succinate) [62]. The authors' explanation for why the viscosity of the 1.2 mol% ionomer was so much lower than for the 0.3 mol% ionomer involved differences in intermolecular and intramolecular associations of the ionic groups in the two ionomers. That explanation does not seem credible, and a more likely reason why the higher sulfonate-containing ionomer had the lower viscosity is that it also had the higher concentration of butanediol, that makes the chain more flexible, which should also lower the melt viscosity. There is no evidence in the literature that shows that intramolecular interactions of the ionic group dominate the viscosity behavior for dense ionomers with very low ionic content. In fact, just the opposite is true. For sulfonated polystyrene ionomers, even at an ionic group concentration of 0.11 mol% (IEC = 0.011 meq $g^{-1}$), there is a significant increase in the melt viscosity compared with the parent polystyrene [64].

For the poly(butylene succinate) there are competing influences when the sulfonate concentration increases. Increasing the SDMF concentration should increase the number of supramolecular crosslinks, which increases viscosity. However, increasing the sulfonate groups also increases the concentration of butanediol residues, which tend to plasticize the ionic associations and increase chain flexibility. So, the higher melt viscosity of the 0.30 mol% SDMF ionomer compared with the 1.2 mol% ionomer may be a consequence of the butanediol concentration dominating the dynamics at low SDMF concentration. Unfortunately, the authors did not report the rheology of ionomers with higher SDMF concentrations to see if the viscosity trend changed at higher sulfonate levels The $T_g$ of those ionomers decreased with increasing SDMF concentration, which is also contrary to the behavior for most ionomers. That trend, however, is consistent with the plasticizing influence of the butanediol residues with increasing concentration of SDMF. The intensity of the tan δ peak increased with increasing SDMF concentration, which the authors concluded was due to an increased volume concentration of ionic aggregates in the ionomer. The crystallization rate of the ionomers was faster than for the non-ionic poly(butylene succinate), but it decreased with increasing SDMF concentration. The authors concluded that the non-aggregated ion-pairs behaved as nucleating sites, which were responsible for producing smaller crystallites in the ionomer than in poly(butylene succinate). The decreasing crystallization rate for the ionomers with increasing SDMF concentration was attributed to a decrease in the concentration of non-aggregated ion-pairs that corresponded to an increase in the concentration of ionic aggregates.

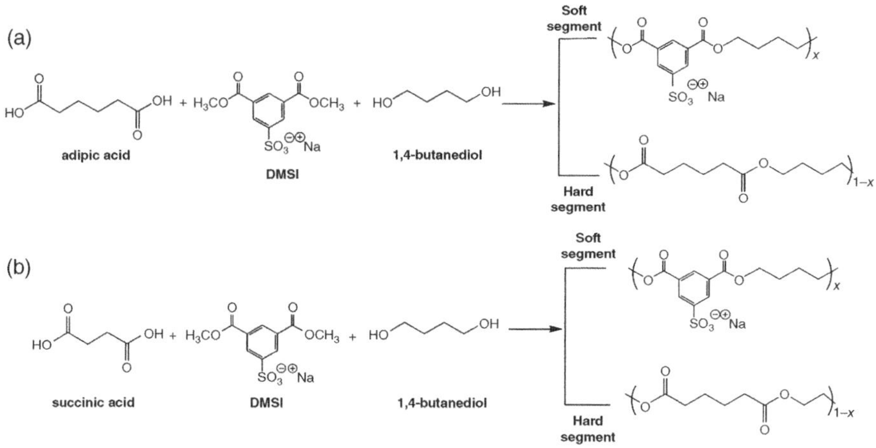


**Scheme 4.17** Synthesis of segmented ionomers of (a) DMSI-adipate and butylene adipate and (b) DMSI-succinate and butylene succinate [65–67].

Im and coworkers [65–67] prepared poly(butylene adipate), PBA, and poly(butylene succinate), PBS, ionomers based on segmented copolymers of DMSI-adipate and butylene adipate [65, 67] and DMSI-succinate and butylene succinate [66, 67], Scheme 4.17. The PBA ionomers [65, 67] had SIP concentrations of 1–5 mol% (IEC = 0.050–0.24 meq $g^{-1}$ for the PBA ionomers). $T_g$ increased and $T_m$ and crystallinity decreased with increasing sulfonate content. The melt viscosity was measured for only one ionomer and it was higher than that of a non-ionic PBA. A microphase separated microstructure of 20–30 nm ionic aggregates was observed by transmission electron microscopy. The room temperature modulus increased and the tensile stress and strain at break decreased with increasing sulfonate concentration. The sulfonate groups nucleated crystallization, which produced much smaller spherulites than observed in the non-ionic PBA and the biodegradation rate of the polymers increased with increasing ionic group content. The ionomers were also biocompatible with human dermal fibroblast cells, which suggested that they had potential for biomedical applications.

The PBS ionomers containing SIP groups [66, 67] also had SIP concentrations of 1–5 mol% (0.058–0.28 meq $g^{-1}$). As with the poly(butylene adipate) ionomers containing SIP groups, $T_g$ increased and $T_m$ and crystallinity decreased with increasing SIP content (mol%), but the modulus, tensile strength and elongation at break did not show any clear correlation with SIP concentration. The PBS ionomers with SIP showed the same anomalous melt viscosity behavior as did the PBS ionomers with SDMF – that is the ionomer with 3 mol% SPI (IEC = 0.17 meq $g^{-1}$) had a higher viscosity than the ionomer with 5 mol% SPI (IEC = 0.28 meq $g^{-1}$), which, again, may be a consequence of the competing influences of the associations of the SPI units and the plasticizing influence of the butanediol units Atomic force microscopy measurements of the surfaces of ionomer samples revealed ionic aggregates with sizes of 40–60 nm. Like the poly(butylene adipate) ionomers containing SIP groups, the ionomers were biocompatible and the sulfonate groups nucleated crystallization and increased the biodegradation rate.

Pekkanen et al. [68] synthesized biocompatible and biodegradable polyester ionomers by transesterification reactions of polycaprolactone diol with DMSI, Scheme 4.18. This reference is an abstract for a talk at a national American Chemical Society meeting, and the work was not otherwise published. The abstract, however, indicated that the ionic groups suppressed the crystallinity of the polycaprolactone and provided physical crosslinks that produced an increase in the tensile and the compression moduli. Biocompatibility was demonstrated by cell attachment to the ionomer and proliferation, and the authors suggested that these PCL-based ionomers may have potential as a biodegradable wound glue.

A hyperbranched polyester ionomer was prepared from a hyperbranched polyester with hydroxyl endgroups obtained from the polymerization of 4,4-bis(4-hydroxyphenyl)valeric acid [69]. The hydroxyl end-groups were then reacted with 1,4-butane sultone in the presence of sodium hydride to produce the hyperbranched polyester with sulfonic acid endgroups. Free-standing membranes were prepared by phase-inversion of the ionomer in a dimethyl formamide solution and assembled into a membrane electrode assembly for testing in a hydrogen fuel cell. The ionomer membranes had poor mechanical properties, but the proton conductivities of the sulfopolyester membranes were comparable to those of Nafion (DuPont). However, at the higher sulfonation levels, the fuel crossover performance was inferior to that of Nafion. Composite membranes composed of titanium dioxide nanoparticles dispersed in the ionomer had poorer membrane performance.

Water dispersions of sulfopolyesters are commercially available as Eastek (Eastman Chemical Co.) and as Hydro-Rez (Lawter, Chicago, IL). Virtanen et al. [70] used Hydro-Rez dispersions to develop printable bioactive paper

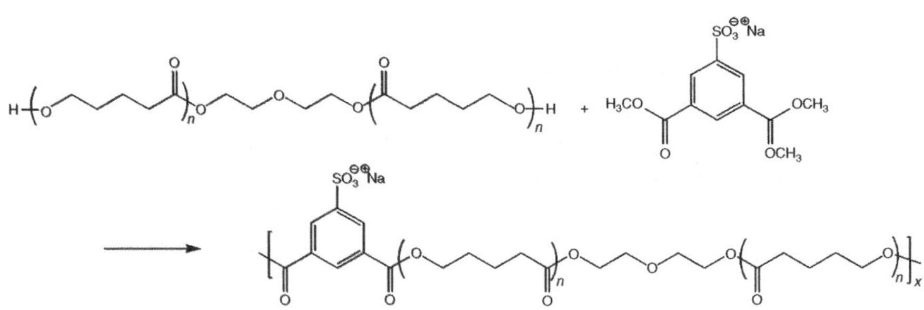


**Scheme 4.18** Synthesis of caprolactone-based ionomers. [68]

containing laccase enzymes that catalyzed oxidation reactions. Al-Ahmed et al. [71] prepared a peroxide biosensor by polymerizing aniline inside the interstitial pores of the sodium salt of an Eastman AQ-55 sulfopolyester ionomer. The nanocomposite biosensor exhibited sensitivity, limit of detection and dynamic linear range characteristics that compared favorably with other peroxide biosensors.

## 4.3 Poly(Ether-Ester) Ionomers

Segmented ionomers based on aromatic polyester hard segments and polyether soft segments, referred to as poly(ether ester) ionomers, are prepared from the polymerization of oligomeric poly(alkylene glycols) and a sulfonate-containing terephthaloyl or isophthaloyl monomer. For example, Scheme 4.19, shows the synthesis of a poly(alkylene oxide-co-sulfoisophthalate) ionomer from the polycondensation of DMSI with a poly(alkylene glycol), PAG-x, where $x$ is the nominal number-average molecular weight (g/mole), $n$ is the degree of polymerization of the alkylene glycol and $p$ is the degree of polymerization of the ionomer.

For the reaction shown in Scheme 4.19, when $x = 2$, the PAG is poly(ethylene glycol) (PEG-x), which is the most common alkylene glycol used in the synthesis of these types of ionomers. Ionomers based on the polymerization of DMSI with PEG, Scheme 4.20, have been synthesized and studied by several research groups [72–81]. In this review, the nomenclature used for these ionomers is M-SIPE-y, where SI and PE refer to the sulfoisophthalate and polyethylene glycol, respectively, M is the metal cation for the sulfonate group and y denotes the sulfonate concentration expressed as ion-exchange capacity (IEC) in milli-equivalents per unit mass (meq/g). Alternatively, sulfo-poly(ether esters) similar to what is shown in Scheme 4.20 can be prepared using sodium bis(2-hydroxyethyl)-5-sulfoisophthalate (SBHEI) instead of DMSI [82]. Note that Ref. [82] referred to the sodium bis(2-hydroxyethyl)-5-sulfoisophthalate monomer as SIPE instead of the SBHEI notation used in this review.

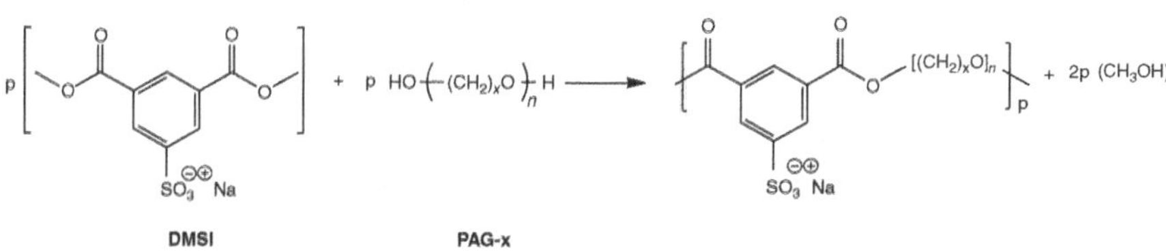


**Scheme 4.19** Example synthesis of a sulfonated poly(ether ester) ionomer from the reaction of DMSI with a PAG-x, where $x$ is the nominal number-averaged molecular weight of the poly(alkylene glycol), $n$ is the degree of polymerization of the alkylene glycol and p is the degree of polymerization of the ionomer.

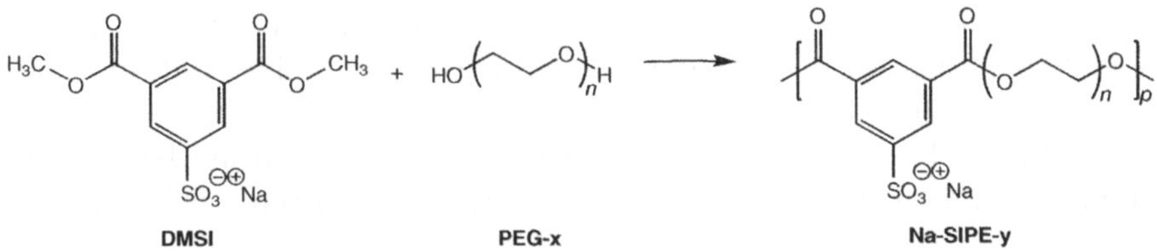


**Scheme 4.20** Synthesis of poly(ether ester) sulfonate ionomer, Na-SIPE-y, from the polycondensation of DMSI and PEG-x, where *x* is the $M_n$ of the PEG, *n* is the degree of polymerization of the PEG-x, *y* is the IEC of the ionomer and *p* is the degree of polymerization of the ionomer. Condensation by-products are not shown [72–81].

The IEC of the M-SIPE-y ionomers can be controlled by varying the degree of polymerization, *n*, of the polyethylene glycol or by diluting the DMSI (or SBHEI) monomer with a non-ionic phthaloyl monomer that produces isophthaloyl or terephthaloyl residues [83–91] Scheme 4.21. The SIPE-y notation is also used herein for the latter ionomers, since the main purpose for the non-ionic monomer was to dilute the SIP concentration.

Dou et al. [72] prepared the Li-, Na- and Cs-salts of three different M-SIPE-y single-ion conductors with PEG-400, PEG-600, PEG-900, which correspond to M-SIPE-1.6, M-SIPE-1.2 and M-SIPE-0.90, respectively. As with most ionomers, $T_g$ increased with increasing IEC, which is a result of the suppressed segmental mobility due to intermolecular association of the ionic species [92]. Unlike most other ionomers, however, none of the M-SIPE-y ionomers exhibited a small-angle X-ray scattering peak, which indicated that those ionomers did not exhibit microphase separation of ion-rich aggregates. The authors attributed that result to the solvating influence of the ether groups, though dielectric spectroscopy analysis suggested that nearly all of the sulfonate groups were paired [73]. Likewise, no relaxation for an ionic phase was observed in rheological measurements. The ionic conductivity increased significantly with decreasing IEC (i.e., increasing PEG spacer length), which the authors attributed to the effect of increasing segmental mobility of the polymer chain from the plasticizing effect of the flexible PEG. An ionic conductivity of $10^{-6}$ S cm$^{-1}$

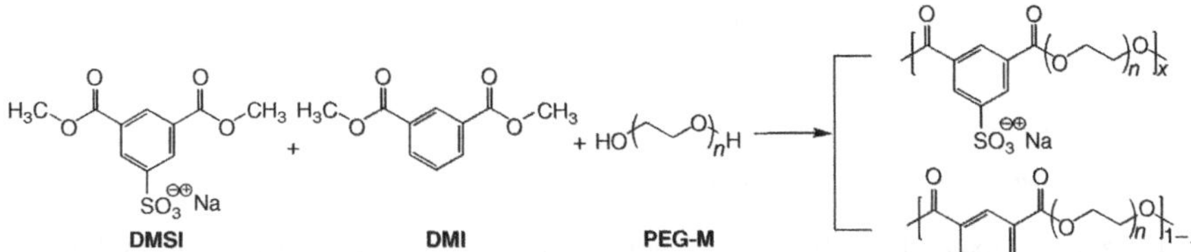


**Scheme 4.21** Synthesis of random copolymer ionomers synthesized from DMSI, DMI and PEG-M, where $x$ = mole fraction of DMSI (based on total amount of dicarboxylate monomer use in the feed composition). Condensation by-products are not shown [83–87, 91].

was achieved at room temperature for the Na-SIPE-0.90, which is comparable to that of other solvent-free single-ion conductors.

Wang et al. [74] measured X-ray diffraction and small angle X-ray scattering (SAXS) of the same M-SIPE-y ionomers used in Refs. [72, 73], but they also included two additional ionomers with lower IECs, 0.76 and 0.28 meq $g^{-1}$. As the PEG spacer length increased (i.e., decreasing IEC), the PEG blocks crystallized into a monoclinic unit cell microstructure identical to that of PEG oligomers. The degree of crystallinity and the crystallization rate of the PEG in the ionomers were lower than for the neat PEG.

As was reported in Ref. [72], no SAXS peak in the region of scattering vector corresponding to a characteristic spacing of 1–6 $nm^{-1}$ was observed for the Na- or Cs-salts of the SIPE ionomers. However, when they analyzed their scattering data by subtracting the normalized scattering of a neutral polymer analog, where the SIP units were replaced with isophthalate, they did observe excess scattering at higher q (smaller characteristic spacing) that they attributed to "closely assembled and strongly interacting ions in various associated states, such as isolated ion pairs, quadrupoles, or larger aggregates". Those authors also reported a SAXS peak for the Li-SIPE-1.2 sample that corresponded to a characteristic distance of 2.3 nm, Figure 4.4, which is similar to the correlation length for separation of ionic aggregates in other ionomers. That correlation persisted

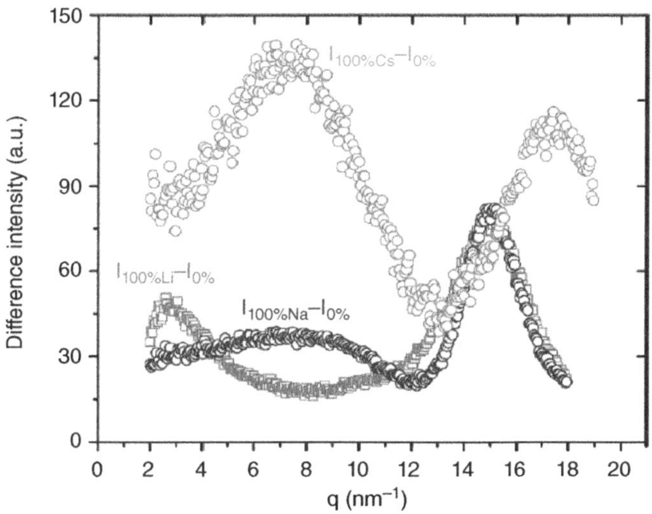


**Figure 4.4** "Difference X-ray scattering" (where the X-ray scattering intensity of PEO-600 was subtracted from the X-ray scattering intensity of the ionomer) at room temperature for M-SIPE-1.2 ionomers, where $M$ = Li, Na or Cs. The scattering vector, $q \equiv 4\pi\sin\theta/\lambda$, where $\lambda$ is the X-ray wavelength and $2\theta$ is the scattering angle. The intensity peak at $q \sim 2.7\,nm^{-1}$ for Li-SIPE-1.2 corresponds to an average spacing between ionic aggregates of ~2.3 nm. *Source:* Wang et al. [74]/with permission from American Chemical Society.

when they subtracted the normalized scattering of the neutral polymer, and the authors concluded that conventional ionic nanodomains did form in the Li-SIPE-1.2 ionomer.

In a separate study by the same research team [75], the authors observed that the microstructure of the Na-SIPE-1.2 and Cs-SIPE-1.2 transformed from small associations of contact ion-pairs to larger scale aggregation upon heating the ionomers to 120 °C, Figure 4.5. They attributed the ion-pair aggregation to a decrease of the dielectric constant of the continuous, non-ionic phase with increasing temperature that correspondingly increased the Coulomb energy of the ion-pair, which presumably drives the phase separation of the ionic groups. The ionic aggregates disappeared when the sample was cooled back to room temperature. Lu et al. [76] analyzed the state of the ionic species in Na-SIPE-1.2 ionomers using FTIR spectroscopy. They detected no free sulfonate ions, but they did resolve FTIR absorbances that they attributed to ion-pairs and aggregates. The absorbance due to ion-aggregates increased in intensity relative to the absorbance due to ion-pairs as the temperature increased, which was consistent with the observations of Wang et al. [75].

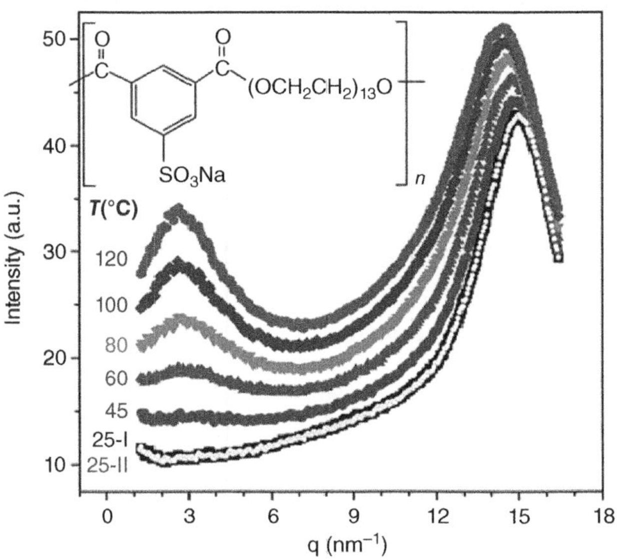


**Figure 4.5** Temperature dependence of the X-ray scattering of Na-SIPE-1.2. The peak that appears at $q \sim 2.8\,nm^{-1}$ with increasing temperature corresponds to the formation of ionic aggregates with an average interdomain spacing of ~2.2 nm. The changes in scattering were thermally reversible. *Source:* Wang et al. [75]/with permission from American Chemical Society.

O'Reilly and Winey [78] compared the effects of adding two types of nanoparticles, bare silica nanoparticles (SNP) and PEG-grafted silica nanoparticles (PEGNP) on the viscosity and ionic conductivity of a Li-SIPE-1.2 ionomer. Better dispersions were achieved using the SNP, but the SNP also increased the viscosity by three orders of magnitude and lowered the ionic conductivity by an order of magnitude. Grafting PEG to the silica nanoparticle suppressed crystallization of the PEG. The addition of PEGNP plasticized the ionomer, lowering $T_g$ and the viscosity. It also decreased the relaxation time for ionic rearrangements and increased the ionic conductivity by as much as an order of magnitude.

Iacob and Runt [79] studied the charge transport properties of Na- SIPE-1.2 and Li-SIPE-1.2 ionomers that were confined in unidirectional silica nanopore membranes with an average pore diameter of 7.5 nm. Nanoconfinement of the ionomer lowered the dc-conductivity and the characteristic dielectric frequency by 1.4 decades. The authors attributed the slower molecular mobility to strong intermolecular interactions between the ionomer and the silica surface of the nanopores, which produced a dynamic adsorption layer at the polymer-silica interface. Silanization of the pore walls suppressed formation of the adsorption layer and increased the dc-conductivity.

Pekkanen et al. [80, 81] synthesized water-soluble M-SIPE-0.062 ionomers (M = Na, K, Ca, Mg and Zn) to use in low temperature (70 °C) extrusion-based additive manufacturing. The choice of the cation had little effect on $T_g$, $T_m$, crystallinity or the tensile properties of the ionomer, but it had a significant effect on the melt viscosity. Ionomers with divalent cations (Ca, Mg and Zn) had higher viscosities than the monovalent Na and K ionomers, and the largest increase was attained for the Ca-SPE-0.062 ionomer. Measurements of the dynamic viscoelastic properties while cooling the melt at a fixed frequency indicated that the monovalent salts did not gel (i.e., create a physically crosslinked network) at the printing temperature, but gelation did occur for the divalent-salt ionomers, as evident from a crossover of G′ and G″ (from $G' < G''$ to $G' > G''$) at ~70 °C. Those data indicated that the intermolecular interactions in the monovalent-cation ionomers were insufficiently strong to produce gelation, but the intermolecular associations in the divalent salt ionomers were sufficient to form a network. The metal sulfonate groups also provided heterogeneous nucleation sites for crystallization. The Ca-salts exhibited the highest nucleation density and smallest spherulites, which produced stronger printed parts with less warpage.

Roach et al. [83] used nuclear magnetic resonance spectroscopy (NMR) to measure the polymer and Li-ion dynamics in Li-SIPE-y ionomers prepared from DMSI, DMI and PE-600 and with IEC = 0.080–1.2 meq $g^{-1}$. Measurements of the temperature dependence of the $^1H$ and $^7Li$ spin-lattice relaxation times ($\tau_1$) indicated that the Li-ion hopping was ten times slower than the segmental motion of the polymer backbone, but the activation energies for both motions were similar and the motions were strongly correlated at low ionic content. The timescale of the motions of the midsection of the PEG spacer was 2–3 orders of magnitude faster than the section of the PEG spacer adjacent to the isophthalate group. Sinha et al. [84] used quasi-elastic neutron scattering (QENS) to estimate the relative concentration of "bridge atoms", which were the PEG groups in the midsection of the spacer, and "anchor atoms" from the PEG ends near the isophthalate groups in similar Li-SIPE-y with IEC = 0.76–1.6 meq $g^{-1}$. The Li-ions associated with the bridge atoms were considered solvated (free) ions, and the authors estimated that the ionomers contained about 8–22% free Li-ions, depending on the length of the PEG spacer that was used. That concentration was considerably higher than the estimates reported from FTIR measurements [76].

Roach et al. [85] also studied the backbone dynamics of Li-SIPE-y and Na-SIPE-y ionomers (IEC = 0.081–1.2 meq $g^{-1}$) over a temperature range of 203–373K using solid state nuclear magnetic resonance with magic angle spinning (MAS-NMR) spectroscopy. For the lower IEC, e.g., M-SIPE-0.081, $T_g$ of the two different salts were similar and about the same as the neutral isophthaloyl-PEG600 copolymer (SIPE-0), $T_g \sim 228$ K. In that case, the polymer backbone was relatively mobile, but as the IEC increased to 1.2, the $T_g$s of the Li-SIPE-1.2 and Na-SIPE-1.2 increased to 253K and 271K, respectively, and the backbone mobility became hindered. For $T_g/T < 0.9$ the segmental motion of the ionomer backbone was similar at all sulfonate concentrations, which indicated a universal behavior where the ionomer motions of the backbone at the microsecond timescale became insensitive to the ionic interactions.

Fragiadakis et al. [86] used dielectric spectroscopy to characterize the segmental and local dynamics and Li-ion transport in the same Li-SIPE-y ionomers used in Refs. [83, 85]. The ionomers were amorphous and exhibited a single $T_g$, with no evidence of ion aggregation. The segmental relaxation of the polymer slowed with SIP concentration above a critical concentration due to an increasing physical crosslink density that decreased the polymer fragility. The Li-ion conductivity was strongly coupled to the segmental relaxation. An additional relaxation process associated with rotation of ion-pairs was also observed, and because of the low dielectric constant and the Li-ion affinity for the ether oxygens of PEG, most of the ions were present as non-associated ion-pairs. However, the ion-pairs rapidly dissociated to release positive triple ions ($Li^+SO_3^-Li^+$), and the authors proposed an ion-hopping mechanism of positive triple ions to explain the observed increase in Li-ion mobility with increasing ion content.

Li-ion self-diffusion and conductivity in Li-SIPE-y ionomers, where the $M_n$ of the PEG spacer was 400–1000 g $mol^{-1}$ and IEC = 0.081–1.6 meq $g^{-1}$, were affected by the ionic state of the Li-ion and the its interactions with the polymer backbone [87]. Ion-pair associations (i.e., physical crosslinks) increased with increasing sulfonate concentration, which hindered Li-ion diffusion. This was especially significant in ionomers with larger-scale ionic aggregation that slowed down Li-ion diffusion. However, when the aggregates overlapped, transfer of Li-ions between ionic aggregates, was facilitated and the diffusion coefficient increased. Only mobile Li-ions and positive triple ions contributed to conductivity. Neutral ion-pairs and Li-ions that comprised the aggregates did not contribute to either Li-diffusion or conductivity. The PEG spacer length had a stronger effect on Li-mobility than the ion concentration, because of the strong interactions between the Li-ions and the ether oxygens of the PEG.

FTIR and $^{1}H$ and $^{7}Li$ NMR diffusion measurements indicated that Li-diffusion was much faster than would be expected from the ionic conductivity data, which led to the conclusion that neutral ion-pairs dominated lithium transport in the Li-SIPE-y ionomers [88]. Segmental motions of the polymer backbone promoted hopping of neutral ion-pairs from one aggregate to another, but only hopping of positive triple ions contributed to the ionic conductivity.

Chen et al. [89] measured the viscoelastic behavior of M-SIPE-y ionomers ($x = 0–1.6\,meq\,g^{-1}$) and M-SIPT-y ionomers ($x = 0–1.2\,meq\,g^{-1}$) that were derived from DMSI, PTMG and, optionally, dimethyl isophthalate (DMI). A "sticky Rouse model" satisfactorily represented the LVE behavior if the $\alpha_2$-relaxation time measured from dielectric relaxation spectroscopy was used as the lifetime of the ionic associations, Figure 4.6. For M-SIPE-y ionomers, the relatively weak associations of the sulfonate groups in a PEG environment (~10 kJ mol$^{-1}$) produced only a short delay of the terminal relaxation beyond that of a neutral polymer analog, but when the weaker solvating PTMG was used in the ionomer formulation, the association energy increased to 58 kJ mol$^{-1}$, which significantly suppressed the terminal relaxation and produced LVE behavior analogous to that of a highly entangled polymer melt.

Chen and Liu [90] prepared water-soluble Na-SIPE-y ionomers by a two-step polymerization of DMSI, PEG-M and phthalic anhydride, with different molecular weight PEG-x ranging from $x = 106–1000\,g\,mol^{-1}$. The nominal sulfonate concentrations of their Na-SIPE-y ionomers (based on the feed composition) ranged from 0.43–1.8 meq g$^{-1}$. Ionomer solutions displayed excellent surface activity, including high surface tension, low-foaming, solubilization and dispersant properties for dyes. The surface tension increased with increasing sulfonation level. The same research team [91] also prepared water-soluble Na-SIPE-y ionomers from the polymerization of DMSI with four different PEGs, with molecular weights of 400–2000 g mol$^{-1}$. The nominal IECs values for those ionomers were 0.23–0.98, and the conclusions reached in that study were essentially the same as those in Ref. [90].

Greener et al. [93] prepared copolymers of DMT, DMSI, EG and PEG-400, Scheme 4.22. The PEG400 acted as an internal plasticizer that lowered the melt viscosity, the $T_g$ and the glassy and rubbery moduli of the ionomer.

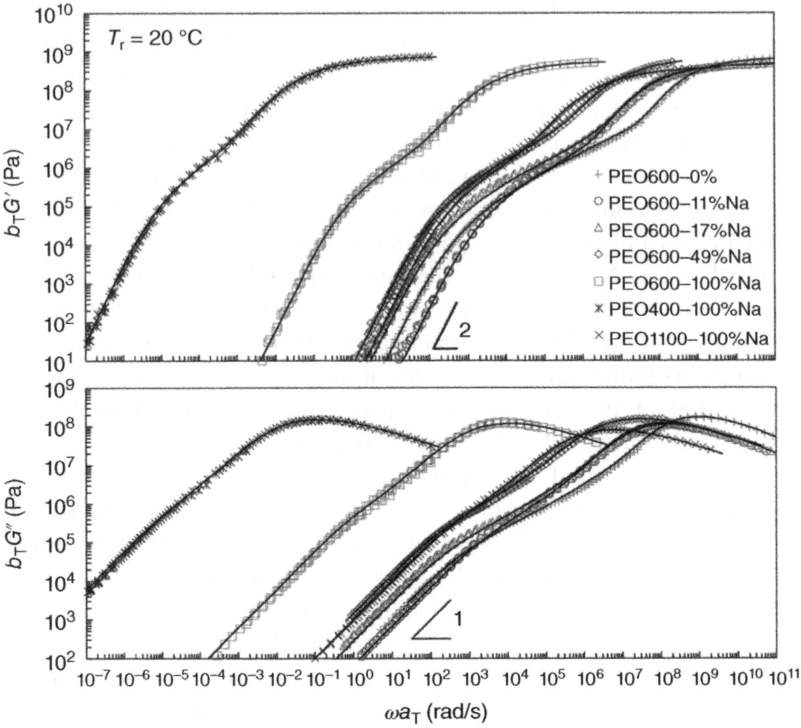


**Figure 4.6** Comparison of master curves dynamic modulus ($G'(w)$) and loss modulus ($G''(w)$) of Na-SIPE-x ionomers at a reference temperature of 20 °C with the predictions of a sticky Rouse model92 (solid curves). SIPE-0 is the non-ionic copolymer of DMI and PEO-600. *Source:* Chen et al. [89]/With permission from AIP Publishing.

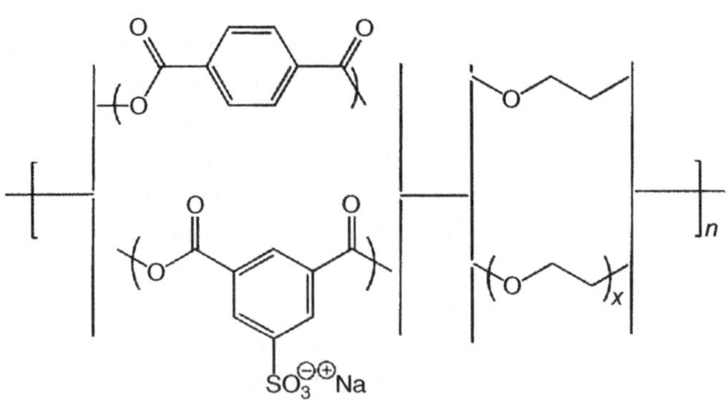


**Scheme 4.22** Poly(ether ester) ionomers prepared from DMT, DMSI, EG and PEG400. In Ref. [93], $x = 9$.

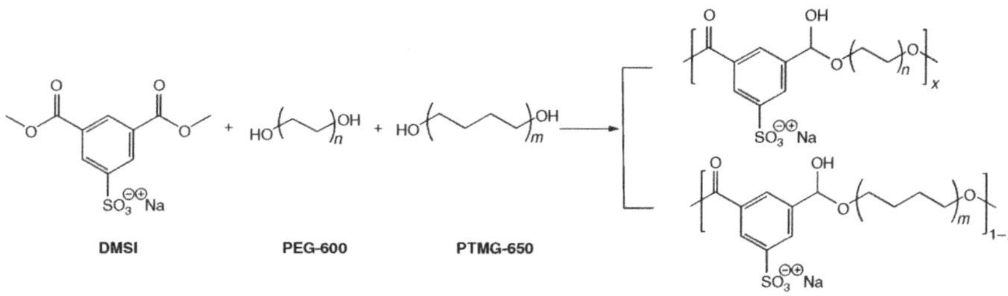


**Scheme 4.23** Synthesis of random copolymer Na-SIPEPT ionomers synthesized from DMSI, PET-600 and PMTG-650, where $x$ = mole fraction of PEG-600 (based on total amount of polyol used in the feed composition). Condensation by-products are not shown [94–96].

Colby and coworkers [94–96] synthesized random copolymers of DMSI with a mixture of poly(ethylene glycol), PEG-600, and poly(tetramethylene glycol), PTMG-650, spacers, Scheme 4.23. Those ionomers are hereafter denoted as M-SIPEPT-y, where M is the sulfonate cation, SI is sulfoisophthalate, PE is poly(ethylene glycol), PT is poly(tetramethylene glycol), and y is the IEC of the ionomer. Note that M-SIPEPT-y is the same as M-SIPE-y when the PEG mole fraction is 1).

Tudryn et al. [94] studied the benefits of using both PEG-600 and PTMG-650 spacers on the ion-transport in M-SIPEPT-1.2 and M-SIPEPT-1.2 ionomers, M = Na or Li, and the mole fraction of PEG-600 (based on total charge of polyol in the reaction) was varied from 0 to 1. The use of PEG provided improved solvation of the cation, while PTMG provided a lower $T_g$ and faster segmental dynamics. Since the molecular weights of the PEG and PTMG were 600 and 650 g mol$^{-1}$, respectively, the IEC of all the ionomers was ~1.2 meq g$^{-1}$. However, the conductivity of the ionomer decreased with increasing PTMG content even though the PTMG had a lower $T_g$, Figure 4.7. That result was attributed to microphase separation of a PTMG-rich phase that confined the ions to the PEG-rich phase. Ion-conductivity and mobility showed Vogel–Fulcher–Tammann (VFT) temperature dependences and increased with increasing PEG concentration. Conducting ion-densities, determined by dielectric spectroscopy, exhibited Arrhenius temperature dependence that was nearly constant for all ionomer compositions.

FTIR spectroscopy and linear viscoelasticity (LVE) measurements were used to study the relaxation behavior and quantify the structure and location of the ions in the same Na- and Li-SIPEPT-1.2 ionomers [95]. Although the ionic groups near the interface of the PEG- and PTMG-rich phases were preferentially in the PEG phase, the Na-sulfonate aggregates formed string-like structure and the Li-sulfonate aggregates formed denser, sheet-like structures. Microphase-separation of PTMG segments improved the mechanical properties of the ionomers, due to physical crosslinking effects, yet the continuous ion-PEG phase provided facile ion-conduction. The relaxation behavior of these ionomers was influenced by the higher segmental friction in the PEG-rich phase due to its higher $T_g$ and a higher

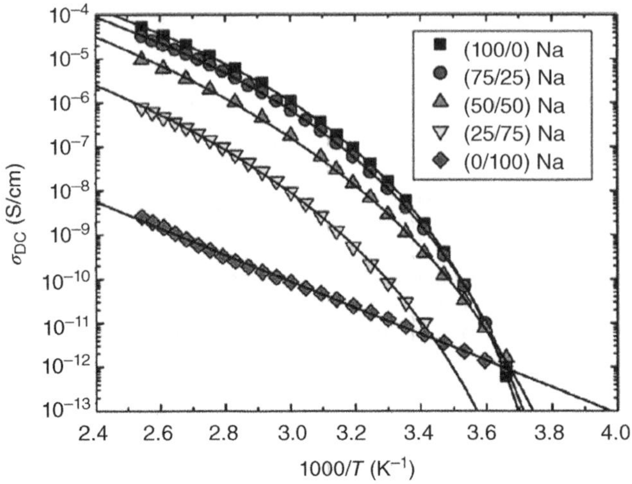


**Figure 4.7** Temperature dependence of the ionic conductivity of Na- SIPEPT-1.2 ionomers for various ratios of PEG-6000/PTMG-650. Conductivity decreased with increasing PTMG-650 concentration and decreasing temperature. The lines correspond to the VFT-dependence of conductivity on temperature, except for the (0/100) ionomer, which had no PEG-600 and exhibited Arrhenius temperature dependence. *Source:* Tudryn et al. [94]/with permission from American Chemical Society.

activation energy for ion-hopping with increasing PTMG concentration. The terminal relaxation of the ionomers was determined by dissociation of the aggregated ion-pairs in the PTMG-rich phase.

Shear and uniaxial extension rheological measurements of the Na- and Li-SIPEPT-1.2 ionomers indicated that the ionomer melts exhibited strain-hardening, which decreased with increasing extension rate [96]. When only PTMG was used in the formulation (i.e., a M-SIPT-y ionomer), the material was brittle in extension, but as the PEG content increased, the ionomer became more ductile, presumably due to the better solvation of the ionic species by PEG. In transient shear experiments, the ionomer melt with only PTMG exhibited flow instabilities, such as wall slip, shear banding and edge fracture, at relatively low strains (~0.1), but when PEG was included in the ionomer formulation, the melt exhibited a stable shear deformation up to a yield stain of ~2.5, though steady shear flow was not achieved.

Roslaniec and coworkers [28, 97–99] synthesized segmented poly(ether ester) elastomeric ionomers. The hard segments that provided the physical crosslinks were composed of copolymers of butylene terephthalate and butylene sulfoisophthalate and the soft segments were copolymers of terephthaloyl-PTMG-1000 and SIP-PTMG-1000, Scheme 4.24. The non-ionic version of this elastomer was developed by DuPont as a thermoplastic elastomer and sold under

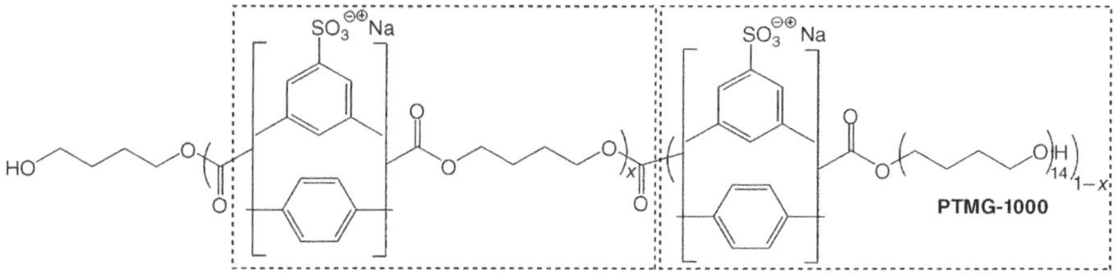


**Scheme 4.24** Poly(ether ester) elastomeric sulfonate ionomers [28, 97–99].

the tradename, Hytrel. The introduction of ionic groups into the hard segments increased the tensile modulus, decreased the stress and elongation at break and increased the melt viscosity, which were a result of the physical crosslinks formed by intermolecular associations of the sulfonate groups. The ionomers had a microphase-separated morphology that consisted of semi-crystalline hard segment nanodomains that contained ionic groups, dispersed in a continuous, homogenous soft phase composed of terephthaloyl and ionic groups dispersed in an amorphous polyether phase. The incorporation of the SIP groups into the hard phase reduced the crystallinity of that phase. The authors suggested that ionic aggregates may be present in the hard phase, especially for the higher sulfonate concentrations, but no direct evidence of this was presented.

Polyamic ester (PAE) ionomers were synthesized, by the reaction of PTMG and/or PEG with 1,2,4,5-benzenetetracarboxylic dianhydride (BTDA) and subsequent neutralization of the carboxylic acid groups in the product polymer with a metal acetate [100–102], or an alkyl amine [103]. An example polymerization based on using only PTMG is shown in Scheme 4.25. Chen et al. [100] synthesized PAE ionomers based on PTMG-650, PTMG-981 or PTMG-1000. They reported that the non-ionic polymers were liquids, but the ionomers were solids, which was a consequence

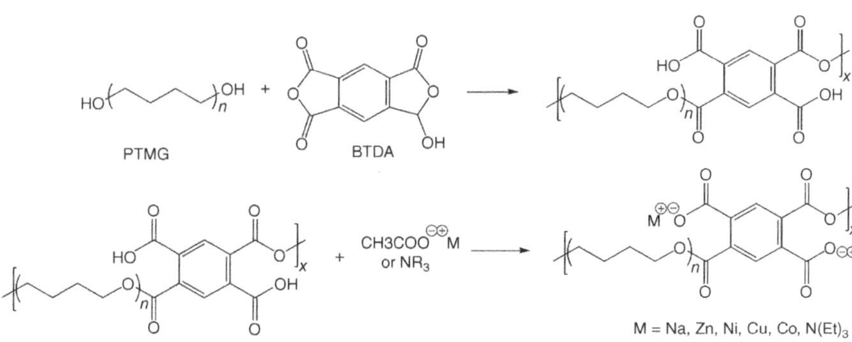


**Scheme 4.25** Synthesis of polyamic ester ionomers using PTMG. [100, 101, 103].

of the physical crosslinked network formed by intermolecular association of the ionic groups. SAXS data indicated that the ionic groups microphase-separated into ion-rich aggregates with a characteristic length of 1–3 nm, which is generally thought to be the average separation of the ionic nanodomains [104]. The characteristic distance for the Cu-salt was larger than for the other three ionomers. These authors also used extended EXAFS measurements to characterize the organization of the metal carboxylate groups in the aggregates. In a subsequent paper, Hu et al. [101] reported photoconductivity measurements for these PAE ionomers.

Yu et al. [102] produced PAE ionomers using PTMG-650 or PTMG-1000 partially or fully neutralized to the Na-, Cu- or Ni-salts. They inferred the existence of microphase separated ion-rich aggregates from dynamic mechanical data. The ionomers behaved as thermoplastic elastomers with good toughness and relatively high extensibility. The choice of cation influenced the extent of the microphase separation and the mechanical properties. For example, the zinc-salts showed less phase separation and poorer tensile properties compared to the other ionomers. Banu et al. [103] made the triethylamine-salt of PEA ionomers using PTMG-2000. Those ionomers dispersed in water without any additional solubilizing agent, and they observed that the ionomer particle size and the viscosity of the dispersion increased with increasing ionic group concentration. Particles were crosslinked with polyaziridine, which increased the $T_g$ of cast-films. Contact angle data revealed a reorganization of the hydrophobic groups at the surface of the crosslinked films that produced a more hydrophobic surface.

Lim and Kim [105] synthesized sulfonated aliphatic polyester ionomers from the polycondensation of adipic acid with mixed monomers of 1,4-butanediol (BD) and sodium sulfonated polyethylene glycol (SPEG), Scheme 4.26. The SPEG was

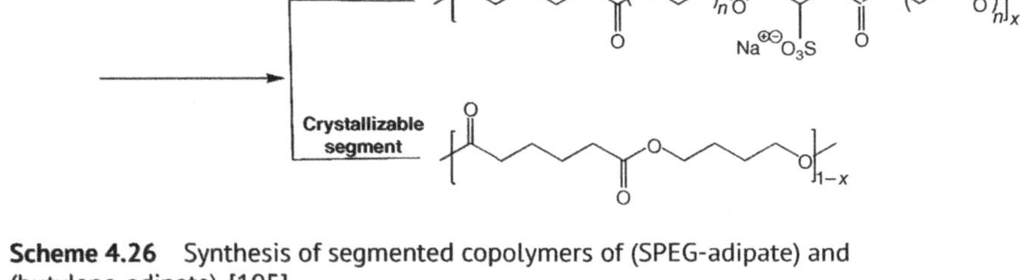


**Scheme 4.26** Synthesis of segmented copolymers of (SPEG-adipate) and (butylene-adipate). [105].

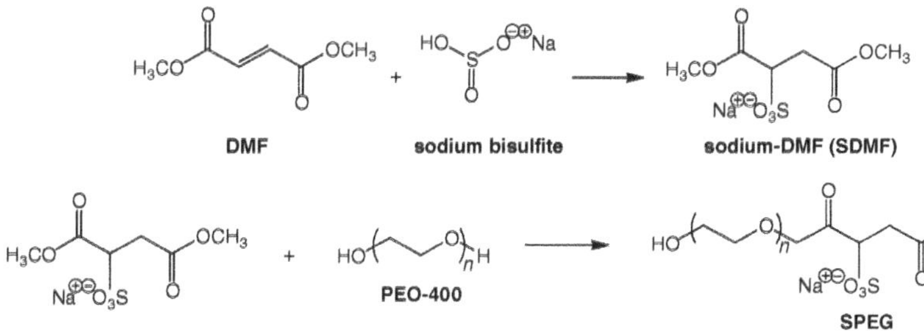


**Scheme 4.27** Synthesis of sulfonated poly(ethylene glycol), SPEG. [106].

synthesized from dimethyl fumerate (DMF) using a procedure described by Wang et al. [106], Scheme 4.27.

The ionomers shown in Scheme 4.26 were segmented copolymers with crystallizable poly(butylene adipate) segments and poly(sulfo-oxyethylene adipate) soft segments. Two sulfonation concentrations were considered, 2.5 and 5.0 mol% (IEC = 0.11 and 0.20 meq $g^{-1}$, respectively). In that study, the direct effect of the ionic groups on the ionomer properties was ambiguous, because the major component of the soft segments was PEG, which acts as an internal plasticizer. The PEG concentration was coupled to the concentration of ionic groups, and in those ionomers, the PEG concentration also had a significant influence on the properties. For example, the introduction of 2.5 mol% ionic groups (IEC = 0.11 meq $g^{-1}$) increased the modulus compared with a non-ionic analog, but the modulus of the 5.0 mol% sample (IEC = 0.20 meq $g^{-1}$) was lower than that of the 2.5 mol% sample, which is in stark contrast to the more common result of increasing modulus with increasing ionic group concentration for ionomers. The results suggest that at the lower ionic groups concentration, the ionic groups had the dominant influence on the modulus, while at higher ionic group concentrations, the softening influence of the PEG appeared to be more significant. The introduction of 2.5 mol% sulfonate groups also decreased the rate of crystallization of the ionomer compared with that of the non-ionic poly(oxyethylene-b-butylene adipate). Only crystallization data for the 2.5 mol% ionomer was included in Ref. [105], so the effects of higher sulfonate concentration (and higher PEG concentration) is not known.

Lee et al. [107] assessed the shape memory behavior of the same segmented copolymer ionomers reported in Ref. [105]. The ionomers with 2.5 and 5.0 mol% (IEC = 0.11 and 0.20 meq $g^{-1}$), sulfonate groups and the non-ionic poly(oxyethylene-b-butylene adipate) exhibited excellent shape fixing behavior, but the ionomers showed better shape recovery than the non-ionic polymer. The authors claim that the shape recovery of the 5.0 mol% ionomer was better than for the 2.5 mol% ionomer, but they did not report any statistics for their data, and it was not clear whether multiple shape memory tests were run on each composition. The difference in the shape recovery of the two ionomers was only ~10%, so any conclusion that one ionomer had better shape memory performance is probably equivocal.

Hwang et al. [108] synthesized similar sulfonate ionomers based on the polycondensation of succinic acid instead of adipic acid with the same mixed monomers of BD and SPEG, Scheme 4.28. Note that the first instance of the structure shown for SPEG and the copolymer ionomer structure in Scheme 4.1 of Ref [108] is incorrect. The ionic concentrations of the ionomers produced in this work were 0.5–2.0 mol%, which corresponded to IEC = 0.028–0.11 meq $g^{-1}$. The same uncertainty with regard to the effect of sulfonate concentration on properties as was discussed in the preceding paragraph for the segmented polyester ionomers based on adipic acid also applies here, as well. The SPEG macromer is composed of two PEG chains for each sulfonate group, so changes in the concentrations of sulfonate groups and PEG were coupled. Whereas physical crosslinks between sulfonate groups tend to stiffen the ionomer and increase the melt viscosity, PEG softens and plasticizes the ionomer. Ref. [108] fails to resolve the individual contributions of these two influences on the properties of the ionomers.

The 0.5 mol% ionomer (IEC = 0.028 meq $g^{-1}$), which had the lowest concentration of sulfonate groups, though also the lowest concentration of PEG, had the highest dynamic melt viscosity and dynamic modulus above the melting point of the PEG crystallites. As the sulfonate concentration increased, so did the mass concentration of PEG, which appears to have dominated the properties, decreasing the viscosity and modulus. The decreasing modulus with increasing concentration of ionic

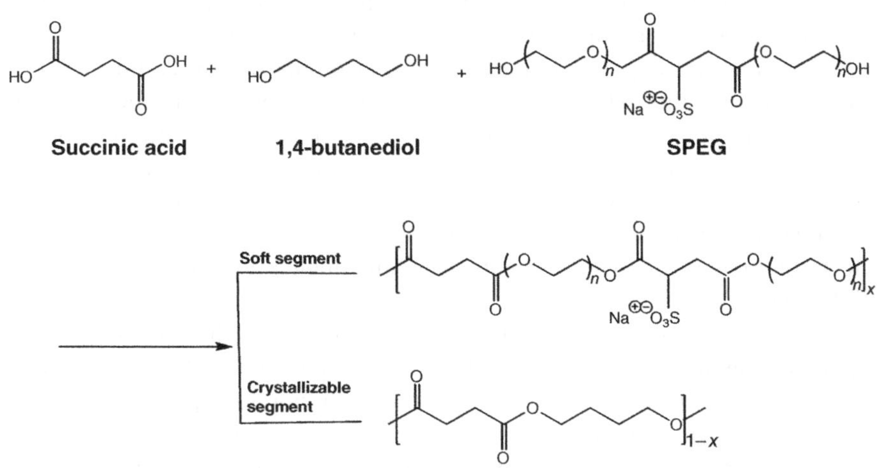


**Scheme 4.28** Synthesis of random copolymers of SPEG-succinate and (butylene succinate). [108].

groups may be explained by the softening effect of the increased PEG concentration as the soft segment concentration increased. The increasing PEG concentration that accompanied increasing sulfonate concentration may also have been responsible for decreasing the $T_{g,}$ $T_{m,}$ and the crystallinity of the hard, crystallizable segments as ionic group concentration increased. The authors also reported that enzymatic hydrolysis of the ionomers occurred more rapidly than for poly(oxyethylene-b-butylene succinate)

Ma et al. [109] synthesized a sulfonated polyester ionomer by the polycondensation of TA, isophthalic acid (IPA), ethylene glycol, SBHEI, and PEG-600. The synthetic procedure was not clearly described, especially what was the feed composition and what the symbols *n* and *m* represent in Figure 4.1 of their paper. Ref. [109] indicated that "the mole ratio of Na-SIP is 8%, IPA is 4% and PEG is 2%". If one assumes that description corresponds to the feed composition, that "mole ratio" meant mol%, that equal molar amounts of diol and carboxylic acid were used, and that the reaction was complete, the molar composition of their polymer would have been 4% sodium sulfoiosphthalic acid (or ester), 2% IPA, 44% TA, 1% PEG-600 and 49% EG, which corresponds to an IEC = 0.41 meq $g^{-1}$. The authors compared the rheological properties of their ionomer, PET and blends of the ionomer with PET. The melt viscosity of the ionomer was greater than that of PET, which would be expected since association of the ionic groups usually increases viscosity. The effects of the blend composition on the melt viscosity and flow activation energy were complicated, which may have been a consequence of the immiscibility of the PET and the ionomer. That could have produced a phase inversion as the composition changed from PET-rich to ionomer-rich.

Sulfopolyester ionomers described as "Easy Cationic Dyeable Copolyester" (ECDC) fibers [110–115] were easier to dye than neat PET. The ionomers were commercial products from the Defulun Chemical Fiber Co., Ltd. (Shanghai, China) and Nirlon Synthetic Fibers and Chemicals Ltd. (Bombay, India), but the references do not clearly describe the composition of the ionomers or how they were synthesized. One paper [112] says that they were made by a copolymerization, while another paper [111] states that they were blends of ionic and non-ionic compounds. From the descriptions of the polymers, it appears that they were poly(ether ester) ionomers with ethylene terephthaloyl, SIP, and PEG residues.

Lee et al. [116] studied the effects of ozone gas treatment of ECDC knitted fabric. Neither the origin, nor the composition of the ECDC, other than that it contained SIP, were provided in the paper. Ozone treatment increased the hydroxyl and carboxyl concentrations in the fiber, which increased moisture regain and water absorption of the fibers and accelerated the dyeing rate, especially for a cationic dye. However, the equilibrium adsorption of dye did not increase.

## 4.4 Random Polylactide Ionomers

Poly(lactic acid), or polylactide, (PLA) is a compostable aliphatic polyester derived from biomass. PLA is a commodity thermoplastic with a global market valued at $567M in 2021 and projected to grow at a compound annual rate of over 26.6% from 2022 to 2030 [117]. High molecular weight PLA is normally synthesized by ring-opening polymerization of lactide, which is the cyclic dimer of lactic acid [118]. PLA can also be prepared by the polycondensation of lactic acid, but that requires removal of water to promote the reaction and high molecular weight polymer is generally difficult to achieve by that route. PLA is competitive in many applications with traditional petroleum-based thermoplastics, such as polystyrene, PET, high-density polyethylene, and polypropylene, and the fact that it is biocompatible, compostable, and non-toxic makes it attractive for biomedical applications [119]. However, PLA has deficiencies, such as low melt strength, a slow crystallization rate, and a relatively low $T_g$ that limit its use temperature. All three of those deficiencies can potentially be addressed by the introduction of ionic groups to produce a PLA ionomer.

Accordingly, Ro et al. [120] synthesized ionomers with random PLA branches with carboxylate end groups by copolymerizing a methacrylate-terminated PLA macromonomer and methyl methacrylate, Scheme 4.29. The macromonomer was prepared by the reaction shown in Scheme 4.29. The macromonomer was then copolymerized with methyl methacrylate to form a polymer with randomly placed branch groups of hydroxyl-terminated PLA oligomers. Some of the hydroxyl groups of the copolymer were then converted to carboxylic acid groups by reaction with succinic anhydride using a 1-methylimidazole catalyst to produce the carboxylic acid derivative of the ionomer with an IEC = 0.69 meq $g^{-1}$, Scheme 4.29. Since not all of the hydroxyl groups were converted to carboxylic acid, the ionomer was actually a terpolymer with MMA units and polymer branches of hydroxyl and carboxylic acid terminated PLA oligomers.

The terpolymer had an $M_n$ = 36,000 g $mol^{-1}$. The carboxylic acid form of the ionomer was converted to an ionomer with Na-, Ca- or Y-carboxylate groups by partial or full neutralization of the carboxylic acid groups with the appropriate metal acetate. Conversion of the carboxylic acid derivatives to metal salts increased $T_g$ and the modulus. A rubber-like plateau region was observed in the viscoelastic behavior of the ionomers to temperatures of 200 °C, which was a direct consequence of the network formed by intermolecular associations of the ionic groups.

Kulkarni et al. [121] synthesized various 6-arm star PLAs with different combinations of carboxylic acid and hydroxyl end-groups. The star polyesters were prepared by ring-opening polymerization of PLA using a dipentaerythritol initiator and a SnOct catalyst, Scheme 4.30. The polymerization produced 6-arm

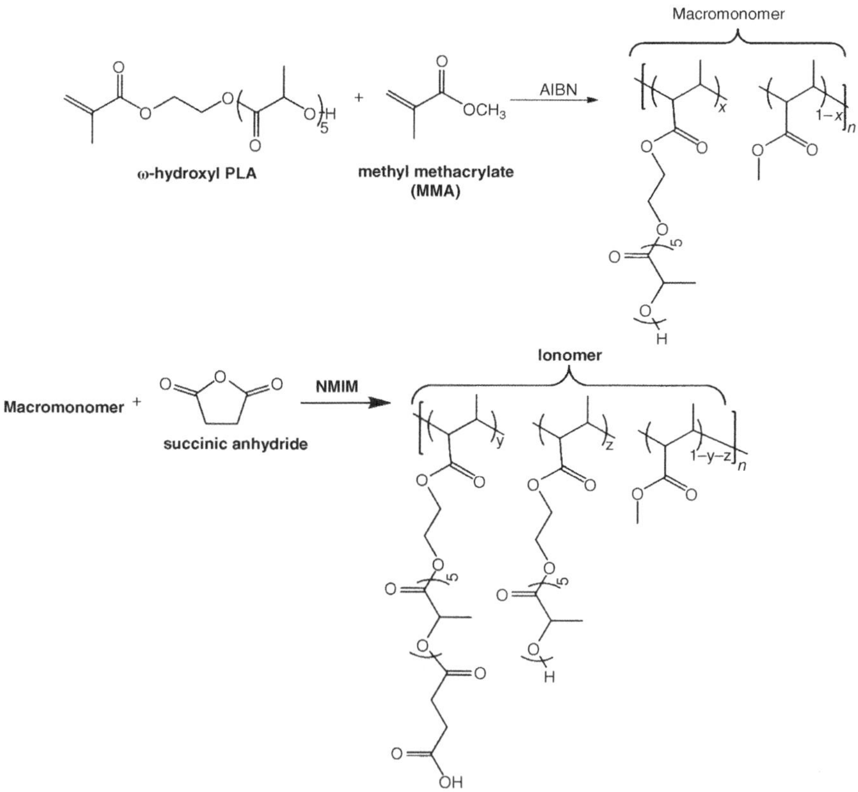


**Scheme 4.29** Synthesis of random methyl methacrylate and ω-carboxylic acid PLA copolymer. [120].

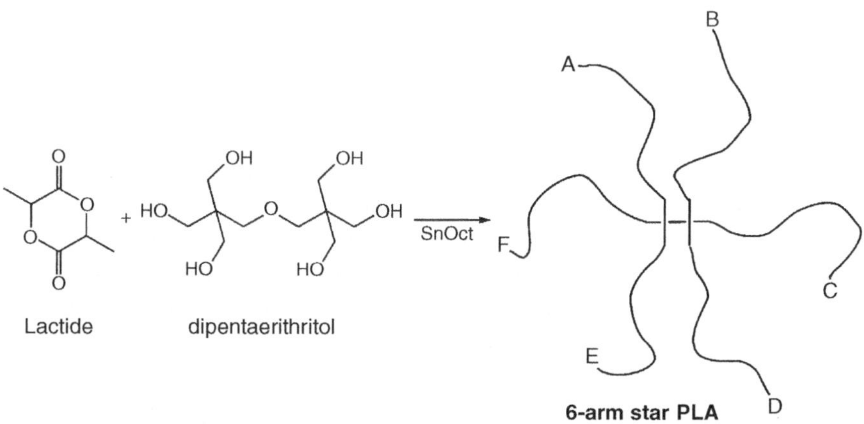


**Scheme 4.30** Synthesis of PLA star polymers [121].

stars of PLA with hydroxyl endgroups (i.e., A = B = C = D= E = F = OH in Scheme 4.30). Some or all of the hydroxyl end-groups were then converted to carboxylic acid by reaction with succinic anhydride, and the carboxylic acid groups were neutralized with sodium hydride to produce Na-carboxylate 6-arm star PLA ionomers with an average of 2, 3, 4 or 5 salt end-groups. The melt viscosity and elasticity increased by one to three orders of magnitude as the concentration of the ionic groups increased. The viscoelastic star ionomer melts were predominantly elastic (G′>>G″) with longer relaxation times compared to the precursor, non-ionic hydroxyl-terminated PLA star polymer melt, which was mostly viscous (G″>>G′). Linear viscoelastic behavior for the ionomer melts persisted to strains of 1–10%, while the non-ionic star PLA exhibited LVE behavior to strains of 100%.

Zhang et al. [122] copolymerized D,L-lactic acid (LA), L-malic acid (MA) and the monomethyl ether of polyethylene glycol with $M_n$ = 1200 g mole$^{-1}$ (mPEG-1200) by a polycondensation reaction. This would appear to produce a fairly complicated and branched microstructure, since the MA has three hydroxyl sites that can react with the carboxylic acid groups from LA or another MA molecule, and the reaction of the mPEG with any carboxylic acid group terminates the growing chain. For all the formulations described in the paper, the stoichiometric ratio of hydroxyl to carboxylic acid groups was at least 2:1, so the polymerization reactions produced only low molecular weight copolymers, $M_n$ = 1100–3100 g mol$^{-1}$. Although the authors claim that they made ionomers, there is little evidence, if any, that their products contained any carboxylic acid groups. On the basis of the large excess of hydroxyl groups used in the syntheses, one would expect all the carboxylic acid groups of LA and MA to have reacted with hydroxyl groups to form ester linkages, so that no carboxyl groups should have remained in the oligomers. The only evidence the authors provided for synthesizing an ionomer was an absorption in the FTIR spectra at 3400 cm$^{-1}$ that they attributed to carboxylic acid group. However, as was discussed above, on the basis of the stoichiometry used it is more likely that the strong 3400 cm$^{-1}$ absorption was due to hydroxyl endgroups and unreacted hydroxyl groups from the monomers, not carboxylic acid groups.

Noga et al. [123] synthesized branched carboxylic acid-functionalized PLA, Scheme 4.31, by first copolymerizing a benzyl-ether substituted monomer with lactide. The polymer product was then debenzylated to achieve pendant hydroxyl groups that were then converted to carboxylic acid by reaction with succinic anhydride. The objective of that work was not to make an ionomer, but rather to use the carboxylic acid groups to graft RGD-containing peptides to make a biofunctional polylactide. Although it was not reported in that paper, the carboxylic acid-polylactide derivatives could be used to prepare PLA ionomers.

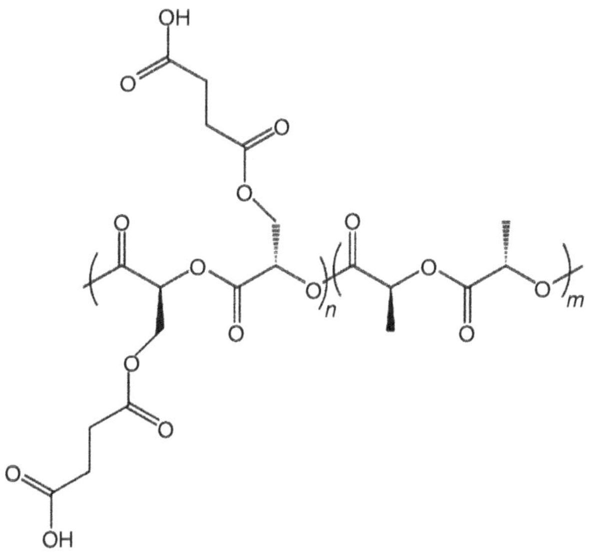

**Scheme 4.31** Carboxylic acid functionalized polylactide. [123].

## 4.5 Telechelic Polyester Ionomers

Poly(ethylene terephthalate) telechelic ionomers with terminal benzyl sulfonate groups, Scheme 4.32a, were prepared by polycondensation of DMT, EG and the sodium salt of 3-sulfobenzoic acid (SSBA) [124]. Ionomers with 1–5 mol% SSBA groups (IEC = 0.085–0.24 meq $g^{-1}$) were prepared. For telechelic ionomers, the IEC and molecular weight are coupled, so that $M_n$ of these PLAs decreased with increasing IEC, from 23,600 g $mol^{-1}$ to 13,100 g $mol^{-1}$. Increasing the IEC had little effect on the $T_g$ of the ionomers, but it increased the crystallization temperature and the rate of crystallization and reduced the $T_m$ and the degree of crystallinity. The ionic end-groups also increased the melt viscosity.

Kang et al. [125] synthesized similar poly(ethylene isophthalate) telechelic ionomers with 1–5 mol% terminal benzyl sulfonate groups (IEC = 0.053–0.24 meq $g^{-1}$) from DMI, EG, and SSBA, Scheme 4.32b. Intermolecular associations of the ionic groups increased the solution viscosity in chloroform at 25 °C compared to that of a non-ionic analog, even though the molecular weight of the non-ionic polymer was about three times greater than that of the ionomer. Similarly, the melt viscosities of the ionomers for $T < 150$ °C were as much as an order of magnitude higher than the non-ionic, higher molecular weight PEI, but above 150 °C the viscosities of the ionomers dropped to about an order of magnitude below that of the PEI. The higher melt viscosities of the ionomers at the lower temperatures

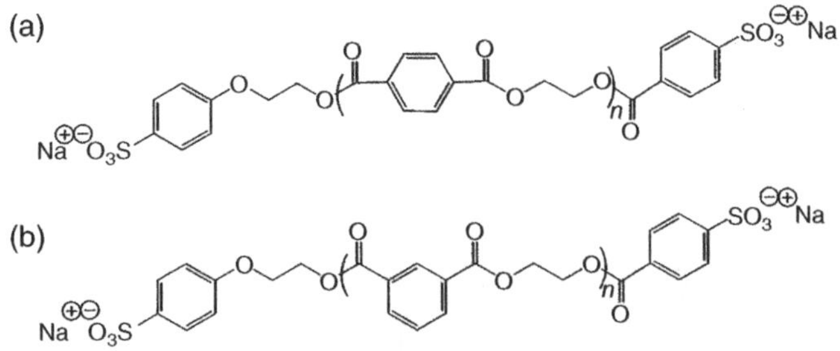


Scheme 4.32 Telechelic ionomers: (a) sulfonated PET; (b) sulfonated PEI [124, 125].

was due to intermolecular ionic group associations that persisted in the melt, but above 150 °C, the aggregates dissociated and the lower viscosities for the ionomers compared to PEI were due to the lower molecular weight of the ionomers.

Storey and Sherman [126] synthesized telechelic α-hydroxyl-ω-(carboxylic acid) poly(ε-caprolactone) oligomers with one carboxylic acid and one hydroxyl end-groups with IEC = 0.36 meq $g^{-1}$ by the reaction of glycolic acid with ε-caprolactone and a stannous octoate catalyst, Scheme 4.33a. That procedure also introduced some glycolic acid groups into the chain. When succinic anhydride was added to the polymerization, an α,ω-carboxylic acid poly(ε-caprolactone) ionomer with carboxylic acid groups at each end of the polymer chain was formed, Scheme 4.33b. The α,ω-carboxylic acid poly(ε-caprolactone) ionomers (IEC = 0.083–0.35 meq $g^{-1}$) were converted to different alkali metal salt ionomers by reaction of the carboxylic acid groups with appropriate metal hydroxides [127].

For the lowest molecular weight ionomer ($M_n$ = 2826 g $mol^{-1}$; IEC = 0.35 meq $g^{-1}$), $T_m$ and the extent of crystallinity increased as the metal counterion was varied in the order of $Cs^+ > Rb^+ > K^+ > Na^+ > Li^+ > H^+$, which corresponded to the same trend as the Coulomb energy of the ion-pair. That correspondence or its

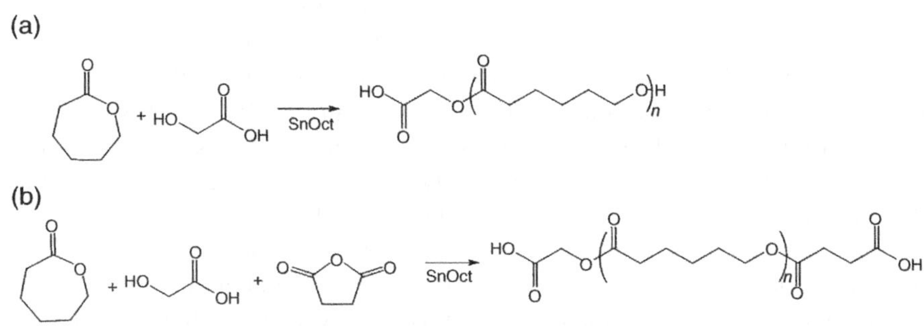


**Scheme 4.33** Synthesis of telechelic poly(ε-caprolactones): (a) **α**-hydroxyl-**ω**-(carboxylic acid) poly(ε-caprolactone) and (b) **α,ω**-carboxylic acid poly(ε-caprolactone). [126, 127].

significance was not noted in the paper, and although it is not apparent why the Coulomb energy of the carboxylate ion-pair should affect the thermal behavior of the polycaprolactone crystallites, it may be related to differences in the ability of the various ionic groups or their supramolecular structures to nucleate crystallization and the kinetics of the crystal growth. Although the authors did not measure melt viscosities of their ionomers, typically ionic groups increase viscosity, and the melt viscosity of ionomers has been shown to scale with the Coulomb energy of the ion-pair [128]. As the molecular weight of the ionomer increased, however, the effect of the counterion on $T_m$ and crystallinity became less distinct.

Essahli et al. [129] prepared zinc-phosphonate telechelic polyester ionomers (M-PETDEGSP), Scheme 4.34a, based on the end-functionalization of a PET glycolysate oligomer. The molecular weights were very low, $M_n$ ~586–800 g mol$^{-1}$ (IEC ~2.5–3.4 meq g$^{-1}$), and the ionomers were amorphous with a $T_g$ = 23 °C, which was about 70 °C higher than the non-ionic PET glycolysate ($T_g$ = −51 °C). SAXS measurements indicated that the ionomer exhibited a microphase-separated morphology of nanoscopic ionic aggregates separated by about 3 nm. The thermal properties of the ionomer were modified by blending it with a zinc-phosphonate telechelic PTMG (Zn-TeraSP), Scheme 4.34b. For example, a 50/50 (w/w) mixture of the Zn-PETDEGSP and Zn-TeraSP ($T_g$ = −68 °C) had a $T_g$ = 3 °C.

Sherman and Storey [130] synthesized a telechelic α-hydroxyl-ω-(carboxylic acid) PLA oligomer by ring-opening polymerization of L,L-lactide, using a tin octoate catalyst and initiated by the secondary hydroxyl group of lactic acid, Scheme 4.35a. The α-hydroxyl-ω-(carboxylic acid) PLA was then converted to a telechelic α,ω-carboxylic acid PLA ($M_n$ = 2000 g mol$^{-1}$; IEC = 1.0 meq g$^{-1}$) by reaction with succinic anhydride, using a 1-methylimidazole catalyst, Scheme 4.35b. Seven different methods were used to neutralize the acid derivatives to the Ca-salt ionomers, including using different bases (calcium acetate or calcium hydride), different solvents (chlorobenzene, a chlorobenzene/toluene mixture and a dichloroethylene/toluene mixture), different temperatures (105–180 °C), different methods for removing the neutralization by-products (reflux with a Dean-Stark trap and

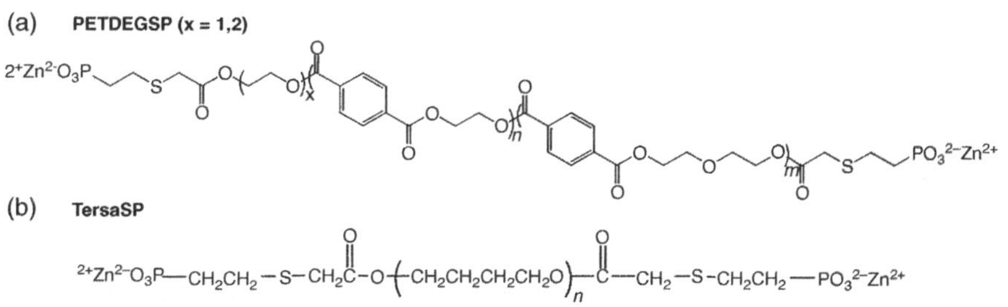


**Scheme 4.34** Telechelic phosphonate polyester ionomers. [129].

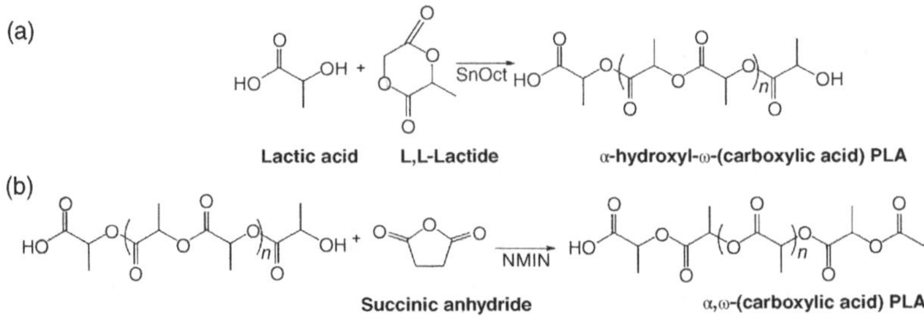


**Scheme 4.35** Two-step synthesis of **α,ω**-carboxylic acid PLA. [130].

vacuum- or azeotropic-distillation) and either a stoichiometric amount or excess neutralizing agent. The different neutralization methods produced ionomers with different properties, which was a consequence of chain degradation occurring during the neutralization process. Nonetheless, the presence of Ca-carboxylate increased the $T_g$ of the ionomer by as much as 18 °C compared to that of the carboxylic acid derivative ($T_g$ = 47 °C), and for the ionomers with the higher $T_g$s (58–65 °C) crystallization of the polymer was suppressed.

Ro and Weiss [131] synthesized telechelic PLA ionomers using a chemical recycling reaction (i.e., transesterification) of a high molecular weight PLA ($M_n$ = 65,000 g mol$^{-1}$; $M_w/M_n$ = 1.9) with 2-hydroxyethyl methacrylate (HEMA) and a tin octoate catalyst, Scheme 4.36a. The resultant ω-hydroxy PLA oligomers were converted to ω-(carboxylic acid) PLA by reaction with itaconic anhydride, Scheme 4.36b, and then neutralized with metal acetates to form ω-metal carboxylate PLA ionomers ($Na^+$-, $Li^+$-, $K^+$-, $Zn^{2+}$-, $Ca^{2+}$- and $Y^{3+}$-salts) with IEC = 0.077 and 0.29 meq g$^{-1}$. α,ω-hydroxyl PLAs were synthesized by transesterification of PLA with EG and a tin octoate catalyst, Scheme 4.37a, and those products were converted to α,ω-carboxylic acid PLAs, (IEC = 0.15, 0.22 and 0.059 meq g$^{-1}$), by reacting the hydroxyl end groups with itaconic anhydride, Scheme 4.37b. The incorporation of ionic groups increased the $T_g$, and $T_g$ increased as the

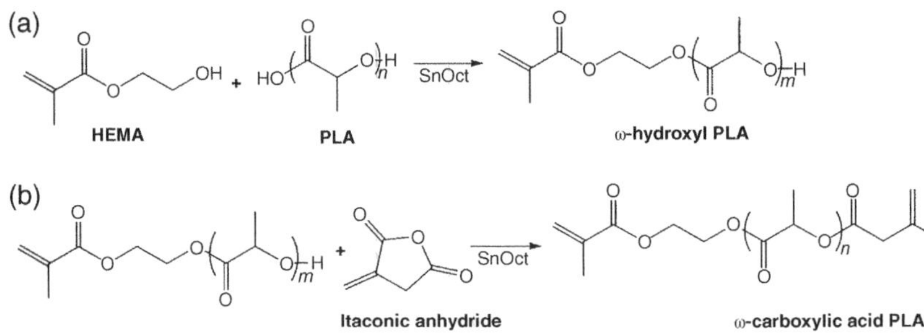


**Scheme 4.36** Two-step synthesis of **ω**-carboxylic acid PLA. [131].

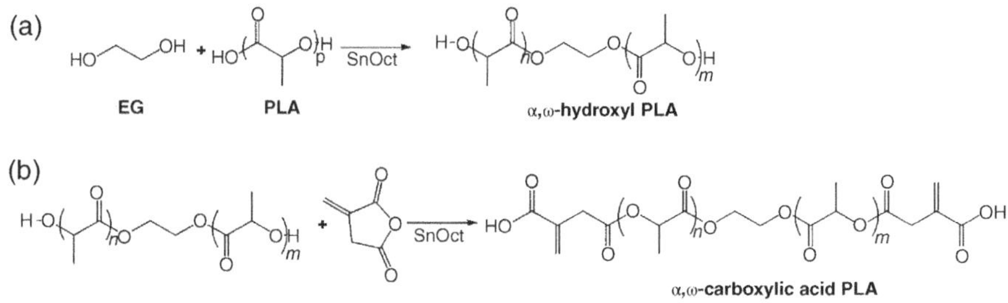


**Scheme 4.37** Two-step synthesis of **α, ω**-carboxylic acid PLA. [131].

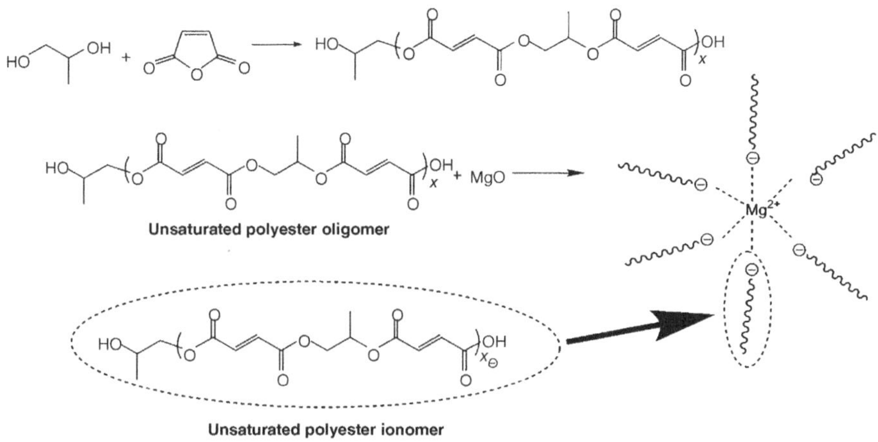


**Scheme 4.38** Synthesis of unsaturated polyester telechelic ionomer. [132].

Coulomb energy of the carboxylate ion-pair increased in the order of cation: $Y^{3+} > Zn^{2+} > Ca^{2+} > Li^{+} > Na^{+} > K^{+}$). The presence of the ionic groups suppressed crystallinity, especially when multivalent cations were used.

Litvinov et al. [132] prepared unsaturated star-like polyester ionomers by polycondensation of propylene glycol and maleic anhydride, Scheme 4.38 The unsaturated polyester had an $M_n \sim 1500\,g\,mol^{-1}$ and a mixture of hydroxyl and carboxylic acid end-groups. The Mg-carboxylate ionomers were formed by reaction of the unsaturated polyester oligomer with MgO, and the authors proposed that the carboxylate oxygens coordinated with the $Mg^{2+}$ cation to form an octahedral – coordinated cluster that was responsible for a $10^3$–$10^4$ increase of the viscosity in a styrene solution.

Telechelic copolymer ionomers of PET and PEI with phosphonium end groups, Scheme 4.39, were synthesized by Unal et al. [133]. Unfortunately, no properties

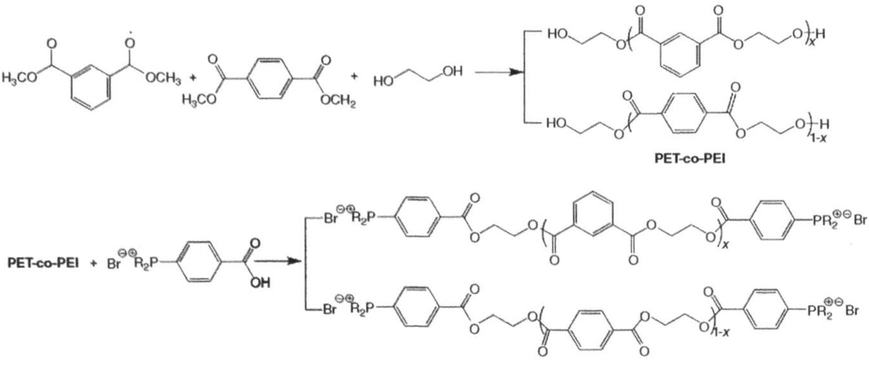

**Scheme 4.39** Synthesis of phosphonium telechelic ionomers based on copolymers of (ethylene terephthalate and ethylene isophthalate). [133].

of these ionomers were reported in that preprint and no subsequent publications that discussed these ionomers were found.

## 4.6 Liquid Crystalline Ionomers (LCI)

Hara and coworkers [134] synthesized wholly aromatic, naphthalene-containing thermotropic (NTP) LCIs with 1–20 mol% SIP (IEC = 0.074–1.4 meq $g^{-1}$) by polycondensation of methyl *p*-hydroxybenzoate (MHBA), methyl 6-hydroxy-2-naphthoate (MHNA) with DMSI, Scheme 4.40a. In addition to the inclusion of the naphthaloyl moieties, the use of DMSI introduced kinks in the linearity of the polymer chain that reduced $T_m$ and rigidity of the LCP. Hydroquinone (HQ) was also added to balance the carboxylic acid and hydroxyl functionalities of the sulfonated monomer when the imbalance was greater than 1 mol%. Xue and Hara [135] also prepared LCIs using the potassium salt of hydroquinone-2,5-disulfonate (K-HQDS) in place of the DMSI, Scheme 4.41b, which eliminated the kinks in the chain from the isophthaloyl groups. Ionomers were made with IEC = 0.075–0.29 meq $g^{-1}$. The HDQS monomer has two sulfonate groups compared to one for the DMSI monomer, so less HQDS was needed to achieve the same IEC. LCIs were also prepared using Ca-HQDS, which was prepared by *in situ* ion-exchange of K-HQDS with Ca-acetate during the polymerization reaction [135].

For the DMSI-NTP ionomers [134], when the IEC ≤ 0.15 meq $g^{-1}$, no $T_g$ was detected by DSC. For IEC < 0.715 meq $g^{-1}$, $T_g$ was nearly constant at ~94 °C, but for IEC = 1.0 meq $g^{-1}$, $T_g$ increased to 127 °C and for IEC = 1.4 meq $g^{-1}$, two $T_g$s were observed at 115 and 153 °C. The multiple $T_g$s were attributed to phase separation of

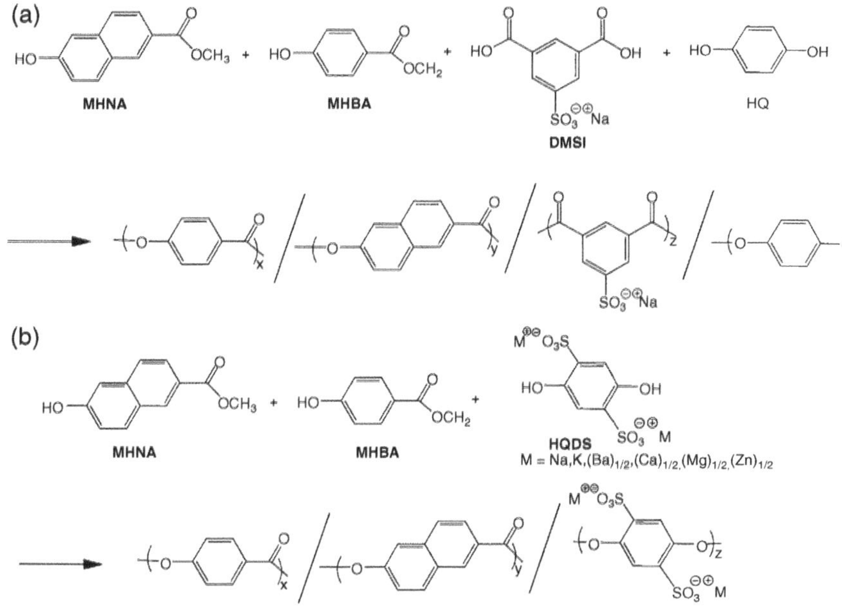


**Scheme 4.40** Wholly aromatic naphthalene-containing polyester LCIs [134–137].

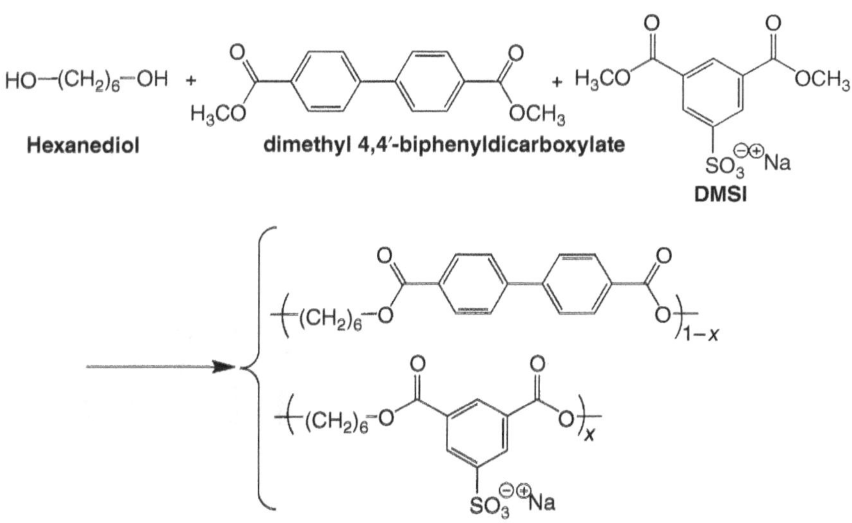


**Scheme 4.41** Polyester LCI synthesized from dimethyl 4, 4′biphenyldicarboxylate, hexane diol, and dimethyl sulfo-isophthalate. [138].

ion-poor and ion-rich phases. The DMSI-NTP ionomer melts exhibited birefringence typical of a liquid crystalline mesophase, and the majority of the ionomers exhibited nematic mesophase textures over a wide temperature range without showing a transition to an isotropic phase to at least 380 °C, above which thermal degradation of the LCI occurred. $T_m$ and the crystallization temperature ($T_c$) of the DMSI-NTP ionomers decreased with increasing ionic group concentration.

The HQDS NTP ionomer melts [135] also exhibited nematic mesophases. $T_g$ of K- and Ca-salts did not show any clear trend as the ionic content increased and $T_m$ and $T_c$ were relatively insensitive to the ionic group concentration, though the range of ionic group concentrations in that paper was small. The tensile modulus and strength of the K-HQDS NTP ionomers increased as the ionic content increased, Figure 4.8(a), despite a decrease in the molecular weight with increasing HQDS. Those results indicated that the physical crosslinks that resulted from intermolecular ionic interactions had a more dominant effect on the mechanical properties, compared with molecular weight.

Xue et al. [136] compared the mechanical properties of the Na-salt and four divalent salts (Ba, Ca, Mg, Zn) of M-DMSI-NTP LCI fibers (IEC = 0.075 meq $g^{-1}$) and a non-ionic NTP LCP. The divalent salts were prepared by ***in situ*** ion-exchange of Na-DMSI with the appropriate metal acetate during the polymerization step. Although all of the ionomers had the same sulfonate concentration (IEC = 0.075 meq $g^{-1}$), the molecular weight of the polymer, as determined by inherent viscosity measurements, decreased in the order of cation: $Na^+ > Ca^{2+} > Zn^{2+} > Mg^{2+} > Ba^{2+}$. For the LCIs with divalent cations, the Ca-HQDS NTP ionomer had the highest molecular weight and the highest modulus and tensile strength, Figure 4.8(b). The Ba-HQDS NTP was brittle and the mechanical

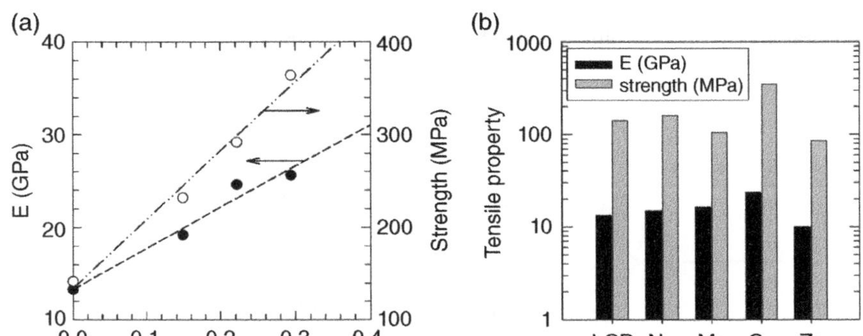


**Figure 4.8** Tensile modulus and strength of NTP ionomers. (a) K-HQDS-NTP; (b) M-DMSI-NTP (IEC = 0.075 meq $g^{-1}$). LCP denotes a non-ionic NTP LCP. *Source:* Data from Xue and Hara [135]. 2005; Xue et al. [136], et al. 1998.

properties of the Mg- and Zn-HQDS ionomers were significantly inferior to those of the Ca-HQDS NTP, which was attributed to their lower molecular weights. The fibers made from Na-HQDS NTP had comparable mechanical properties as the non-ionic NTP LCP, but better mechanical properties compared to the Mg- and Zn-salts. The modulus and strength values for the Ca-salt were about twice those for the Na-salt and the non-ionic NTP LCP, even though the molecular weights of the Na-salt and the non-ionic LCP were higher.

The ionic interactions also influenced the tensile fracture surface morphology [136]. Fracture surfaces of DMSI-NTP ionomer fibers showed brittle fracture [134], and the spontaneous fibrillation generally observed in fibers of non-ionic liquid crystalline polymers [139] was increasingly suppressed with increasing IEC. That result was attributed to the decreasing rigidity of the polymer backbone due to incorporation of the ionic monomer and increasing cohesion of the material that resulted from association of ionic groups. Flat fracture surfaces observed for the high ionic content NTP fibers reflected the brittleness arising from easier crack propagation [134].

Whereas a non-ionic NTP polymer film showed a relatively uniform fracture surface with a hairline texture pleated along the transverse direction, the Na-DMSI-NTP (IEC $= 0.75\,\text{meq g}^{-1}$) ionomer films were more resistant to fracture along the transverse direction. In contrast, fracture of the Ca-DMSI-NTP (IEC $= 0.75\,\text{meq g}^{-1}$) ionomer film split the film into continuous parallel strips along the extension direction with some fibrils connecting the strips [136], Figure 4.9. That result was attributed to the stronger physical crosslinks between the polymer chains for the Ca-salt.

Kink band formation during compression in DMSI-NTP fibers with IEC $= 0.075\,\text{meq g}^{-1}$ suggested that the compressive properties of the ionomers were better than for non-ionic NTP fibers [137]. That result was attributed to intermolecular links from ionic group associations. Tensile strengths of the ionic NTP fibers were also slightly higher than for non-ionic NTP fibers, even though the molecular orientation was lower for the LCI fibers.

Sulfonate-LCIs based on dimethyl 4,4′-biphenyldicarboxylate mesogens and flexible spacers of hexane diol containing 1–20 mol% of SIP (IEC $=0.031$–$0.62\,\text{meq g}^{-1}$) were reported by Lin et al. [138], Scheme 4.41. The inherent viscosities of the polymers decreased with increasing SIP concentration of the LCI, which the authors attributed to an increase in the melt viscosity of the reaction melt that suppressed the molecular weight. For a SIP concentration of 13 mol% or less (IEC $\leq 0.40\,\text{meq g}^{-1}$), the LCI melt formed a smectic A morphology, but no liquid crystalline phase was observed for the ionomer with 20 mol% SPI (IEC $= 0.62\,\text{meq g}^{-1}$). The melting point to a smectic liquid phase ($T_{ms}$) of the LCIs varied between 200 and 210 °C and the clearing transition from a smectic liquid to an isotropic liquid ($T_{si}$) varied between 221 and 241 °C as the concentration of SIP increased

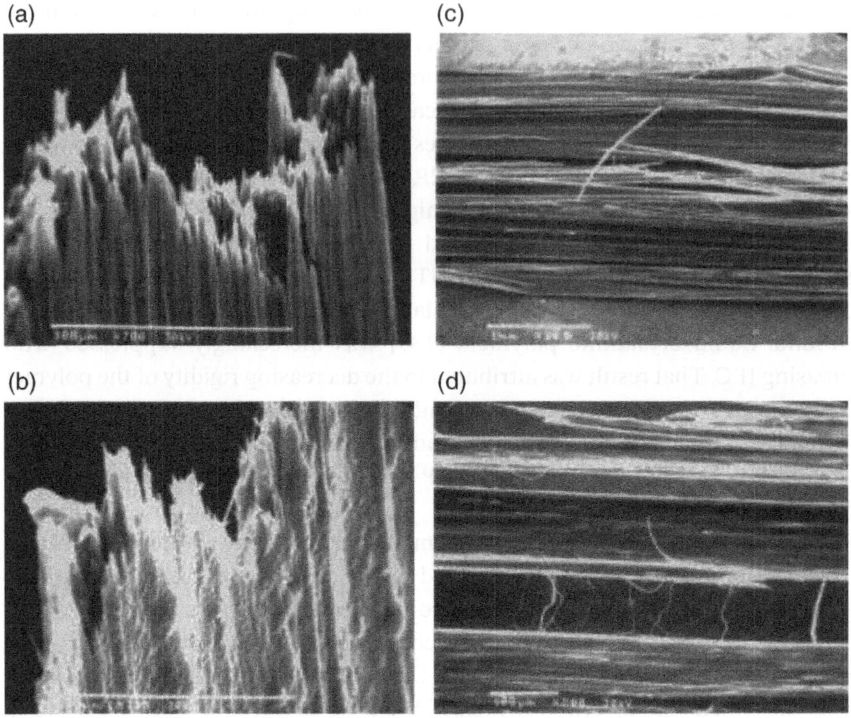

**Figure 4.9** Scanning electron micrographs of tensile fracture surfaces: (a) nonionic NTP; (b) Na-DMSI-NTP (IEC = 0.75 meq g$^{-1}$); (c, d) Ca-DMSI-NTP (IEC = 0.75 meq g$^{-1}$). *Source:* Reproduced with permission from Ref. [136], American Chemical Society.

from 0 to 20 mol%. Actually, the reported clearing points first decreased with increasing SIP between 1 and 5 mol% SIP (IEC = 0.031 and 0.093 meq g$^{-1}$, respectively), then increased between 5 and 10 mol% (IEC = 0.15 and 0.31 meq g$^{-1}$, respectively) and then decreased again with the incorporation of more SIP. The authors noted that these $T_{si}$ values were reproducible, and explained that the unusual behavior of the clearing point was a result of a competition between the effect of the kinks from meta linkages of the SIP that tended to decrease $T_{si}$ and the ionic associations that tended to stabilize the mesophase and increase $T_{si}$.

Song et al. [140] prepared a polyester LCI by sulfonating a preformed polyester LCP based on a terephthaloyl bis(4-oxybenzoyl) mesogen called a "triad aromatic ester" mesogen [141] and a butylene terephthalate flexible spacer previously reported by Ignatious et al. [142], Scheme 4.42. The liquid crystalline polyester was sulfonated with acetyl sulfate in homogeneous solution following the procedure of Makowski et al. [143]. The LCI was then formed by neutralizing the sulfonic acid groups with NaOH. The schematic sulfonation reaction provided by

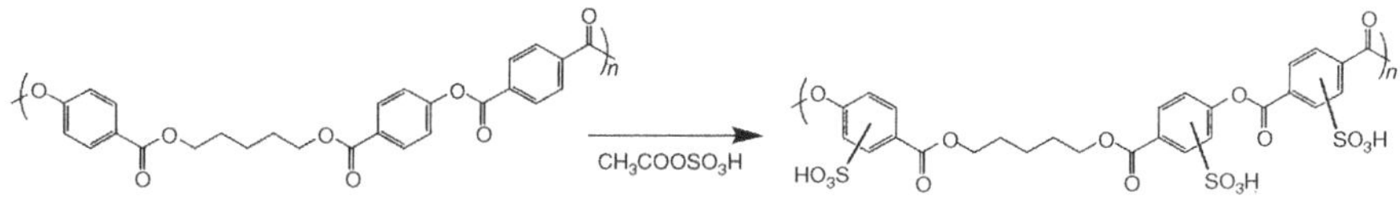

**Scheme 4.42** Synthesis of polyester LCI by sulfonating a liquid crystalline polymer composed of a Triad mesogen and butylene terephthalate flexible spacer. [140].

the authors indicated that any of the aromatic groups may be sulfonated, as is shown in Scheme 4.42, though not every aromatic groups was sulfonated. The degree of sulfonation was controlled by the amount of acetyl sulfate used, and LCIs were produced with IEC = 0.279–0.457 meq $g^{-1}$. The acid derivatives and the Na-sulfonate ionomers exhibited nematic melts and melting transition from a crystal to a nematic liquid, $T_{mn}$ ~ 250 °C, was unaffected by either the sulfonate concentration or whether the polymer was the sulfonic acid or the Na-sulfonate derivative. The clearing point, in this case the nematic to isotropic phase transition ($T_{ni}$), for the acid derivatives was insensitive to the IEC ($T_{ni}$ ~ 287 °C), but $T_{ni}$ decreased when the acid derivative was converted to the Na-salt and with increasing Na-sulfonate concentration. For the ionomers, a new thermal transition was observed at 166 °C, which the authors attributed to melting of a phase-separated polybutylene terephthalate phase.

Lin et al. [144] prepared wholly aromatic polyester LCPs containing sulfonate pendent groups by an interfacial polycondensation of the potassium salt of hydroquinone sulfonic acid with equimolar amounts of either terephthaloyl chloride (TC), 2-bromoterephthaloyl chloride (BTC) or nitroterephthaloyl chloride (NTC). The nominal composition of each of the three sulfo-LCIs was 50 mol% of the sulfonate monomer (IEC = 3.4, 2.5, and 2.7 meq $g^{-1}$ for the ionomers prepared with TC, BTC, and NTC, respectively). The ionomers did not exhibit thermotropic behavior, but the NTC-containing polyester ionomer formed lyotropic liquid crystalline solutions in a mixed solvent of dimethyl sulfoxide and water when the ionomer concentration was ~15–20 wt. %.

Zhang and coworkers [145–149] synthesized a number of main-chain polyester LCIs using the diol brilliant yellow (BY) to provide the ionic functionality, Schemes 4.43–4.47. For a random reaction of two dihydroxyl monomers (A and B) with two diacid chloride monomers (C and D), the probabilities for all the possible ester repeat units are:

$xy$ for A—C

$x(1-y)$ for A—D

$(1-x)y$ for B—C

$(1-x)(1-y)$ for B—D

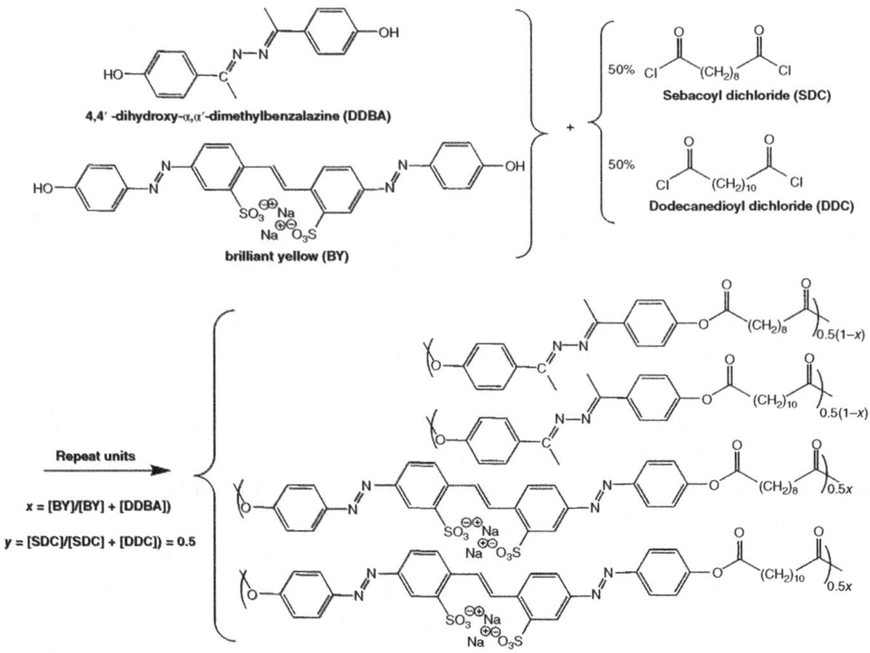


**Scheme 4.43** Synthesis of an azine main-chain polyester LCI containing brilliant yellow. [145].

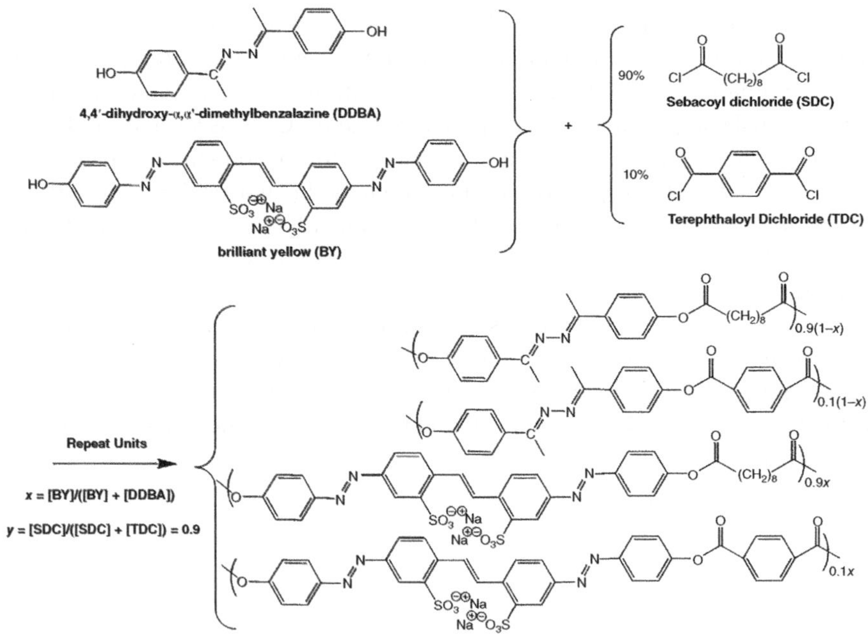


**Scheme 4.44** Synthesis of main-chain polyester LCI containing brilliant yellow [146].

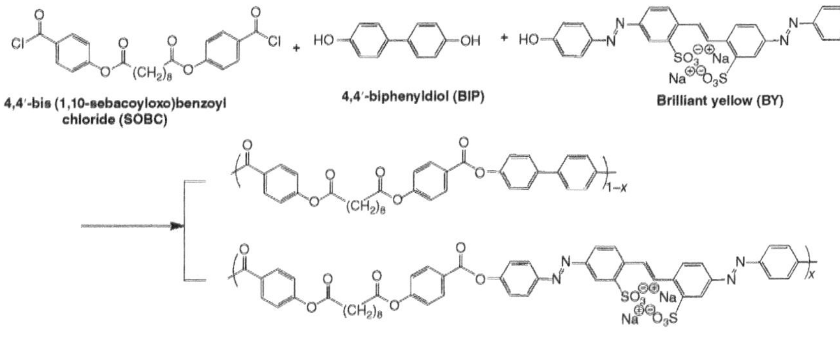


**Scheme 4.45** Main-chain thermotropic LCI containing brilliant yellow [147].

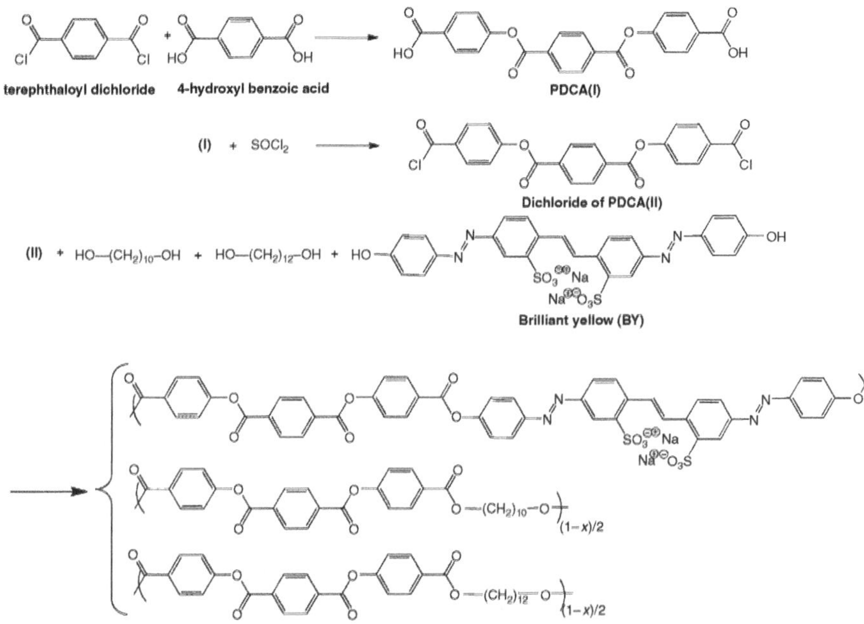


**Scheme 4.46** Main-chain thermotropic LCI containing brilliant yellow [148].

where $x \equiv ([A]/[A] + [B])$ = mole fraction of A based on the total amount of dihydroxyl monomers in the feed composition and $y \equiv ([C]/[C] + [D])$ = mole fraction of C based on the total amount of diacid chloride monomers in the feed composition.

Sulfonate LCIs with IEC = 0.034–0.31 meq $g^{-1}$ were prepared by an interfacial condensation reaction of BY, 4,4′-dihydroxy-α,α′-dimethylbenzalazine (DDBA), and a 50/50 molar mixture of sebacoyl and dodecanedioyl dichlorides [145],

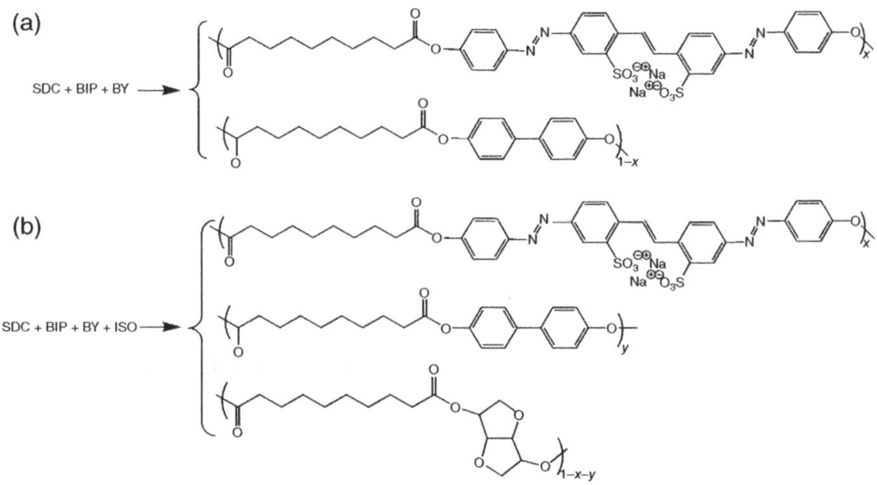


**Scheme 4.47** Main-chain thermotropic polyester LCI prepared from sebacoyl dichloride (SD), 4,4′-biphenyldiol (BIP), isorbide (IS), and brilliant yellow (BY). [149]

Scheme 4.43. The sulfonate concentration was controlled by varying the ratio of BY/DDBA in the feed composition. The authors reported that the addition of the BY monomer appeared to limit the polymerization, which may indicate that the reactivity of the BY monomer was less than that of the DDBA. In that case, once a BY monomer reacted with the growing chain, further polymerization of that chain was suppressed, which would leave the BY units primarily at the chain ends. Their explanation for that hypothesis was that the addition of the hydrophilic BY pulls the growing chain into the aqueous phase, which removes it from the interface with the oil phase containing the acid chlorides. That explanation was supported by the decreasing polymer yields and inherent viscosity with increasing BY feed concentration. However, no characterization evidence for that conclusion was presented.

The BY-azine ionomers shown in Scheme 4.43 were thermally stable to about 300 °C and exhibited a broad nematic mesophase region of 70–100 °C between $T_{mn}$ (126–151 °C) and $T_{ni}$ (200–240 °C). $T_g$ decreased with increasing BY concentration, and although there was scatter in the data, $T_{mn}$ and $T_{ni}$ appeared to decrease with increasing BY. Plots of the reduced viscosity ($\eta_{red}$) versus concentration showed similar anomalous behavior as that reported for flexible ionomers by Lundberg and Phillips [150], in that intramolecular associations of the ionic groups were predominant at low polymer concentrations and lowered the viscosity in chloroform below that of a solution of the non-ionic analogous polymer and intermolecular ionic associations were predominant at higher polymer concentrations and produced a viscosity greater than that of the non-ionic polymer solution.

Similar sulfonate LCIs with IEC = 0.034–0.313 meq $g^{-1}$ were prepared by an interfacial condensation reaction of BY, DDBA and a 10/90 molar mixture of terephthaloyl and sebacoyl dichlorides [146], Scheme 4.44. The nominal compositions of the ionomers were 1–12 mol% BY based on the moles of dihydroxyl monomer in the feed, which corresponded to IEC = 0.093–1.1 meq $g^{-1}$. The semicrystalline ionomers exhibited thermotropic behavior with nematic mesophase textures. $T_g$ and $T_{mn}$ decreased with increasing sulfonate concentration, but no clearing point was observed before the polymer degraded. The solution viscosity behavior ($\eta_{red}$ vs. c) in *N*-methyl-2-pyrrolidone for those ionomers was unusual in that as the concentration decreased for dilute solutions, the reduced viscosity increased, Figure 4.10. Such behavior has been previously been reported for sulfonated polystyrene ionomers in dilute polar solutions [150], where the unshielded negative charges of the anions fixed to the polymer backbone at low polymer concentrations repel one another and produce an expansion of the chain and an increase in its hydrodynamic volume and the solution viscosity. That *polyelectrolyte effect* [151] is commonly observed with high charge density polymers in water, but it is unusual to see such an effect with relatively low charge density ionomers. At higher polymer concentrations, the ion-pair is condensed and the usual increase in viscosity with increasing polymer concentration occurs. The ionomers

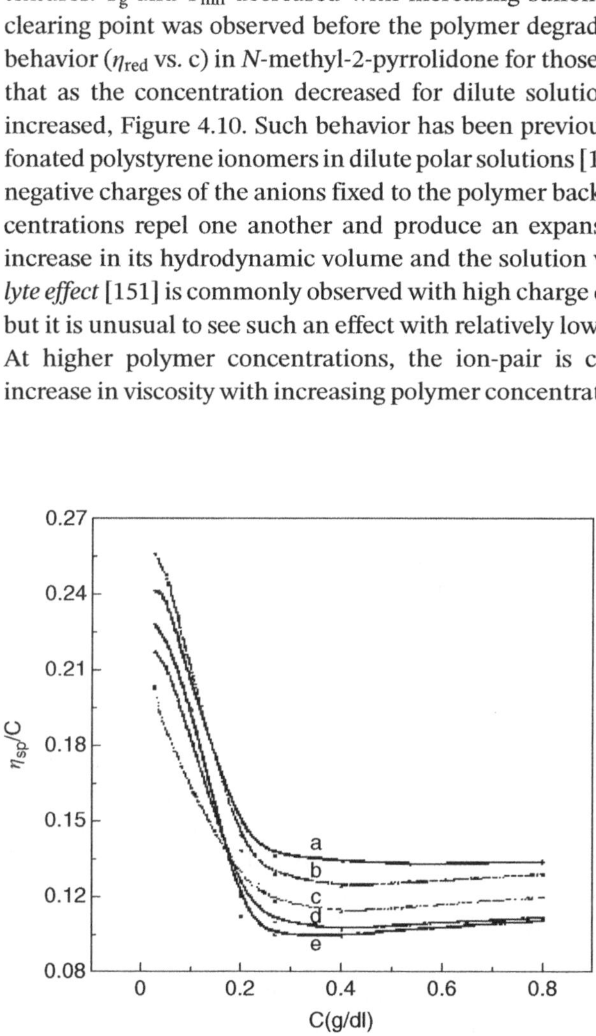


**Figure 4.10** Reduced solution viscosities at 80 °C in *N*-methyl-2-pyrrolidone for non-ionic LCP and BY-azine LCIs: IEC = (a) 0, (b) 0.18, (c) 0.37, (d) 0.72, (e) 1.1. *Source:* Adapted from [145].

shown in Scheme 4.44 also exhibited lyotropic behavior in sulfuric acid solutions with willowy leaf-shaped textures.

Scheme 4.45 shows the synthesis of a thermotropic LCI synthesized by an interfacial condensation of 4,4′-bis(1,10-sebacyloxo)benzoic acid chloride (SOBC), BY and 4,4′-biphenyldiol. The nominal compositions of the ionomers, based on the feed compositions, ranged from 1 to 5 mol% BY (IEC = 0.063–0.31 meq $g^{-1}$). The LCIs were thermally stable to ~270 °C and the melts exhibited a nematic mesophase texture. $T_g$ and $T_{mn}$ of the ionomers were lower than those of an analogous non-ionic LCP, but the transitions for the ionomers were relatively independent of the sulfonate concentration. No clearing point was observed before the ionomers degraded. The reduced viscosity of the ionomers in dimethylformamide exhibited a polyelectrolyte effect, similar to the LCIs shown in Scheme 4.44.

LCIs based on interfacial polycondensation of BY, the dichloride of 1,4-phenylene-dicarbonic acid bis(4-carboxyl phenyl ester), PDCA, and a 50/50 mixture of decanediol and dodecanediol are shown in Scheme 4.46 [148]. The compositions of the LCIs were 0.35–8 mol% BY (IEC = 0.0072–0.55 meq $g^{-1}$). The LCIs were thermally stable to 320 °C and the melts exhibited a nematic mesophase texture. $T_g$ increased slightly and $T_{mn}$ decreased with increasing BY concentration. No clearing point was observed before the ionomers degraded.

Two other LCIs based on BY, shown in Scheme 4.47, were synthesized by an interfacial polycondensation from sebacoyl dichloride (SDC), 4,4′-biphenyldiol (BIP), isosorbide (ISO), and BY [149]. The ionomer compositions for the LCIs denoted by (a) in Scheme 4.47 were 0.5–3.0 mol% BY (IEC = 0.050–0.29 meq $g^{-1}$). All of the formulations for that series of LCIs used an excess of SDC, so the IEC calculations were made assuming that the amount of SDC that was incorporated into the polymers was a stoichiometric amount based on the total hydroxyl groups in the feed composition. For the series of LCIs denoted by (b) in Scheme 4.47 the compositions ranged from 0.5 to 3.0 mol% (IEC = 0.052–0.30 meq $g^{-1}$). As with the A-series, the IEC for the B-series was calculated assuming that the amount of SDC incorporated into the ionomers was the stoichiometric amount based on the total hydroxyl group concentration in the feed. No explanation was provided in the paper for why an excess of SDC was used in the polymerizations. The IECs for the two series of ionomers were similar, and the only difference in the composition of the two series of LCIs was the replacement of some of the BIP with ISO in the B-series LCIs. The objective of the research was to understand how ISO affected the properties of the LCI.

The series A and B LCIs melted into chiral smectic C mesophases ($S_mC^*$), while the introduction of ISO into the structure, series B, produced a mesophase-mesophase transition ($T_{CB}$) from SmC* to a smectic B mesophase. The A-series ionomers had a crystal to crystal transition near ~60 °C and a $T_g$ at ~135 °C that decreased weakly with increasing BY concentration. The $T_{ms}$, where the crystals melted to a SmC* phase, decreased from 164 to 151 °C as the ionic concentration increased from 0.050 to 0.29 meq $g^{-1}$. The clearing point, $T_{si}$ was more sensitive to the ionic content, decreasing from 242 to 184 °C as IEC increased from 0.50 to 0.29 meq $g^{-1}$. For the B-series, $T_g$, $T_{ms}$, $T_{CB,}$ and the $T_{si}$ decreased as IEC increased from 0.052 to 0.30 meq $g^{-1}$ (note that the BIP and ISO concentrations also decreased about 5% each as the ionic concentration increased over that range).

Zhang and coworkers [152, 153] synthesized an LCI similar to that shown in Scheme 4.43, except that they used only a single acid chloride, sebacoyl dichloride. Those LCIs were used to dope a solid electrolyte formed from a blend of PEG and PLA [152] and to intercalate a montmorillonite nanoclay into the PEG/PLA electrolyte [153]. Conductivities of ~$10^{-4}$ S $cm^{-1}$ were achieved, which was four orders of magnitude higher than that of the neat PEG-PLA solid polymer electrolyte.

Meng et al. [154] synthesized a main-chain potassium sulfonate LCI by polycondensation of bis(4-chlorocarbonylphenyl)hexanedioate (CCPH), hydroquinone (HQ) and potassium bis(2-hydroxy-5-ethyleneoxycarbony)benzenesulfonate (KBHEBS), Scheme 4.48. Bis(4-carboxyphenyl) hexanedioate (COPD) was prepared from the reaction of 4-hydroxy benzoic acid (HBA) and hexanedioyl dichloride (HDC), and CCPD was synthesized from the reaction of COPD with thionyl chloride. KBHEBS was prepared from the reaction of EG and 2-hydroxy-5-carboxybenzenesulfonic acid (HCBSA). The paper reported the ionic group concentration as the mass concentration of potassium ions, and the corresponding. IEC values were 0.21–1.0 meq $g^{-1}$. For ionomers with IEC ≤ 0.43 meq $g^{-1}$, the melts exhibited a nematic mesophase texture, but for the ionomers with higher sulfonate

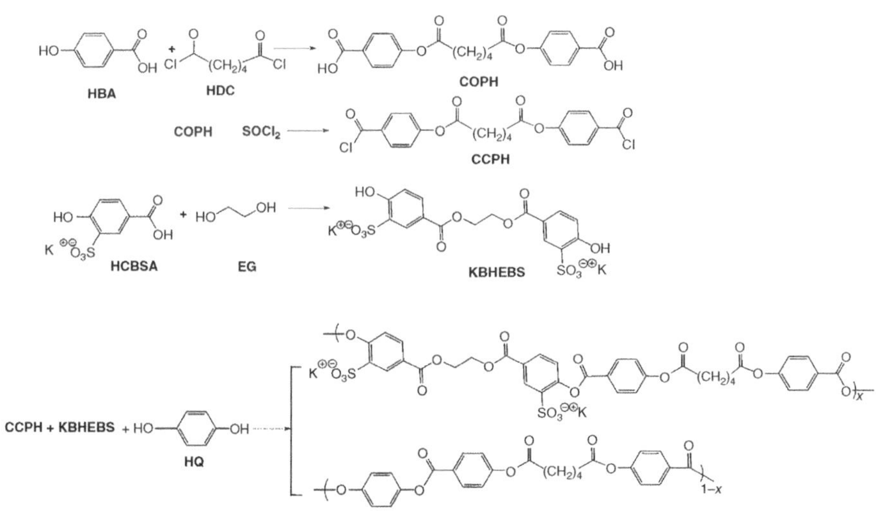


**Scheme 4.48** LCI synthesized from CCPH, HQ, and KBHEBS. [154].

concentrations the crystalline polymer melted to a smectic A mesophase ($T_{ms}$) and a mesophase-mesophase transition to a nematic phase occurred at a higher temperature ($T_{sn}$). The $T_g$, $T_{ms,}$ and $T_{ni}$ of the ionomers were relatively insensitive to the sulfonate concentration, but $T_{sn}$ increased with increasing IEC.

Meng and coworkers [155] also synthesized a main-chain sulfonate LCI by the reaction of BIP, CCPD, and 6,7-dihydroxynaphthalene-2-sulfonic acid (DNSA), Scheme 4.49. Bis(4-(carboxyl)phenyl) decanedioate (COPD) was prepared from the reaction of 4-hydroxy benzoic acid (HBA) and dodecanedioyl dichloride (DDC) and COPD was reacted with thionyl chloride to form CCPD. The sulfonate concentrations were reported as 0.54–2.26 wt.% in that paper, which correspond to IEC = 0.067–0.28 meq $g^{-1}$. The LCIs were complexed with polyaniline (PAN) emeraldine base, and the complexes were dispersed in silicone oil to form electrorheological fluids. The ionic conductivity and the relative permittivity of the LCI-PAN complexes increased with increasing sulfonate concentration. When an electric field was applied to the LCI-PAN dispersions in silicone oil, an electrorheological effect [156] was observed, i.e., the shear stress increased and the rheological behavior was that of a Bingham fluid [157] with a large yield stress. The shear stress achieved at a fixed electric field increased with increasing ionic group concentration.

Telechelic LCIs have also been reported [158, 159], Schemes 4.50 and 4.51. Zhang and Weiss [158] prepared telechelic LCIs that were terminated at each end

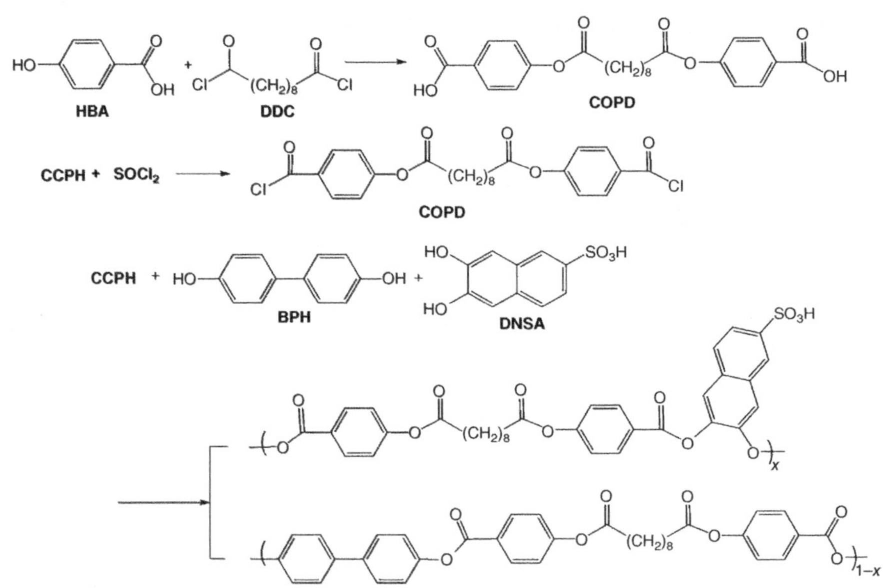


**Scheme 4.49** LCI synthesized from BIP, CCPD and DNSA. [155].

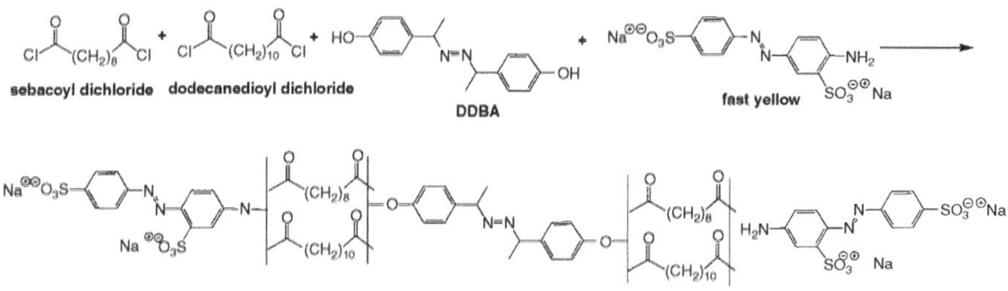


**Scheme 4.50** Telechelic LCI synthesized from sebacoyl dichloride, dodecanedioyl dichloride, 4,4′-dihydroxy-α,α′-dimethylbenzalazine (DDBA) and 4-amino-1,1′-azobenzene-3,4′-disodium sulfonate (fast yellow). [158].

with the ionic dye, fast yellow. (IEC = 0.0078–0.18 meq $g^{-1}$) by an interfacial polymerization of DDBA and a 50/50 mixture of sebacoyl and dodecanedioyl dichlorides, Scheme 4.50. All the LCIs formed nematic melts and $T_g$ (~29 °C), $T_{mn}$ (~148 °C) and $T_{ni}$ (~230 °C) did not change with ionic concentration. The ionic nature of the LCIs was demonstrated by the solution viscosity behavior in mixed solvents of chloroform and methanol, Figure 4.11. The solution viscosity of the ionomer in chloroform was greater than that of a non-ionic LCP analog. However, the addition of small amounts of methanol to the LCI/chloroform solution significantly decreased the solution viscosity due to solvation of the ionic associations by the polar co-solvent. That type of behavior for ionomer solutions, where polar

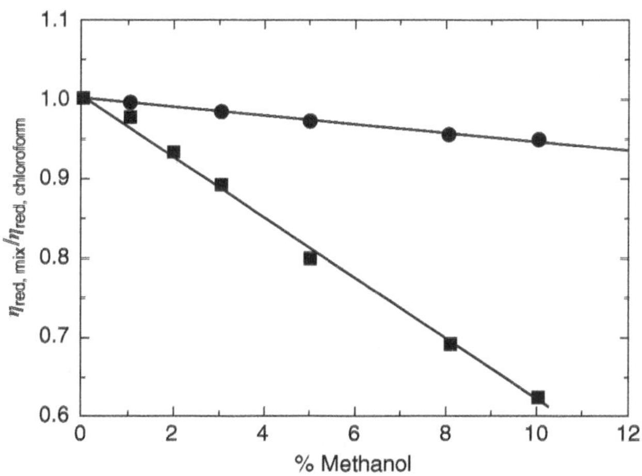


**Figure 4.11** Normalized reduced viscosities of: (●) non-ionic LCP and (■)telechelic LCI (IEC = 0.18 meq $g^{-1}$) at 25 °C in chloroform/methanol mixed solvents as a function of methanol concentration vs. methanol. The viscosities were normalized by the reduced viscosity in chloroform). *Source:* Zhang and Weiss [158]/with permission from John Wiley & Sons.

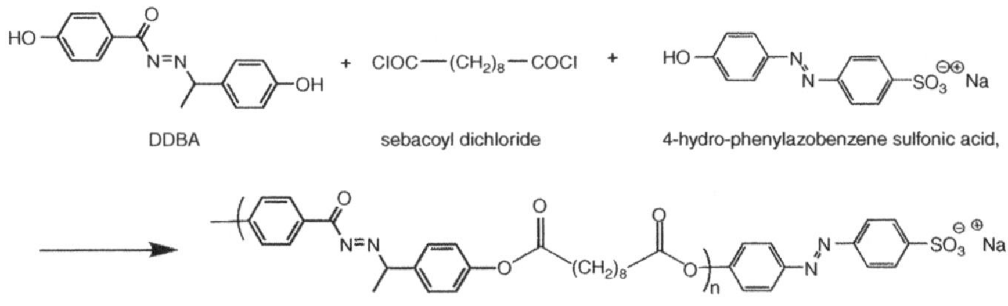


**Scheme 4.51** Telechelic LCI synthesized from sebacoyl dichloride, 4,4′-dihydroxy-**α**,**α**′-dimethylbenzalazine (DDBA) and 4-hydro-phenylazobenzene sulfonic acid. [159].

cosolvents were added to a nonpolar solvents to solvate the ionic interactions was previously reported for flexible ionomers [160].

Zhang et al. [159] synthesized a telechelic LCI by the polymerization of DDBA with sebacoyl dichloride and 4-hydro-phenylazobenzene sulfonic acid. It is not clear from the paper whether both ends of the chain were terminated with 4-hydro-phenylazobenzene sulfonic acid groups, which would be expected, but the mono-telechelic structure shown in Scheme 4.51 is what is shown in Ref. [159]. The LCI was used as a compatibilizer for blends of PBT and polypropylene (PP). The addition of 1 wt.% of the LCI to a 50/50 (w/w) blend of PP and PBT improved the tensile strength and ultimate elongation, but those properties decreased significantly when additional LCI was added.

## 4.7 Polyester Ionomer Patents

Because of the extraordinary number of patents and patent applications where polyester ionomers, mainly sulfopolyesters, are mentioned, this section is not intended to be a comprehensive record of all the intellectual property involving polyester ionomers. Instead, the objective of this section is to provide the reader with a sampling of the compositions and applications of polyester ionomers. The discussion of these patents/applications is purposely brief, since as indicated earlier in the *Introduction* section patents are not peer-reviewed and are usually less precise as to what exactly was accomplished and scientifically verified. Although patents are important with regard technology, they often are composed of shopping lists of possibilities that obfuscate, rather than illuminate reality.

### 4.7.1 Ionomer Synthesis

#### 4.7.1.1 Polyester Ionomers

Larson et al. [161, 162] described the synthesis of an assortment of sulfonate-containing polyester ionomers, including the SPET shown in Scheme 4.1. The synthesis of SPET-type sulfopolyester ionomers were also disclosed by Fragerburg and McFarlane [163] and O'Neill [164]. Duldner et al. [165] reported a process for preparing SPET oligomers by a transesterification process of PET waste with diesters of sodium-5-sulfonate.isophthalic acid.

Water-dispersible, sulfopolyester ionomers were disclosed by Schade [166]. Hartman et al. [167] developed water-dispersible polyester ionomers using cycloalkylsulfonate monomers that are more resistant to ion-exchange and scum formation. Those ionomers had applications as water-developable, dye imbibition and photo-hardening imaging materials with positive color-proofing applications [168]. George [169] synthesized sulfopolyester containing 4,4'-biphenyldicarboxyloyl residues, and George and Cannon [170] described the use of 1,4-cyclohexanedimethanol (CHDM) to synthesize sulfonate-containing copolyesters. Teasley [171] prepared aliphatic-aromatic copolyester ionomers.

Larson [172] described low molecular weight, film-forming, ethylenically-unsaturated sulfonate-containing polyester ionomers that can be crosslinked with actinic radiation. Chan et al. [173] described the synthesis of a variety of sulfonated aromatic monomers for use in the polymerization of sulfopolyesters, and Hayes [174] described the synthesis of sulfonated aromatic copolyesters, including some compositions that were biocompostable. Sulfopolyesters that were biodegradable were disclosed by Hayes and Kosinski [175] and Hayes [176]. Loos et al. [177] reported biodegradable sulfopolyesters based on 2-methylsuccinic acid (or ester). Gupta et al. [178] and Parker and Kirk [179, 180] described processes for recovering sulfopolyesters from an aqueous dispersion or concentrate, which is important with regard to preparing solid sulfopolyester articles. Storey et al. [181] described the synthesis of bioerodable, mono-, and bis-carboxylic acid telechelic polyester ionomers for use in medical applications, such as implants for controlled release of bioactive substances and tissue repair. An example of an $\alpha,\omega$-carboxylic acid poly($\varepsilon$-caprolactone) ionomer was shown in Scheme 4.34.

#### 4.7.1.2 Poly(Ether Ester) Ionomers

A variety of sulfo-poly(ether ester) ionomers were described by Kibler and Lappin [182, 183], and water-dispersible poly(ether esters) were disclosed by Bayless [184], Heiberger [185], Sublett [186], and George [187]. O'Neill and Pacifici [188] described the synthesis of crosslinkable poly(ether ester) ionomers with a variety of applications, such as textile sizing materials, adhesives, coatings, inks, packaging materials, and sequestering agents [189].

### 4.7.2 Polyester Ionomer Applications

#### 4.7.2.1 Adhesives

Noonan and McConkey [190–192] described heat-activatable, water-dispersible polyester adhesives for bonding hydrophilic and hydrophobic surfaces for laminated structures in photographic materials. The ionic species was an iminodisulfonyl compound, such as shown in Schemes 4.11–4.13. A variety of sulfopolyesters have been proposed as water-born adhesive compositions [193, 194] and as hot-melt adhesives [195–199].

Sulfopolyesters have also been used as adhesives in medical and personal care products. For example, Chatelier and Rylatt [200] described the use of a sulfopolyester ionomer as an adhesive in blood contacting devices, such as an immunosensor. O'Brien and Atkins [201] developed switchable pressure sensitive sulfopolyester ionomer adhesives suitable for attachment to skin and for use in medical articles, such as wound and surgical dressings, tapes, and patches, including transdermal drug delivery. Kerins [202] disclosed body-activated adhesive compositions for personal care products, where the adhesive properties increase with exposure to temperature and humidity of the body. Williams and Oldfield [203] reported adhesive and personal care product applications of pourable mixtures of a water-dispersible, branched sulfopolyester and an alcohol.

#### 4.7.2.2 Films, Fibers, and Textiles

Griffing and Remington [204] and Gallagher et al. [205] described film- and fiber-forming sulfopolyester and sulfo-poly(ether esters) that can be processed into shapes by extrusion, molding or casting. Polyester fibers based on ionomers with imino sulfonyl groups (see Schemes 4.11–4.13) with improved dyeability were described by Caldwell and Jones [206]. They reported that fibers made with those ionomers exhibited improved soil or stain release characteristics.

Multicomponent fibers [207–210] containing sulfopolyester have a variety of applications, including microdenier fibers [211], nonwoven fabrics or articles [212–219], flushable personal care products and medical products [220, 221], high strength specialty paper [222, 223], high efficiency filters [224, 225], paperboard or cardboard [226] and battery separators [227]. Sulfopolyester surface treatments have been used to improve soil resistance [228–230], to improve durability of fibers to washing [231, 232], for provide anti-static properties and as fiber sizings [233–237] and as leveling agents in the dyeing of polyester fibers [238]. Other fiber/textile applications of sulfopolyesters include coatings that reduce pilling of fabrics [239], odor preventing textile treatments [240–242], and water-soluble binders [243]. Sulfopolyesters are also used in detergent and fabric conditioning compositions [244–256].

#### 4.7.2.3 Imaging and Inks

Sulfopolyester ionomers are used as imaging elements, receiver films and dispersants in image-forming compositions for photographic, lithographic, and electrographic processes, such as photolithographic plates [257], conductive or antistatic layers [258, 259], binders [260], photo conductors in xerography [261], developer compositions [262], thermal dye printing [263–272], ink jet printing [273], toners [274–276], and color filter arrays [277].

Improved ink compositions with improved quality [278–280], colorant stability [281], antibacterial properties [282] or scents [283] using sulfopolyester ionomers have been described. Krutak et al. [284] and Escano et al. [285] disclosed inks based on sulfopolyesters that contain a near-infrared fluorophore. Such inks can be used in applications that require intelligible markings that are virtually invisible to the human eye on the surface of articles for identification, authentication or sorting. The use of water-dispersible sulfopolyesters in ink-jet printing inks can improve properties, such as drying, water fastness, abrasion resistance, and image quality [286–289]. From an environmental perspective, Langston et al. [290] described a waste treatment for water-based printing operations, specifically to isolate sulfopolyesters used in aqueous ink dispersions.

#### 4.7.2.4 Coatings and Packaging

Malotky et al. [291] disclosed sulfopolyester ionomer coating compositions and methods for achieving aqueous dispersions for applying coatings. Jenkins et al. [292] described using a sulfopolyester ionomer to produce high solids aqueous dispersions for coatings or adhesives. Wu [293] described water vapor barrier and chemically resistant polyester or poly(ether ester) ionomers films for applications as protective fabric and clothing, gloves, shoes, hats, tents, and sleeping bags resistant to noxious liquids and gases. Nair et al. [294] disclosed the use of the imino sulfonyl polyester ionomers shown in Schemes 4.11–4.13 as protective overcoats for photograph that resist fingerprints, common stains, spills and wet abrasion. Carrie et al. [295] described protective gas barrier composite coatings composed of a water-emulsifiable sulfopolyester ionomer and high aspect ratio platelet fillers, e.g., mica or clay. George et al. [296] disclosed water-dispersible sulfopolyester compositions with high concentrations of pigments or fillers that are useful as packaging materials.

Erneta and Priewe [297] described antimicrobial sulfopolyester ionomer coatings. Podhajny [298] described an inexpensive process to make antimicrobial packaging materials by printing an inert sulfopolyester dispersion containing zeolites that release antimicrobial metal ions, such as silver, upon exposure to moisture.

Moore [299] described rinseable primers that allow for the removal of temporary functional coatings for applications where coatings need to be removed to achieve the final product. An example is the removal of temporary adhesives used in the manufacture of semiconductors and flat panel displays. Sulfopolyester coatings can also be used to remove organic substances from a substrate, such as microelectronic wafers or flat panel displays [300]. Kuo et al. [301] described the use of a sulfopolyester ionomer as a stabilizer in a process for preparing a latex-modified alkyd composition.

Buchanan et al. [302] described the use of sulfopolyester ionomers, specifically Eastman's AQ-ionomers, as components of a topically-applied skin and face coating/painting compositions that protect the skin from damage due to intense external heat or thermal flux. Stockl et al. [303] described the use of sulfopolyesters as a low VOC additive to an aqueous coating that extended the coating's wet-edge time (time period over which the edge of the coating is workable) and open time (time period over which main area of coating is workable). George and Sublet [304] disclosed sulfopolyester compositions useful as adhesives, coating materials, sizes, laminated products, inks, and films, particularly where increased abrasion resistance, inherent higher temperature conditions and improved blocking resistance are required. Agarwal et al. [305] described oxygen-scavenging polyester compositions containing a cobalt-neutralized sulfopolyester.

Li et al. [306] described sulfopolyester coating compositions that can hide, but later reveal a hidden layer. One example application is to mask a substrate and later expose the substrate by applying an eradicator fluid that reacts with the coating to render it clear. Another example is a coating that covers artwork or printing that can be mechanically scratched off. Sharma et al. [307] described a waterborne correction fluid containing a water-dispersible sulfopolyester with a high solids content that dried quickly and bonded to paper surfaces for correcting laser-jet printed and photocopied markings.

#### 4.7.2.5 Antistatic and Conductive Compositions

Sulfonate-containing polyester ionomers can be used as antistatic films [308] or as binders for an antistatic agent, such as vanadium oxide, in various applications, e.g., photographic or cinematographic or photographic film [309, 310], abrasives [311], delaminating coatings for use in thermal transfer systems [312, 313] and antistatic coatings for self-supporting articles [314, 315]. Matsubayashi [316] described a method for producing an antistatic molded article by coating the article with a conductive composite containing a π-conjugated conductive polymer and a sulfopolyester binder. Polyester ionomers containing iminodisulfonyl groups are also effective for stabilizing superconductive ceramic oxides in electrically conductive materials that exhibit superconductivity at relatively high temperatures [317, 318].

#### 4.7.2.6 Polymer Blends, Compatibilizers, and Nanocomposites

Sulfopolyester can compatibilize polyamide/polyester blends used in bottles and containers [319–322]. The compatibilized blends exhibited improved color and clarity compared to blends without the ionomer. Similarly, sulfopolyesters can compatibilize polyamide/poly(arylene ether) blends and reduce the moisture absorption and improve paint adhesion [323]. Sanjay [324] disclosed thermoplastic blends of a sulfopolyester and poly(glycolic acid) for improving the gas barrier properties of clear articles, containers, bottle or films. Sulfopolyester have also been used to improve the dispersion and intercalation of orgnoclays in polyester-based nanocomposites [325–327]. Farrugia et al. [328] described composites of silver nanoparticles, dispersed in a sulfopolyester matrix for antimicrobial applications, such as biosensors, fibers, cosmetics, and electronic components.

#### 4.7.2.7 Cosmetics and Personal Care Products

Crane and Fonolla-Moreno [329] described mascara formulations containing a water-dispersible sulfopolyester, such as Eastman's AQ55 ionomer. Piot et al. [330] disclosed mascara compositions made from hard waxes and a film forming sulfopolyester. Hsieh et al. [331] prepared nail enamel compositions that contained low molecular weight sulfopolyester to improve adhesion and removal of the enamel. Myers et al. [332] and Dobbs [333] reported cosmetic formulations based on sulfopolyester compositions that are quick drying, exhibit a glossy finish, have good vapor transmission, excellent adhesion, and excellent durability. Hapern-Chirch et al. [334] disclosed cosmetic compositions that incorporate a complex of a sulfopolyester and an ammonium methacrylate copolymer that improves the water resistance and reduces the stickiness of the cosmetic composition. Pays et al. [335] describe the use of sulfopolyesters, such as Eastman's AQ ionomers, as a component of artificial eye lashes.

Bebot and Pasquet [336] used a branched sulfopolyester ionomer as a component in cosmetic compositions for styling and shaping hair. Bhatt and Galleguillos [337] described using sulfopolyester ionomers as a component of an aqueous aerosol or non-aerosol hair styling aid or mousse composition. Pays et al. [338] described water-resistant sulfopolyester cosmetic compositions for skin, lips, and keratin fibers. Those compositions can be used for makeup products for the face, the body and/or the lips and for artificial eyelashes, eyebrows and hair. Sulfopolyester ionomers also have application in hair coloring [339] and hair spray formulations [340–343]. McEntire et al. [344] disclosed a sulfopolyester film for masking skin fissures and imperfections, such as wrinkles, cracks and abrasions. Spina and Maletsky [345] described a depilatory for removal of body hair that uses a hot melt sulfopolyester-based adhesive.

Catalan et al. [346] disclosed the use of sulfopolyester ionomer films for durable hydrophilic nanocomposite coating compositions. These coatings may be used to produce absorbent, disposable articles, such as diapers, training pants, adult incontinence products and feminine care products. McEntire et al. [347] described a water-dispersing film using a sulfopolyester for delivery of a cosmetic, dermatological or pharmaceutical active ingredient to the skin. Similarly, a water-dispersible dermal patch can be achieved produced from a sulfopolyester ionomer containing an active agent for release [348]. McEntire and McCaulley [349] disclosed sulfopolyester sunscreen compositions that are stable in oil-in-water emulsions and required less phosphate surfactant than formulas without the sulfopolyester.

#### 4.7.2.8 Miscellaneous Applications

Nagarajan [350] described methods for increasing the enzymatic hydrolytic degradation by microorganisms of aliphatic-aromatic copolyesters, including sulfo-copolyesters. Hahn [351] disclosed sulfo-poly(ether ester) compositions used to treat crude oil during production and refinery processing. Campbell [352] disclosed the use of sulfopolyester ionomers, e.g., SPET and SPBT, as nucleating agents to promote crystallization of polyester thermoplastics, such as PET. Ghosh and Majumdar [353] described a process using a polyester ionomer-intercalated or exfoliated smectic clay to produce a ceramic. An ionomer-clay solution was mixed with zirconia and stabilized with at least one rare earth oxide element (Mg, Ca, Y, Sc, or Ce), and the mixture was sintered to form a ceramic.

## Abbreviations

| | |
|---|---|
| BD | 1,4-butanediol |
| BHEC | 1,4-bis(2-hydroxyethoxy) cyclohexane diol |
| BIP | 4,4′-biphenyldiol |
| BTC | 2-bromoterephthaloyl chloride |
| BTDA | 1,2,4,5-benzenetetracarboxylic dianhydride |
| BY | brilliant yellow |
| Ca-HQDS | calcium salt of hydroquinone-2,5-disulfonate |
| CHDM | 1,4-cyclohexanedimethanol |
| CCPH | bis(4-chlorocarbonylphenyl)hexanedioate |
| COPD | bis(4-carboxyphenyl) hexanedioate |
| DDBA | 4,4′-dihydroxy-α,α′-dimethylbenzalazine |
| DEG | diethylene glycol |
| DMCD | dimethyl-1,4-cyclohexanedicarboxylate |
| DMI | dimethyl isophthalate |
| DMISDBNa | sodium- dimethyl-3,3′-bis(imino)-disulfonyl dibenzoate |
| DMPI | sodium-dimethyl-5-phosphoisophthalate |
| DMSI | dimethyl sodium-5-sulfoisophthalate |
| DMSI-NTP | naphthalene-containing thermotropic LCI prepared using DMSI |
| DMST | dimethyl sodium sulfoterephthalate |
| DMT | dimethyl terephthalate |
| DMF | dimethyl fumerate |
| DNSA | 6,7-dihydroxynaphthalene-2-sulfonic acid |
| DPG | dipropylene glycol |
| DSC | differential scanning calorimetry |
| ECDC | easy cationic dyeable copolyester |
| EG | ethylene glycol |

| | |
|---|---|
| EXAFS | extended X-ray absorption fine structure |
| FTIR | Fourier transform infrared spectroscopy |
| HBA | 4-hydroxy benzoic acid |
| HCBSA | 2-hydroxy-5-carboxy-benzenesulfonic acid |
| HDC | Hexanedioyl dichloride |
| HEMA | 2-hydroxyethyl methacrylate |
| HQ | hydroquinone |
| IEC | ion exchange capacity |
| IPA | isophthalic acid |
| ISO | isosorbide |
| KBHEBS | potassium bis(2-hydroxy-5-ethyleneoxycarbony) benzenesulfonate |
| K-HQDS | potassium salt of hydroquinone-2,5-disulfonate |
| LA | D,L-lactic acid |
| LCI | liquid crystal ionomer |
| LCP | liquid crystalline polymers |
| LVE | linear viscoelastic |
| MA | L-malic acid |
| meq/g | milliequivalents per gram (measure of IEC), M denotes the metal cation |
| MHBA | methyl *p*-hydroxybenzoate |
| MHNA | methyl 6-hydroxy-2-naphthoate |
| M-HQDS NTP | naphthalene-containing thermotropic LCI prepared using M-HQDS, M denotes the metal cation |
| MPD | 2-methyl-1,3-propanediol |
| mPEG-1200 | monomethyl ether of polyethylene glycol with $M_n = 1200\,\text{g mole}^{-1}$ |
| M-PETDEGSP | M-phosphonate telechelic polyester ionomer shown in Scheme 4.34 |
| M-SIPEPT-y | sulfo-poly(ether ester) copolymers of SIP and a mixture of PEG-x and PTMG-x, where M denotes a metal cation and y is the IEC (meq/g) |
| M-SIPE-y | sulfo-poly(ether ester) copolymers of SIP and PEG, where M denotes a metal cation and y is the IEC (meq/g) |
| M-SIPT-y | sulfo-poly(ether ester) copolymers of SIP and PTMG, where M denotes a metal cation and y is the IEC (meq/g) |
| NMR | nuclear magnetic resonance |
| NTP | naphthalene-containing thermotropic |
| NPG | 2,2-dimethyl-1,3-propanediol |
| NTC | nitroterephthaloyl chloride |
| OMC | organically-modified montmorillonite clay |

| | |
|---|---|
| PAE | polyamic ester |
| PAG-x | poly(alkylene glycol), where $x$ = number average molecular weight |
| PBI | poly(butylene isophthalate) |
| PBT | poly(butylene terephthalate), |
| PD | 1,3-propanediol |
| PDCA | 1,4-phenylene-dicarbonic acid bis(4-carboxyl phenyl ester) |
| PEGNP | peg-grafted silica nanoparticles |
| PEG-x | poly(ethylene glycol), where $x$ = number average molecular weight |
| PET | poly(ethylene terephthalate) |
| PG | propylene glycol or 1,2-propanediol |
| PTMG-x | poly(tetramethylene glycol), where $x$ = number average molecular weight |
| PVA | poly(vinyl alcohol) |
| SAXS | small angle x-ray scattering |
| SBHEI | sodium bis(2-hydroxyethyl) 5-sulfoisophthalate |
| SDC | sebacoyl dichloride |
| SIP | sulfoisophthaloyl groups |
| SNP | silica nanoparticles |
| SOBC | 4,4'-bis(1,10-sebacyloxo)benzoic acid chloride |
| SPBI | random copolymers of butylene isophthalate and butylene sulfoisophthalate |
| SPBT | random copolymers of butylene terephthalate and butylene sulfoisophthalate |
| SPEG | sodium sulfonated polyethylene glycol |
| SPET | random copolymers of ethylene terephthalate and ethylene sulfoisophthalate |
| SPETT | random copolymers of ethylene terephthalate and ethylene sulfoterephthalate |
| SSBA | 3-sulfobenzoic acid |
| TA | terephthalic acid |
| TC | terephthaloyl chloride |
| TGA | thermogravimetric analysis |
| VFT | Vogel-Fulcher-Tammann |
| Zn-TeraSP | zinc-phosphonate telechelic PTMG (see Scheme 4.34) |
| $G'$ | dynamic modulus |
| $G''$ | loss modulus |
| $M_n$ | number-average molecular weight |
| $M_v$ | viscosity-average molecular weight |
| $S_{mC*}$ | chiral smectic C mesophase |
| $T_{CB}$ | Transition temperature where a smectic C mesophase becomes a smectic B mesophase |
| $T_g$ | glass transition temperature |

$T_m$ melting point, defined as the maximum rate of the enthalpy change during the melting transition measured by differential scanning calorimetry
$T_{mn}$ melting transition of crystal to a nematic liquid
$T_{ms}$ melting transition of crystal to a smectic liquid
$T_{ni}$ clearing transition of a nematic liquid to an isotropic liquid
$T_{si}$ clearing transition of a smectic liquid to an isotropic liquid
$[\eta]$ intrinsic viscosity
$\eta_{red}$ reduced viscosity

## References


1 Tant, M.R., Mauritz, K.A., and Wilkes, G.L. (ed.) (1997). *Ionomers: Synthesis, Structure, Properties and Application*. London: Blackie Academic & Professional.
2 Rees, R.W. (1966). Ionic hydrocarbon polymer. US Patent 3,264,272, 2 August 1966.
3 Peyser, P. (1989). Glass transition temperatures of polymers. In: *Polymer Handbook* (ed. E.H. Immergut). New York: Wiley.
4 The Business Research Company (2023). *Polyethylene Terephthalate Global Market Report 2023*. Dublin, Ireland: The Business Research Company https://www.thebusinessresearchcompany.com/report/polyethylene-terephthalate-global-market-report.
5 Greener, J., Gillmor, J.R., and Daly, R.C. (1993). Melt rheology of a class of polyester ionomers. *Macromolecules* 26: 6416–6424.
6 Guo, X.-Y., Gu, L.-X., and Feng, X.-X. (2002). The glass transition, crystallization, and melting characteristics of a class of polyester ionomers. *J. Appl. Polym. Sci.* 86: 3660–3666.
7 Timm, D.A. and Hsieh, Y.-L. (1993). The microstructure and macrostructure of sulfonated poly(ethylene terethphalate) fibers. *J. Polym. Sci, Part B: Polym. Phys.* 31: 1873–1883.
8 Timm, D.A. and Hsieh, Y.-L. (1994). Solvent-induced structural changes in sulfonated poly(ethylene terephthalate) (SPET) fibers. *J. Appl. Polym. Sci.* 51: 1291–1301.
9 Ingamells, W., Lilou, S.H., and Peters, R.H. (1981). The accessibility of sulphonic acid groups in basic hydrolysis during dyeing dyeable polyester fibers. I. The prevention of fiber hydrolysis during dyeing. *J. Appl. Polym. Sci.* 26: 4087–4094.
10 Ingamells, W., Lilou, S.H., and Peters, R.H. (1981). The accessibility of sulphonic acid groups in basic dyeable polyester fibers. 1I. The influence of plasticization on the ion exchange process. *J. Appl. Polym. Sci.* 26: 4095–4101.
11 Ostrowska-Gumkowska, B. and Ostrowska-Czubenko, J. (1994). Property-structure relationships in partially sulfonated poly(ethylene terephthalate). 1. Infrared studies. *Eur. Polym. J.* 30: 869–874.

**12** Register, R.A. and Prud'homme, R.K. (1997). Melt rheology. In: *Ionomers: Synthesis,Structure, Properties and Applications* (ed. M.R. Tant, K.A. Mauritz, and G.L. Wilkes), 208–260. London: Chapman and Hall.

**13** Weiss, R.A. (1994). Glass transition(s) of ionomers. In: *Assignment of the Glass Transition*, Standard Tech. Publ. 1264 (ed. R.J. Seyler), 214–225. Philadelphis: American Society for Testing Materials.

**14** Lee, C.H., Inoue, T., and Nah, J.-W. (2002). Morphology development in a range of nanometer to micrometer in sulfonated poly(ethylene terephthalate) ionomer. *Bull. Korean Chem. Soc. 23*: 580–586.

**15** Zhang, H., Li, S., and Luo, H. (1987). Melt flow and fiber properties of copolyesters. *J. Appl. Polym. Sci.* 34: 1353–1366.

**16** Hsiao, K., Shu, Y.-C., and Tsen, W.-C. (2003). Thermal and mechanical properties of polyblends of polycaprolactam with cationic dyeable poly(ethylene terephthalate). *J Polym. Res.* 10: 161–169.

**17** Shu, Y.-C. and Hsiao, K.-J. (2007). Flow and thermal characteristics of cationic dyeable nylon 6/cationic dyeable poly(ethylene terephthalate) polyblended polymers and filaments. *J. Appl. Polym. Sci.* 103: 2049–2056.

**18** Hsiao, K.-J., Jen, Z.-F., and Lu, C.-L. (2002). Analysis of physical properties of PET/CD-PET polyblend hollow fiber and its kinetics of alkaline hydrolysis. *J. Appl. Polym. Sci.* 86: 3601–3610.

**19** Hsiao, K.-J., Kuo, J.-L., Tang, J.-W., and Chen, L.-T. (2005). Physics and kinetics of alkaline hydrolysis of cationic dyeable poly(ethylene terephthalate) (CDPET) and polyethylene glycol (PEG)-modified CDPET polymers: effects of dimethyl 5-sulfoisophthalate sodium salt/peg content and the number-average molecular weight of the PEG. *J. Appl. Polym. Sci.* 98: 550–556.

**20** Iyer, S. and Schiraldi, D.A. (2006). Role of ionic interactions in the compatibility of polyester ionomers with poly(ethylene terephthalate) and nylon 6. *J. Polym. Sci., Part B: Polym Phys.* 44: 2091–2103.

**21** Prattipati, V., Hu, Y.S., Bandi, S. et al. (2005). Effect of compatibilization on the oxygen-barrier properties of poly(ethylene terephthalate)/poly(*m*-xylylene adipamide) blends. *J. Appl. Polym. Sci.* 97: 1361–1370.

**22** Barber, G.D., Calhoun, B.H., and Moore, R.B. (2005). Poly(ethylene terephthalate) ionomer based clay nanocomposites produced via melt extrusion. *Polymer* 46: 6706–6714.

**23** Vidotti, S.E., Chinellato, A.C., Hu, G.-H., and Pessan, L.A. (2007). Preparation of poly(ethylene terephthalate)/organoclay nanocomposites using a polyester ionomer as a compatibilizer. *J. Polym. Sci., Part B: Polym. Phys.* 45: 3084–3091.

**24** Paranhos, C.M., Soares, B.G., Machado, J.C. et al. (2007). Microstructure and free volume evaluation of poly(vinyl alcohol) nanocomposite hydrogels. *Eur. Polym. J.* 43: 4882–4890.

**25** Gorda, K.R. and Peiffer, D.G. (1992). Properties of sulfonated poly (butylene terephthalate). *J. Polym. Sci., Part B: Polym. Phys.* 30: 281–292.

**26** Nicholson, L.K., Mahmood, K., Gabrys, B. et al. (1999). An investigation into the behaviour of SPBT and SPBT/PC blends. *Polymer* 40: 6099–6111.

**27** Kalhoroa, M.S., Gabrys, B.J., Zajacc, W. et al. (2001). Small angle neutron scattering study of SPBT/PC blends. *Polymer* 42: 1679–1690.

**28** Szymczyk, A. and Roslaniec, Z. (1999). Sulfonated poly(ether-block-ester) ionomers with anions in the polyester hard segments. *Polym. Adv. Technol.* 10: 579–587.

**29** Stein, R.S. (1973). Optical studies of the morphology of polymer films. In: *Structure and Properties of Polymer Films*, Polym. Sci. Tech. 1973, vol. 1 (ed. R.W. Lenz and R.S. Stein), 1–24. Boston, MA: Springer.

**30** Nishi, T. and Wang, T.T. (1975). Melting point depression and kinetic effects of cooling on crystallization in poly(vinylidene fluoride)-poly(methyl methacrylate) mixtures. *Macromolecules.* 8: 909–915.

**31** Chisholm, B.J., Moore, R.B., Barber, G. et al. (2002). Nanocomposites derived from sulfonated poly(butylene terephthalate). *Macromolecules* 35: 5508–5516.

**32** Pilati, F., Manaresi, P., Ruperto, M.G. et al. (1993). Ion-containing polymers: 1. Synthesis and properties of poly(1,4-butylene isophthalate) containing sodium sulfonate groups. *Polymer* 34: 2413–2421.

**33** Lotti, N., Siracusa, V., Finelli, L., and Munari, A. (2002). Melting behavior and crystallization kinetics of sulfonated poly(butylene isophthalate). *Polym. Eng. Sci.* 42: 1590–1599.

**34** Munari, A., Fiorini, M., Pezzin, G., and Pilati, F. (1993). Rheological behavior of poly(1,4-butylene isophthalate) containing sodium sulfonate groups. *J. App. Polym. Sci.* 50: 159–164.

**35** Chen, B., Zhong, L., and Gu, L. (2010). Thermal properties and chemical changes in blend melt spinning of cellulose acetate butyrate and a novel cationic dyeable copolyester. *J. Appl. Polym. Sci.* 116: 2487–2495.

**36** Wang, J., Li, X., Cai, Z., and Gu, L. (2015). Absorption kinetics and thermodynamics of cationic dyeing on easily dyeable copolyester modified by 2-methyl-1,3-propanediol. *Fiber Polym.* 16: 2384–2390.

**37** Fu, C. and Gu, L. (2013). Structures and properties of easily dyeable copolyesters and their fibers respectively modified by three kinds of diols. *J. Appl. Polym. Sci.* 128: 3964–3973.

**38** Wang, Q., Wang, C., Wang, H. et al. (2012). Preparation and characterization of the hydrophilic copolyester. *Adv. Mater, Res-Switz.* 535–537: 1499–1502.

**39** Subramani, S., Lee, S.J., Lee, J.M. et al. (2009). A facile preparation of microparticles from sulfonated polyester nanoparticles via emulsion–aggregation process. *Colloid Surf. A* 340: 40–49.

**40** Nikolić, D., Moffat, A., Farrugia, V.M. et al. (2013). Multi-scale modeling and synthesis of polyester ionomers. *Phys. Chem. Chem. Phys.* 15: 6128–6138.

**41** Baik, D.H. and Kim, G.L. (2001). Synthesis and characterization of copolyester sizing agents. *Fibers Polym.* 2: 26–29.

**42** Ju, L., Pretelt, J., Chen, T. et al. (2018). Synthesis and characterization of phosphonated poly(ethylene terephthalate) ionomers. *Polymer* 151: 154–163.

**43** Ju, L., Dennis, J.M., Heifferon, V. et al. (2019). Compatibilization of polyester/polyamide blends with a phosphonated poly(ethylene terephthalate) ionomer: comparison of monovalent and divalent pendant ions. *ACS Appl. Polym. Mater.* 1: 1071–1080.

**44** Turner, S.R., Seymour, R.W., and Dombroski, J.R. (2003). Amorphous and crystalline polyesters based on 1,4-cyclohexanedimethanol. In: *Modern Polyesters: Chemistry and Technology of Polyesters and Copolyesters* (ed. J. Scheirs and T.E. Long). West Sussex, England: Wiley; Chapter 7.

**45** Turner, S.R. (2004). Development of amorphous copolyesters based on 1,4-cyclohexanedimethanol. *J. Polym. Sci., Part A: Polym. Chem.* 42: 5847–5852.

**46** Grady, B.P., Floyd, J., Genetti, W.B., and Moore, R.B. (1996). EXAFS Investigations of transition metal ionomers. *ACS Polym. Preprints* 37 (1): 418–419.

**47** Islam, S., Inglefield, D.L., and Velev, O.D. (2018). Revisiting the colloidal fundamentals of water dispersible polyesters: interactions and selfassembly of polymer nanoaggregates in water. *Soft Matter* 14: 2118–2130.

**48** Boykin, T.L. and Moore, R.B. (1997). Compatibility of polyester ionomers with poly(ethylene terephthalate) and nylon 6,6: melt versus solution blending. *ACS Polym. Preprints* 38 (2): 179–180.

**49** Boykin, T.L. and Moore, R.B. (1998). The role of specific interactions and transreactions on the compatibility of polyester ionomers with poly(ethyene terephthalate) and nylon 6,6. *Polym. Eng. Sci.* 38: 1658–1665.

**50** Boykin, T.L., Ciacco, J., and Moore, R.B. (1996). The utilization of ionomers as compatibilizers for immiscible blends of polar polymers. *ACS Polym. Preprint* 37 (1): 717–718.

**51** Moore, R.B. and Boykin, T.L. (2001). Polyester ionomers in binary and compatibilized blends with poly(ethylene terephthalate), poly(butylene terephthalate) and nylon 6,6. *Proc. An. Tech. Conf., Soc. Plast. Eng.* 59 (2): 2249–2253.

**52** Boykin, T.L. and Moore, R.B. (1998). Extent of interchange reactions for blends of polyester ionomers with poly(ethylene terephthalate). *Polym. Preprints* 39: 393–394.

**53** Ammala, A., Bell, C., and Dean, K. (2008). Poly(ethylene terephthalate) clay nanocomposites: improved dispersion based on an aqueous ionomer. *Compos. Sci. Technol.* 68: 1328–1337.

**54** Tudryn, G.J., Liu, W., Wang, S.-W., and Colby, R.H. (2011). Counterion dynamics in polyester sulfonate ionomers with ionic liquid counterions. *Macromolecules* 44: 3572–3582.

**55** Yeom, B.Y. and Pourdeyhimi, B. (2011). Web fabrication and characterization of unique winged shaped, area-enhanced fibers via a bicomponent spunbond process. *J. Mater. Sci.* 46: 3252–3257.

**56** Noonan, J.M. and McConkey, R.C. (1978). Water-soluble polyester adhesives. *ACS Polym. Preprints.* 19 (2): 407–413.

**57** Noonan, J.M. and McConkey, R.C. (1980). Water-soluble polyester adhesives. In: *Ions in Polymers*, Adv. Chem. Ser., vol. 187, 185–194. American Chemical Society.

**58** Connelly, R.W., McConkey, R.C., Noonan, J.M., and Pearson, G.H. (1982). Melt rheology of ion-containing polymers. I. Effect of ionic content in a model polyesterionomer. *J. Polym. Sci., Polym. Phys. Ed.* 20: 259–268.

**59** Bernstein, B., Kearsley, E., and Zapas, L.J. (1963). A study of stress relaxation with finite strain. *Trans. Soc. Rheol.* 7: 391–410.

**60** Kaye, A. and Kennett, A.J. (1974). Constrained elastic recovery of a polymeric liquid after various shear flow histories. *Rheol. Acta* 13: 916–923.

**61** Ferrar, W.T., Molaire, M.F., Cowdery, J.R. et al. (2004). Electron-transport agents incorporated experimental Into polyester ionomers. *ACS Polym. Prepr.* 45 (1): 232–233.

**62** Ishida, K., Han, S.-I., Inoue, Y., and Im, S.-S. (2005). Novel poly(butylene succinate)-based ionomers with sulfonated succinate units: synthesis, morphology, and the unique nucleation effect on crystallization. *Macromol. Chem. Phys.* 206: 1028–1034.

**63** Lim, J.S., Lee, Y., and Im, S.S. (2008). Influence of ionic association on the nonisothermal crystallization kinetics of sodium sulfonate poly(butylene succinate) ionomers. *J. Polym. Sci., Part B: Polym. Phys.* 46: 925–937.

**64** Weiss, R.A. and Yu, W.-C. (2007). Viscoelastic behavior of very lightly sulfonated polystyrene ionomers. *Macromolecules* 40: 3640–3643.

**65** Yoo, Y.T., Lee, B.J., Han, S.-I. et al. (2003). Physical properties and biodegradation of poly(butylene adipate) ionomers. *Polym. Degrad. Stabil.* 79: 257–264.

**66** Han, S.-I., Im, S.S., and Kim, D.K. (2003). Dynamic mechanical and melt rheological properties of sulfonated poly(butylene succinate) ionomers. *Polymer* 44: 7165–7173.

**67** Han, S.-I., Yoo, Y., Kim, D.K., and Im, S.S. (2004). Biodegradable aliphatic polyester ionomers. *Macromol. Biosci.* 4: 199–207.

**68** Pekkanen, A., Horatz, K., Drummey, K., et al. (2017). Charged polycaprolactone copolymers as bioadhesive would glue. Abstract 254th ACS Meeting.

**69** Ghanem, A.F., El-Gendi, A., Abdel Rehim, M.H., and El-Khatib, K.M. (2016). Hyperbranched polyester and its sodium titanate nanocomposites as proton exchange membranes for fuel cells. *RSC Adv.* 6: 32245–32257.

**70** Virtanen, H., Orelma, H., Erho, T., and Smolander, M. (2012). Development of printable bioactive paper containing laccase. *Process Biochem.* 47: 1496–1502.

**71** Al-Ahmed, A., Ndangili, P.M., Jahed, N. et al. (2009). Polyester sulphonic acid interstitial nanocomposite platform for peroxide biosensor. *Sensors* 9: 9965–9976.

**72** Dou, S., Zhang, S., Klein, R.J. et al. (2006). Synthesis and characterization of poly(ethylene glycol)-based single-ion conductors. *Chem. Mater.* 18: 4288–4295.

**73** Klein, R.J., Zhang, S., Dou, S. et al. (2006). Modeling electrode polarization in dielectric spectroscopy: Ion mobility and mobile ion concentration of single-ion polymer electrolytes. *J. Chem. Phys.* 124: 144903.

**74** Wang, W., Liu, W., Tudryn, G.J. et al. (2010). Multi-length scale morphology of poly(ethylene oxide)-based sulfonate ionomers with alkali cations at room temperature. *Macromolecules* 43: 4223–4229.

**75** Wang, W., Tudryn, G.J., Colby, R.H., and Winey, K.I. (2011). Thermally driven ionic aggregation in poly(ethylene oxide)-based sulfonate ionomers. *J. Am. Chem. Soc.* 133: 10826–10831.

**76** Lu, M., Runt, J., and Painter, P. (2009). An infrared spectrocopic study of a polyester copolymer ionomer based on poly(ethylene oxide). *Macromolecules* 42: 6581–6587.

**77** Rudzinski, J.F., Lu, K., Milner, S.T. et al. (2017). Extended ensemble approach to transferable potentials for low-resolution coarse-grained models of ionomers. *J. Chem. Theory Comput.* 13: 2185–2201.

**78** O'Reilly, M.V. and Winey, K.I. (2015). Silica nanoparticles densely grafted with peg for ionomer plasticization. *RSC Adv.* 5: 19570–19580.

**79** Iacob, C. and Runt, J. (2016). Charge transport of polyester ether ionomers in unidirectional silica nanopores. *ACS Macro Lett.* 5: 476–480.

**80** Pekkanen, A.M., Zawaski, C., Stevenson, A.T. et al. (2017). Poly(ether ester) ionomers as water-soluble polymers for material extrusion additive manufacturing processes. *ACS Appl. Mater. Interfaces* 9: 12324–12333.

**81** Zawaski, C.E., Wilts, E.M., Chatham, C.A. et al. (2019). Tuning the material properties of a water-soluble ionic polymer using different counterions for material extrusion additive manufacturing. *Polymer* 176: 283–292.

**82** Zhao, M.-L., Li, F.-X., Yu, J.-Y., and Wang, X.-L. (2014). Preparation and characterization of poly(ethylene terephthalate) copolyesters modified with sodium-5-sulfo-bis-(hydroxyethyl)-isophthalate and poly(ethylene glycol). *J. Appl. Polym. Sci.* 131: 39823.

**83** Roach, D.J., Dou, S., Colby, R.H., and Mueller, K.T. (2012). Nuclear magnetic resonance investigation of dynamics in poly(ethylene oxide)-based lithium polyether-ester-sulfonate ionomers. *J. Chem. Phys.* 136: 014510.

**84** Sinha, K., Wang, W., Winey, K.I., and Maranas, J.K. (2012). Dynamic patterning in peg-based single ion conductors for li ion batteries. *Macromolecules* 45: 4354–4362.

**85** Roach, D.J., Dou, S., Colby, R.H., and Mueller, K.T. (2013). Solid state nuclear magnetic resonance investigation of polymer backbone dynamics in poly(ethylene oxide) based lithium and sodium polyether-ester-sulfonate ionomers. *J. Chem. Phys.* 145: 194907.

**86** Fragiadakis, D., Dou, S., Colby, R.H., and Runt, J. (2009). Molecular mobility and $Li^+$ conduction in polyester copolymer ionomers based on polyethylene oxide. *J. Chem. Phys.* 130: 064907.

**87** LaFemina, N.H., Chen, Q., Colby, R.H., and Mueller, K.T. (2016). The diffusion and conduction of lithium in poly(ethylene oxide)-based sulfonate ionomers. *J. Chem. Phys.* 145: 114903.

**88** LaFemina, N.H., Chen, Q., Mueller, K.T., and Colby, R.H. (2016). Diffusive flux as a new metric for ion-conducting soft materials. *ACS Energy Lett.* 1: 1179–1183.

**89** Chen, Q., Tudryn, J., and Colby, R.H. (2013). Ionomer dynamics and the sticky Rouse model. *J. Rheol.* 57: 1441–1462.

**90** Chen, K.M. and Liu, H.J. (1987). Preparation and surface activity of water-soluble polyesters. *J. Appl. Polym. Sci.* 34: 1879–1888.

**91** Lin, L.-H., Chiang, C.-Y., Liu, H.-J., and Chen, K.-M. (2002). Preparation and properties of water-soluble polyester surfactants. I. Preparation and surface activity of poly(ethylene glycol)-dimethyl 5-sulfo-isophthalate sodium salt polyester surfactants. *J. Appl. Polym. Sci.* 86: 2727–2731.

**92** Ward, T.C. and Tobolsky, A.V. (1967). Viscoelastic study of ionomers. *J. Appl. Polym. Sci.* 11: 2403–2415.

**93** Greener, J., Gillmor, J.R., and Daly, R.C. (1995). Effect of internal plasticization on the melt rheology of a polyester ionomer. *Proc. An. Tech. Conf., Soc. Plast. Eng.* 53 (2): 1963–1967.

**94** Tudryn, G.J., O'Reilly, M.V., Dou, S. et al. (2012). Molecular mobility and cation conduction in polyether—ester—sulfonate copolymer ionomers. *Macromolecules* 45: 3962–3973.

**95** Chen, Q., Masser, H., Shiau, H.-S. et al. (2014). Linear viscoelasticity and fourier transform infrared spectroscopy of polyether—ester—sulfonate copolymer ionomers. *Macromolecules* 47: 3635–3644.

**96** Shabbir, A., Huang, Q., Baeza, G.P. et al. (2017). Nonlinear shear and uniaxial extensional rheology of polyether-ester-sulfonate copolymer ionomer melts. *J. Rheol.* 61: 1279–1289.

**97** Szymczyk, A., Ezquerra, T.A., and Roslaniec, Z. (2001). Poly(ether-block-sulfonated ester) copolymers. I. Phase structure and physical properties. *J. Macromol. Sci., Part B: Physics* 40: 669–684.

**98** Szymczyk, A., Ezquerra, T.A., and Roslaniec, Z. (2001). Poly(ether-block-sulfonated ester) copolymers. II. Mechanical and dielectric relaxation. *J. Macromol. Sci., Part B: Physics* 40: 685–708.

**99** Szymczyk, A., Baltá Calleja, F.J., Roslaniec, Z. et al. (2002). Poly(ether-*block*-sulfonated ester) copolymers. III. Morphology and ionic aggregation in PESE. *J. Macromol. Sci., Part B: Physics* 41: 507–528.

**100** Chen, L., Yu, X.-H., Yang, C.-Z. et al. (1997). Synthesis and characterization of model polyether-ester ionomers. *J. Polym. Sci., Part B: Polym. Phys.* 35: 799–806.

**101** Hu, T.D., Xien, Y.N., Liu, T. et al. (1999). Photoconductivity measurement of polymers by x-ray absorption fine structure. *J. Appl. Phys.* 85: 7755–7758.

**102** Yu, X., Grady, B.P., Reiner, R.S., and Cooper, S.J. (1993). Properties of polyamic acid lonomers. *J. Appl. Polym. Sci.* 47: 1673–1683.

**103** Banu, P., Sundar, S., Dhathathreyan, A., and Radhakrishnan, G. (2004). Synthesis and surface properties of aqueous dispersions of poly(ester-imide) prepared from anhydride terminated polyester prepolymer and diisocyanate. *J. Colloid Interf. Sci.* 277: 304–308.

**104** Grady, B.P. and Cooper, S.L. (1997). Morphological structure characterization. In: *Ionomers: Synthesis, Structure, Properties and Applications* (ed. M.R. Tant, K.A. Mauritz, and G.L. Wilkes), 41–92. London: Chapman and Hall.

**105** Lim, J.S. and Kim, J.H. (2010). Effects of ionic substitution on physical properties, and crystallization rates of, poly(oxyethylene-b-butylene adipate) ionomers. *Mat. Sci. Eng. A* 527: 4641–4645.

**106** Wang, X., Wang, L., Li, H. et al. (2000). Syntheses of poly(ethylene oxide) polyurethane ionomers. *J. Appl. Polym. Sci.* 77: 184–188.

**107** Lee, C.I., Hwang, S.Y., Yoo, E.S., and Im, S.S. (2011). Phase transition behavior of a poly(oxyethylene-b-butylene adipate) ionomer with shape memory properties. *Polym. Degrad. Stabil.* 96: 190–196.

**108** Hwang, S.Y., Jin, X.Y., Yoo, E.S., and Im, S.S. (2011). Synthesis, physical properties and enzymatic degradation of poly (oxyethylene-b-butylene succinate) ionomers. *Polymer* 52: 2784–2791.

**109** Ma, X.-G., Guo, X.-Y., and Gu, L. (2007). Rheological behavior in blends of PET with ionomeric polyester. *Eur. Polym. J.* 43: 3613–3620.

**110** Zhao, G.L., Wu, R., Curiskis, J., and Deboos, A. (2004). Dyeing properties of blends of wool/modified polyester fibers-effects of temperature. *Text. Res. J.* 74: 27–33.

**111** Pal, S.K., Gandhi, R.S., and Kothariz, V.K. (1996). Effect of comonomer on structure and properties of textured cationic dyeable polyester. *J. Appl. Polym. Sci.* 61: 401–406.

**112** Liu, L., Cheng, L., Yu, J., and Xie, H. (2011). Evaluation of the availability of easy cationic dyeable copolyester fibers as electrostatic flocking piles. *J.Appl. Polym. Sci.* 120: 195–201.

**113** Kim, T.-K., Son, Y.-A., and Lim, Y.-J. (2005). Thermodynamic parameters of disperse dyeing on several polyester fibers having different molecular structures. *Dyes Pigm.* 67: 229–234.

**114** Pal, S.K., Mehta, Y.C., and Gandhi, R.S. (1989). Effects of heat setting on tensile and dye sorption characteristics of anionic modified PET-A comparison with normal PET. *Text. Res. J.* 5: 734–738.

**115** Teli, M.D. and Rao, B.R. (1995). Electrokinetic properties of modified polyester fibers. *Text. Res. J.* 65: 693–696.

**116** Lee, M.S., Lee, M., Wakida, T. et al. (2007). Ozone-gas treatment of cationic dyeable polyester and poly(butylene terephthalate) fibers. *J. Appl. Polym. Sci.* 104: 2423–2429.

**117** (2018). *Global Polylactic Acid Market Forecast 2019–2027*. Market Research Report. Boston, MA: Inkwood Research https://inkwoodresearch.com/reports/polylactic-acid-market/.

**118** Mehta, R., Kumar, V., Bhunia, H., and Upadhyay, S.N. (2005). Synthesis of poly(lactic acid): a review. *J. Macromol. Sci-Pol R* 45: 325–349.

**119** Nofar, M., Sacligil, D., Carreau, P.J. et al. (2019). Poly (lactic acid) blends: processing, properties and applications. *Int J Biol Macromol* 125: 307–360.

**120** Ro, A.J., Huang, S.J., and Weiss, R.A. (2009). Synthesis and properties of random poly(lactic acid)-based ionomers. *Polymer* 50: 1134–1143.

**121** Kulkarni, A., Lele, A., Sivaram, S. et al. (2015). Star telechelic poly(l-lactide) ionomers. *Macromolecules* 48: 6580–6588.

**122** Zhang, Y., Ni, C., Shi, G. et al. (2015). The polyion complex nano-prodrug of doxorubicin (DOX) with poly(lactic acid-co-malic acid)-block-polyethylene glycol: preparation and drug controlled release. *Med. Chem. Res.* 24: 1189–1195.

**123** Noga, D.E., Petrie, T.A., Kumar, A. et al. (2008). Synthesis and modification of functional poly(lactide) copolymers: toward biofunctional materials. *Biomacromolecules* 9: 2056–2062.

**124** Kang, H., Lin, Q., Armentrout, R.S., and Long, T.E. (2002). Synthesis and characterization of telechelic poly(ethylene terephthalate) sodiosulfonate ionomers. *Macromolecules* 35: 8738–8744.

**125** Kang, H., Armentrout, R.S., Wang, J.L., and Long, T.E. (2002). Synthesis and characterization of ionic and non-ionic terminated amorphous poly(ethylene isophthalate). In: *Functional Condensation Polymers* (ed. C.C. Carraher Jr. and G.G. Swift), 249–262. Dordrecht: Kluwer Academic.

**126** Storey, R.F. and Sherman, J.W. (1996). Novel synthesis of (carboxylic acid)-telechelic poly(e-caprolactone). *ACS Polym. Prepr.* 37: 624–625.

**127** Storey, R.F. and Sherman, J.W. (1998). Synthesis and characterization of halatotelechelic poly(ε-caprolactone) ionomers. *ACS Polym. Prepr.* 39: 602–603.

**128** Weiss, R.A. and Zhao, H.-Y. (2009). Viscoelastic behavior of sulfonated polystyrene oligomers. *J. Rheol.* 53: 191–212.

**129** Essahli, M., Colomines, G., Monge, S. et al. (2008). Synthesis and characterization of ionomers based on telechelic phosphonic polyether or aromatic polyesters. *Polymer* 49: 4510–4518.

**130** Sherman, J.W. and Storey, R.F. (1999). Synthesis of calcium-ion containing telechelic poly(L,L-lactide) ionomers. *ACS Polym. Prepr.* 40: 952–953.

**131** Ro, A.J., Huang, S.J., and Weiss, R.A. (2008). Synthesis and thermal properties of telechelic poly(lactic acid) ionomers. *Polymer* 49: 422–431.

**132** Litvinov, V.M., Braam, A.W.M., and van deer Ploeg, A.F.M.J. (2001). Telechelic ionomers: molecular structure and kinetics of physical gelation of unsaturated polyester as studied by solid state NMR and X-ray. *Macromolecules* 34: 489–502.

**133** Unal, S., McKee, M.G., Massa, D.J., and Long, T.E. (2005). Synthesis and characterization of phosphonium based telechelic polyester ionomers. *ACS Polym. Prepr.* 46 (2): 1028–1029.

**134** Xue, Y. and Hara, M. (1997). Ionic naphthalene thermotropic copolyesters: effect of ionic content. *Macromolecules* 30: 3803–3812.

**135** Xue, Y. and Hara, M. (2005). Ionic naphthalene thermotropic copolyesters with para-linked ion-containing units. *Polymer* 46: 7293–7300.

**136** Xue, Y., Hara, M., and Yoon, H.N. (1998). Ionic naphthalene thermotropic copolyesters: divalent salts and tensile mechanical properties. *Macromolecules* 31: 7806–7813.

**137** Xue, Y., Hara, M., and Yoon, H.N. (2001). Ionic naphthalene thermotropic copolyesters: enhanced compressive properties of fibers. *Macromolecules* 34: 844–851.

**138** Lin, Q., Pasatta, J., Wang, Z.-H. et al. (2002). Synthesis and characterization of sulfonated liquid crystalline polyesters. *Polym. Int.* 51: 540–546.

**139** Plate, N.A. (1993). Conclusion. In: *Liquid-Crystal Polymers* (ed. N.A. Plate), 422. New York: Plenum Press.

**140** Song, J.-M., Ko, K.-H., Choi, Y.-S. et al. (2009). Randomly sulfonated liquid crystalline polymers containing triad mesogen of butylene terephthalate. *Mol. Cryst. Liq. Cryst.* 511: 255–264.

**141** Ober, C., Jin, J.-I., and Lenx, R.W. (1982). Liquid crystal polymers. V. Thermotropic polyesters with either dyad or triad aromatic ester mesogenic units and flexible polymethylene spacers in the main chain. *Polym J.* 14: 9–17.

**142** Ignatious, F., Lu, C., Kantor, S.W., and Lenz, R.W. (1994). Alternating, block, and random copolymers of a triad mesogen with alkylene terephthalate flexible segments. *Macromolecules* 27: 7785–7793.

**143** Makowski, H.S., Lundberg, R.D., and Singhal, G.L. (1975). Flexible polymeric compositions comprising a normally plastic polymer sulfonated to about 0.2 to about 10 mole% sulfonate. US Patent 3,870,841.

**144** Jian, L., Sherrington, D.C., Nield, E., and Richards, R.W. (1992). Novel wholly aromatic lyotropic liquid crystalline polyesters. Synthesis, characterization, and solution properties. *Macromolecules* 25: 7107–7113.

**145** Zhang, B. and Weiss, R.A. (1992). Liquid crystalline lonomers. 1.Main-chain liquid crystalline polymer containing pendant sulfonate groups. *J. Polym. Sci., Part A: Polym. Chem.* 30: 91–97.

**146** Zhi, J., Zhang, B., Wu, Y., and Feng, Z. (2001). Study on a series of main-chain liquid–crystalline ionomers containing sulfonate groups. *J. Appl. Polym. Sci.* 2001 (81): 2210–2218.

**147** Zhang, B.-Y., Tian, M., Meng, F.-B., and He, X.-Z. (2005). Synthesis and characterization of main chain liquid crystalline ionomers containing sulphonate groups. *Liquid Crystals* 32: 997–1003.

**148** Xu, X.Y., Zhang, B.Y., and Gu, W.M. (2009). Synthesis, structure, and mesomorphic properties of main-chain liquid-crystalline ionomers containing sulfonate groups. *J. Mater. Sci.* 44: 1787–1792.

**149** Tian, M., Zhang, B.-Y., Meng, F.-B., and Zang, B.-L. (2006). Main-chain chiral smectic liquid-crystalline ionomers*J. Appl. Polym. Sci.* 2006 (99): 254–1263.

**150** Lundberg, R.D. and Phillips, R.R. (1982). Solution behavior of metal sulfonate ionomers. II. Effects of solvents. *J Polym. Sci., Polym. Phys. Ed.* 20: 1143–1154.

**151** Fuoss, R.M. and Strauss, U.P. (1948). Electrostatic interaction of polyelectrolytes and simple electrolytes. *J. Polym. Sci.* 3: 602–603.

**152** Zhang, M., Zhang, A.-L., Li, Q. et al. (2019). Conductivity of PEG/PLA doped liquid crystal ionomer solid polymer electrolyte in mesomorphic range. *J. Polym. Environ.* 27: 2369–2379.

**153** Zhang, N., Zhang, A.-L., Liu, Q.-F. et al. (2018). Effect of liquid crystal ionomer intercalated montmorillonite nanocomposites on PEG/PLA solid polymer electrolytes. *Ionics* 24: 3805–3813.

**154** Meng, F.-B., Zhang, B.-Y., Xu, Y., and Lium, J.-Y. (2005). Main-chain liquid-crystalline ionomers bearing potassium sulfonate groups. *J. Appl. Polym. Sci.* 96: 2021–2026.

**155** Meng, F.-B., Zhou, N.-Y., Du, C. et al. (2013). Synthesis, characterization and electrorheological effect of sulfosalt-type liquid-crystalline ionomers containing polyaniline units. *J. Appl. Polym. Sci.* 130: 3395–3403.

**156** Winslow, W.M. (1949). Induced fibration of suspensions. *J. Appl. Phys.* 20: 1137–1140.

**157** Bingham, E.C. (1916). An investigation of the laws of plastic flow. *J Res Natl Inst Stan.* 13: 309–353.

**158** Zhang, B. and Weiss, R.A. (1992). Liquid crystalline ionomers. II. Main chain liquid crystalline polymers with terminal sulfonate groups. *J. Polym Sci., Part A: Polym. Chem.* 30: 989–996.

**159** Zhang, A., Zhang, B., and Reng, Z. (2002). Compatibilization by main-chain thermotropic liquid crystalline ionomer of blends of PBT/PP. *J. Appl. Polym Sci* 85: 1110–1117.

**160** Chakrabarty, K., Shao, L.-Y., and Weiss, R.A. (1997). Solution properties. In: *Ionomers: Synthesis,Structure, Properties and Applications* (ed. M.R. Tant, K.A. Mauritz, and G.L. Wilkes), 158–207. London: Chapman and Hall.

**161** Larson, W.K., Lynn, M.M., and McAllister, E.S. (1982). Process for modifying the surfaces of polyester fibers. US Patent 4,330,588.

**162** Larson, W.K. (1984). Amorphous sulfopolyesters. US Patent 4,480,085.

**163** Fragerburg, D.R. and McFarlane, R.E. (1985). Process for the preparation of sulfo-modified polyesters. US Patent 4,499,262, February 12.
**164** O'Neill, G.J. (1986). Ethylene terephthalate copolymers. US Patent. 4,579,936.
**165** Duldner, M.M., Iancu, S., Vuluga, Z., et al. (2011). Oligoester ionomers from polyethylene therephthalate wastes, to be used as dispersing agents for nano composites, and process for preparing the same. Romanian Patent 128907.
**166** Schade, G. (1978). Water-dispersible ester resin containing a moiety of polyacid or bivalent alcohol containing a sulfo group. US Patent 4,104,262.
**167** Hartman, S.E., Allen, M.E., Pascoe, W.E. (1984). Cycloalkylsulfonates, polymers and processes relating to same. US Patent 4,433,130.
**168** McGuckin, H.G., Hartman, S.E., and Specht, D.P. (1983). Dye imbibition photohardenable imaging material and process for forming positive dye images. US Patent 4,374,194.
**169** George, S.E. (2000). Essentially linear crystalline water-dispersible polyesters comprising residues of 4,4'-biphenyldicarboxylic acid. US Patent 6020420.
**170** George. S.E. and Cannon, J.S. (2012). Copolyesters from 1,4-cyclohexanedimethanol synthesis byproducts. US Patent Appl. 20120329980.
**171** Teasley, M.F. (2011) Polymerization of aliphatic-aromatic copolyetheresters. US Patent Appl. 20110229674.
**172** Larson, W.K. (1985). Sulfonate-containing photopolymer systems. US Patent 4,558,149.
**173** Chan, J.W., Cobb, M.W., and Williams, S.R. (2015). Sulfonated aromatic compounds. US Patent 9,150,506.
**174** Hayes, R.A. (2008). Sulfonated aromatic copolyesters containing hydroxyalkanoic acid groups and shaped articles produced therefrom. US Patent 7,358,325.
**175** Hayes, R.A. and Kosinski, L.E. (2004). Sulfonated aliphatic-aromatic copolyesters. US Patent 6,746,779.
**176** Hayes, R.A. (2002). Sulfonated aliphatic-aromatic copolyesters with good biodegradability. US Patent 6,368,710.
**177** Loos, R., Mijolovic, D., Heimann, J., and Szarka, Z.J. (2013). Biodegradable polyester based on 2-methylsuccinic acid, composition, use, and process for preparing thereof. US Patent 8,546,472.
**178** Gupta, R.K., Crain, A.L., Klosiewicz, D.W. et al. (2019). Sulfopolyester recovery and article fabrication. US Patent 7,635,745.
**179** Parker, K.R. and Kirk, S.K. (2017). Ultrafiltration process for producing a sulfopolyester concentrate. US Patent 9,598,802.
**180** Parker, K.R. and Kirk, S.K. (2017). Ultrafiltration process for producing a sulfopolyester concentrate. US Patent 9,605,126.
**181** Storey, R.F., Deng, Z.D., Glancy, T.P., and Peterson, D.R. (1997). Polyester ionomers for implant Fabrication. US Patent 5,668,288.

**182** Kibler, C.J. and Lappin, G.R. (1973). Polyesters and polyesteramides containing ether groups and sulfonate groups in the form of a metallic salt. US Patent 3,734,874.
**183** Kibler, C.J. and Lappin, G.R. (1973). Polyesters and of polyyesteramides containing ether groups and sulfonates groups in the form of a metallc salt. US Patent 3,779,993.
**184** Bayless, J.H. (1980). Modified polyester films. US Patent 4,217,441.
**185** Heiberger, P. (1971). Aqueous disperson of copolyesters modified with a sulfonated aromatic compound. US Patent 3,563,942.
**186** Sublett, B.J. (1980). Polyester and polyesteramide compositions. US Patent 4,233,196.
**187** George, S. E. (2000). Water-dispersible polyesters. US Patent 6,020,420.
**188** O'Neill, G.J. and Pacifici, J.G. (1978). Radiation crosslinkable polyester and polyesteramide compositions containing sulfonate groups in the form of a metallc salt and unsaturated groups. US Patent 4,073,777.
**189** Coney, C.H. (1982). Sequestering agents and compositions produced therefrom. US Patent 4,335,220.
**190** Noonan, J.M., McConkey, C., and Hanrahan, M.J. (1978). Anonic imino-containing polymeric adhesives for photographic materials. US Patent 4,097,282.
**191** Noonan, J.M. and McConkey, R.C. (1981). Water-dispersible polyester adhesives for photographc materials. US Patent 4,307,174.
**192** Noonan, J.M. and McConkey, R.C. (1981). Water-dispersible polyester adhesives for photographic materials. US Patent 4,291,153.
**193** Choi, Y.H., Kim, J.B., and Lim, T.J. (2004). Water-based adhesive compositions using waterborne polyesters and multilayer material. US Patent. Appl. 20040072003.
**194** Ng, C.C.L. and Ulman, J.T. (2011). Water dissolvable medical adhesive article for topical application. US Patent Appl. 20110311611.
**195** Larson, W.K. (1977). Water-dispellable hot melt polyester adhesives. US Patent 4,052,368.
**196** Arendt, W.D., Makarand, J., Berry-Walker, Y.A. et al. (2009). Structural isomers of 1,4-cyclohexanedimethanol dibenzoate for use in adhesives. US Patent Appl. 20090142981.
**197** Miller, R.A., George, S.E., Parsons, T.E., and Montgomery, M.A. (1996). Water-dspersible adhesive composition and process. US Patent 5543488.
**198** Blumenthal, M., Paul, C.W., and Sharak, M.L. (1999). Use of anonic or cationc plasticizers with sulfonated polyesters in hot melt formulations. US Patent 6001910.
**199** O'Neill, G.J. and Belote, S.N. (1981). Water dissipatable polyesters. US Patent 4,304,901.

**200** Chatelier, R.C. and Rylatt, D. (2012). Adhesive composition for use in an immunosensor. US Patent 8,221,994.

**201** O'Brien, E.P. and Atkins, D.G. (2011). Switchable pressure-sensitive adhesive articles suitable for attachment to skin. US. Patent 7,879,942.

**202** Kerins, J., Soerens, D.A., Tanzer, R.W., and Pomplun, W.S. (2003). Body-activated adhesive compositions, films, personal care articles, and pantiliners. US Patent Appl. 2003007855.

**203** Williams, T.H. and Oldfield, T.A. (2006). Fluid sulfopolyester formulations and products made therefrom. US Patent Appl. 20060058438.

**204** Griffing, J.M. and Remington, W.R. (1962). Sulfonate containing polyesters. US Patent 3,018,272.

**205** Gallagher, F.G., Hamilton, C.J., Hansen, S.M. et al. (1992). Polyesters and their use in compostable products such as disposable diapers. US Patent 5171308.

**206** Caldwell, J.R. and Jones, G.C. (1970). Polyesters containing dsulfonamdo compounds having improved dyeing properties. US Patent 3,546,180.

**207** Rudisill, E.N., Everett, C.S., Holbert, R.M. et al. (2019. Multicomponent fibers. US Patent Appl. 20190309445.

**208** Rudisill, E.N., Holbert, R.M., Holbrook, L.P. et al. (2019). Method for production of multicomponent fiber with shaped cross-section. US Patent Appl. 20190309447.

**209** Rudisill, E.N., Everett, C.S., Holbert, R.M. et al. (2019). Multicomponent fiber with shaped cross-section. US Patent Appl. 20190309446.

**210** Gupta, R.K., George, S.E., Klosiewicz, D.W. et al. (2011). Water-dispersible and multicomponent fibers from sulfopolyesters. US Patent 7,892,993.

**211** Gupta, R.K., Klosiewicz, D.W., and George, S.E. (2008). A process for manufacture of nonwoven fabrics from multicomponent fibers comprising sulfopolyesters. US Patent Appl. 20080160859.

**212** Clark, M.D., Mitchell, M.G., and Mitchell, M L. (2012). Short cut microfibers. US Patent Appl. 20120302119.

**213** Gupta, R.K., Mitchell, M.G., Klosiewicz, D.W., et al. (2016). Wet-laid process to produce a bound nonwoven article. US Patent 9.273,417.

**214** Gupta, R.K., Mitchell, M.G., Kosiewicz, D.W. et al. (2012). Sulfopolyester binders. US Patent Appl. 20120183861.

**215** Gupta, R.K., Mitchell, M.G., Kosiewicz, D.W. et al. (2012). Sulfopolyester binders. US Patent Appl. 20120180968.

**216** Gupta, R.K., Mitchell, M.G., Kosiewicz, D.W. et al. (2012). Nonwoven article with ribbon fibers. US Patent Appl. 20120177996.

**217** Gupta, R.K., Mitchell, M.G., Kosiewicz, D.W. et al. (2012). Nonwoven article with ribbon fibers. US Patent Appl. 20120175074.

**218** Gupta, R.K., Mitchell, M.G., Kosiewicz, D.W. et al. (2012). Nonwoven article with ribbon fibers. US Patent Appl. 20120178331.

**219** Clark, M.D., Mitchell, M.G., and Mitchell, M.L. (2012). Short-cut microfiber-containing mixture, process for forming same and fibrous article made from same. US Patent Appl. 20120302120.

**220** Gupta, R.K., Kosiewicz, D.M., Mitchell, M. G. et al. (2012). Fibrous end products incorporating short-cut microfibers. US Patent Appl. 20120251597.

**221** Gupta, R.K., Klosiewicz, D.W., and Mitchell, M.G. (2013). Nonwovens produced from multicomponent fibers. US Patent 8,513,147.

**222** Gupta, R.K., Mitchell, M.G., Kosiewicz, D.W. et al. (2012). High strength specialty paper incorporating thermoplastic fibers. US Patent Appl. 20120219766.

**223** Gupta, R.K., Klosiewicz, D.W., Mitchell, M. G., and Mitchell, M. L. (2013). Sulfopolyesters for paper strength and process. US Patent 8,512,519.

**224** Gupta, R.K., Mitchell, M.G., Kosiewicz, D.W. et al. (2012). High efficiency filter. US Patent Appl. 20120175298.

**225** Parker, I. (2017). High efficiency and high capacity glass-free fuel filtration media and fuel filters and methods employing the same. US Patent 20170296953.

**226** Gupta, R.K., Mitchell, M.G., Kosiewicz, D.W. et al. (2012). Paperboard or cardboard. US Patent Appl. 20120184164.

**227** Gupta, R.K., Mitchell, M.G., Kosiewicz, D.W. et al. (2012). Battery separator. US Patent Appl. 20120183862.

**228** Stockburger, G.J. (1984). Anionic textile treating compositions. US Patent 4,427,557.

**229** Hayes, H.J., Fang, X., Lock, S.S. (2006). Fluorochemical-containing textile finishes exhibiting wash-durable soil release and moisture wicking properties. US Patent 7,012,033.

**230** Wo, S. (2003). Preparation of sulfonated polyesters as soil releasing and textile sizing agents. US Patent 6576716.

**231** McIntryre, J. E.; Robertson, M. M. (1968). Surface modifying treatment of shaped articles made from polyesters. US Patent 3,416,952.

**232** McIntyre, J.E. and Robertson, M.M. (1971). Aqueous dispersion of block or graft polymer useful in surface modifying treatment of polyester shaped articles. US Patent 3,557,039.

**233** Lesley, D.J., Lesley, G.D., Nasser, D. S et al. (2006). Formaldehyde-free binder compositions for simultaneous warp sizing and coloration of filament yarns. US Patent 6,984,675.

**234** Fleury, E. and Girardeau, Y. (1994). Hydrosoluble/hydrodispersible polyesters and sizing of textile threads therewith. US Patent 5,290,631.

**235** George, S.E., Lefler, E.W., and Spain, J.M. (1997). Water-dispersible sulfo-containing copolyester-ether compositions for sizing. U. S. Pat. 5646237.

**236** George, S.E., Lefler, E.W., and Spain, J.M. (1996). Water-dispersible copolyester-ether compositions for sizing. US Patent 5571620.

237 Login, R.B. (1979). Fibrous article sized with à branched water-dispersible polyester. US Patent 4,158,083.
238 Engelhardt, F., Hintermeier, K., Ribka, J., and Beutler, H. (1979). Water-soluble branched polyesters. US Patent 4,167,395.
239 Frazier, D.E. (1991). Process for producing zero pilling polyester. US Patent 5,039,467.
240 Lan, T. and Aylward, B.P. (2019). Odor control composition and method of using. US Patent Appl. 20190145045.
241 Wempe, M.F. (2009). Sulfo-polymer powder and sulfo-polymer powder blends with carriers and/or additives. US Patent Appl. 20090163449.
242 Wempe, M. F. (2009). Sulfo-polymer powder and sulfo-polymer powder blends with reduced volatility. US Patent Appl. 20090163603.
243 Pomplun, W.S., Mumick, P.S., Jackson, D.M. and Chang, Y. (1999). Ion sensitive polymeric materials as binders for water dispersible fabrics. US Patent 5972805.
244 Borcher, T.A., Delgado, R. and Trinh, T. (1989). Soil release polymer compositions having improved processability. US Patent 4,863,619.
245 Borcher, T.A., Delgado, R., and Trinh, T. (1990). Soil release polymer compositions having improved processability. US Patent 4,925,577.
246 Borcher, T.A., Delgado, R., and Trinh, T. (1991). Soil release polymer compositions having improved processability. US Patent 5,041,230.
247 Eppelsheim, A.K., Boeckh, D., Trieselt, W. et al. (1992). Polyesters containing nonionic surfactants as cocondensed units, preparation thereof and use thereof in detergents. US Patent 5,142,020.
248 Gosselink, E.P. (1987). Block poly esters and like compounds useful as soil release agents in detergent compositions. US Patent 4,702,857.
249 Pan, Y.L., Gooselink, E.P., Pancheri, E.J., and Morrall, S.W. (1997). Soil release agents for granular laundry detergents. US Patent 5,599,782.
250 Rohrbaugh, R.H. and Gosselink, E.P. (1998). Soil release polymers with fluorescent whitening properties. US Patent 5,728,671.
251 Blokzijl, W., Creeth, A.M., Elmes, A.R. et al. (1998). Soil release polymers and detergent compositions containing them. US Patent 5,786,318.
252 Blokzijl, W., Creeth, A.M., Falou, M. et al. (1998). Detergent compositions containing soil release polymers. US Patent 5,789,365.
253 Blokzijl, W., Creeth, A.M., Falou, M. et al. (1998). Detergent compositions containing soil release polymers. US Patent 5,789,366.
254 Blokzijl, W., Creeth, A.M., Green, A.D., and Hull, M. (1998). Detergent compositions containing soil release polymers. US Patent 5,789,367.
255 Bakaert, K., Sciacca, A., Depoot, K.J.M., and Boutoille, A.M. (2019). Liquid polymer composition, and method for making liquid laundry detergent composition. US Patent Appl. 20190225743.

256 George, S.E. (2009). Manufacturing of polyester textile as fabric softener agent. US Patent Appl. 20090163402.
257 Arcesi, J.A. and Rauner, F.J. (1975). Lithographic plates having radiation sensitive elements developable with aqueous alcohol. US Patent 3,929,489.
258 Anderson, C.C., Leszyk, G.M., and Tingler, K.L. (1994). Imaging element. US Patent 5,360,706.
259 Anderson, C.C. and DeLaura, D. (1996). Imaging element having a process surviving electrically conductive layer with polyesterionomer binder. US Patent 5,576,163.
260 Molaire, M.F. and Sorriero, L.J. (1998). Electrophotographic elements with generating layers contaning polyester ionomers. US Patent 5,733,695.
261 Noonan, J.M. and Perlstein, J.H. (1983). Condensation polymeric photoconductors containing pendant arylamines. US Patent 4,395,475.
262 Merrill, S.H., Hartman, S.E., Noonan, J. M. et al. (1981). Polyesterionomers having utility in liquid electrographic developer compositions. US Patent 4,252,921.
263 Majumdar, D., Kung, T.-M., Falkner, C.A. et al. (2013). Image receiver elements with aqueous dye receiving layer. US Patent 8.404,332.
264 Majumdar, D., Kung, T.-M., Falkner, C.A. et al. (2013). Image receiver elements with aqueous dye receiving layer. US Patent 8,501,666.
265 Kung, T.-M., Bonsignore, K., Daniels, R.L. et al. (2017). Conductive thermal imaging receiving layer with receiver overcoat layer. US Patent 9,707,788.
266 Bowman, W.A., Shuttleworth, L., and Weber, H. (1996). Thermal dye transfer system with primary polyester ionomer receiver. US Patent 5534478.
267 Light, W.A. (1992). Ink-jet recording sheet. US Patent 5,126,193.
268 Light, W.A. (1992). Ink jet transparency. US Patent 5,126,194.
269 Evans, S., Kung, T.-M., Lawrence, K.B., and Van Hanehaem, R.C. (1998). Thermal dye transfer system with polyester ionomer receiver. US Patent 5,804,531.
270 Kung, T.-M. (2014). Thermal image receiver elements having release agents for image printing photographs. US Patent 8,673,535.
271 Kung, T.-M., Ghyzel, P.J., and Muehlbauer, J.L. (2014). Thermal image receiver elements prepared using aqueous formulations. US Patent 8,691.489.
272 Kung, T.-M., Ghyzel, P.J., and Muehlvauer, J.L. (2014). Thermal image receiver elements prepared using aqueous formulations. US Patent 8,691.489.
273 Missell, G.E. and Merkel, P.B. (2004). Ink jet recording element. US Patent 6,815,019.
274 Sorriero, L.J. and Tyagi, D. (2002). Electrophotographic toner binders containing polyester ionomers. US Patent 6,369,136.
275 Farrugia, V.M., Guerinao, G.G., Hawkins, M.S., and Zhou, K. (2008). Ultra low melt toners having surface crosslinking. US Patent 7416,827.

276 Anderson, C.C., Jennings, D.F., Leszyk, G.M., and Niemeyer, D.A. (1995). Imaging element for use in electrostatography. US Patent 5,380,584.
277 Noonan, J.N., Ryan, R.W., and Houle, J.F. (1983). Image-forming compositions and elements containing ionic polyester despersing agents. US Patent 4,419,437.
278 Keoshkerian, B., Vanbesien, D.W., Chretien, M.N. et al. (2015). Inkjet ink containing sub 100 nm latexes. US Patent 8991992.
279 Foucher, D.A., Patel, R.D., Burns, P.A. et al. (1999). Ink jet inks. US Patent 5958998.
280 Sharma, M.K. (1994). Process for preparing sulfo-polyester/acrylic resin blends without volatile organic compounds and surfactants. US Patent 5294650.
281 Barlow, F.D. and Chamberlin, K.S. (1991). Stabilization of ink compositions. US Patent 5,028,262.
282 Chopra, N., Sacripante, G.G., and Farrugia, V.M. (2019). Antibacterial aqueous ink compositions comprising water soluble sodio-sulfonated polyester. US Patent Appl. 20190233665.
283 Vernardakis, T.G. and Nasilevich, N. (2002). Scented ink composition and method of preparation. US Patent 6454842.
284 Krutak, J.J., Cushman, M.R., Parham, W.W. et al. (1994). Water-dissipatable polyesters and amides containing near infrared fluorescent compounds copolymerized. US Patent 5292855.
285 Escano, N.Z. and Krutak, J.J. (1997). Water based inks containing polymeric near infrared fluorophores. US Patent 5614008.
286 Sorriero, L.J. and Santilli, D. (1998). Inkjet inks containing polyester ionomers. US Patent 5,716,436.
287 Wong, R.W., Breton, M.P., Sacripante, G.G. et al. (1998). Fast drying inks with polyhydroxy alcohol surfactants. US Patent 5762695.
288 Breton, M.P., Bedford, C.E., and Sacripante, G. G. (2007). Heterogeneous low-energy gel ink composition. US Patent 7,172.276.
289 Burns, P.A., Breton, M.P., Gagnon, Y., and MacKinnon, D.N. (1999). Smear-resistant ink jet inks. US Patent 5973026.
290 Langston, J.R., Tortorici, F.J., and Barlow, F.D. (1988). Waste treatment process for printing operations employing water dispersible inks. US Patent 4,738,785.
291 Malotky, D.L., Kainz, B., Diehl, C.F. et al. (2013). Coating composition and articles made therefrom. US Patent 8,357,749.
292 Jenkins, W.L., Barton, K.R., and Jaffe, H.L. (1990). Aqueous polymer dispersions and their preparation. US Patent 4,939,233.
293 Wu, H. (2012). Chemical resistant ionomers and protective coverings. US Patent 8,299,147.
294 Nair, M., Jones, T.K., Seita, J.S. et al. (2001). Protective overcoat comprising polyester ionomers for photographic elements. US Patent US 6,274,298.

**295** Carrie, F., Goldberg, H.A., and Farrell, M. (2012). Barrier coatings post-formation treated with multi-valent metal cations. US Patent Appl. 20120053281.
**296** George, S.E., Miller, R.A., Trotter, J. R. et al. (2001). Water-dispersible films and fibers based on sulfopolyesters. US Patent 6171685.
**297** Erneta, M. and Priewe, J. (2014). Antimicrobial polymer compositions for coated medical articles. US Patent 8,691,899.
**298** Podhajny, R.M. (2003). Anti-microbial packaging materials, their manufacture, and coating application. US Patent Appl. 20030091767.
**299** Moore, J.C. (2015). Compositions and methods of rinsable primers used to aid in the removal of temporary functional coatings. US Patent Appl. 20150279666.
**300** Wuillen, M.W., O'Dell, D.E., Lee, Z.P. et al. (2012). Polymeric or monomeric compositions comprising at least one mono-amide and/or at least one diamide for removing substances from substrates and application and use of the compositions. US Patent Appl. 20120073607.
**301** Kuo, T., Moncier, J.D., Jones, A.S., and Murray, D.L. (2005). Modified alkyd compositions comprising polyol latex compositions and processes of making them. US Patent 6,844,390.
**302** Buchanan, J.P., Anderson, L.L., Mackey, N.M., and Lochhead, R.Y. (2016). Compositions and methods for protection of skin against thermal insult. US Patent 9.271,907.
**303** Stockl, R.R., Kuo, T., and McCreight, K.W. (2013). Low VOCadditives for extending the wet edge and open time of aqueous coatings. US Patent 8.444,758.
**304** George, S.E. and Sublett, B.J. (1994). Heat-resistant water-dispersible sulfopolyester compositions. US Patent 5369210.
**305** Agarwal, U.S., Venkatarishnam, B.B., Jalan, R. et al. (2016). Oxygen-scavenging polyesters. US Patent 9381709.
**306** Li, J., Drieger, C., Urbanski, A.N. et al. (2009). Coated art material system. US Patent 7,521,112.
**307** Sharma, M.K., Singleton, A.H., and Hiller, J.J. (1995). Waterborne correction fluids for correcting laser-jet and photocopied markings. US Patent 5464883.
**308** Yuan, Z. and Jacober, D. (2015). Anti-powdering and anti-static polymer film for digital printing. US Patent Appl. 20150140318.
**309** Morrison, E.D. (1995). Vanadium oxide colloidal dispersons and antistatic coatings. US Patent 5,407,603.
**310** Boston, D.R., Kausch, W.L, Martino, E. et al. (1995). Photographic elements comprising antistatic layers of vanadium pentoxide, epoxy-silane, and sulfopolymer. US Patent 5,439,785.
**311** Buchanan, S.J., Morrison, E.D., Boston, D.R. et al. (1993). Abrasive article having vanadium oxde incorporated therein. US Patent 5,203,88.
**312** Chang, J.C. and Morrison, E.D. (1994). Thermal transfer systems having delaminating coatings. US Patent 5,372,985.

**313** Chang, J.C. and Morrison, E.P. (1995). Thermal transfer systems having delaminating coatings. US Patent 5,424,269.
**314** Morrison, E.D., Boston, D.R., Hedrick, S.T. et al. (1995). Sulfopolymer/vanadium oxde antistatic compositions. US Patent 5,427,835.
**315** Morrison, E.D., Boston, D.R., Hedrick, S.T. et al. (1995). Sulfopolymer vanadumoxde antistatic compositions. US Patent 5,468,498.
**316** Matsubayashi, S. (2017). Method for producing antistatic molded article. US Patent Appl. 20170058167.
**317** Hung, L.-S. and Webster (1992). Electrically conductive article (I). US Patent 5,086,035.
**318** Hung, L.-S. and Chatterjee, D.K. (1991). Electrically conductive article. US Patent 5,073,537.
**319** Liu, Z., Mehta, S., Huang, X., and Schiraldi, D.A. (2011). Method to make single-layer pet bottles with high barrer improved clarity. US Patent 7,943.216.
**320** Liu, Z.; Mehta, S.; Huang, X., Schiraldi, D. A. (2011). Method to make single-layer PET bottles with high barrier and improved clarity. US Patent 7,919,159.
**321** Heater, P.L. and Elliott, G.A. (2013). Polyamides and polyesters blended with a lithium salt interfacial tension reducing agent. US Patent 8.465,818.
**322** Elliott, G.A., Heater, P., Knudsen, R. et al. (2017). Composition for maintaining good color when thermally treating polyester-polyamide blends. US Patent 9,540,507.
**323** Bastiaens, J.H.P. and Carminati, A. (2002). Thermoplastic composition based on a blend of poly(arylene ether) and polyamide. US Patent 6500895.
**324** Sanjay, M. (2013). Articles having improved gas barrier properties. US Patent 8609783.
**325** Trexler, J.W., Piner, R.L., Turner, S.R., and Barbee, R.B. (2000). Process for making polyester/platelet particle compositions displaying improved dispersion. US Patent 6,162,857.
**326** Trexler, J.W., Piner, R.L., Turner, S.R., and Barbee, R.B. (2002). Polyester/platelet particle compositions displaying improved dispersion. US Patent US 6,359,052.
**327** Chisholm, B.J. and Khouri, F.F. (2004). Ionomeric polyester compositions containing organoclays, and method. US Patent 6,831,123.
**328** Farrugia, V.M., Desouza, A., and Gardner, S.J. (2016). Metal nanoparticle-sulfonated polyester composites and green methods of making the same. US Patent 9.458,305.
**329** Crane, C.M. and Fonolla-Moreno (2019). Mascara compositions comprising a bimodal acrylic polymer and anionic, water – dispersible polyester and an aliphatic tackifier. US Patent Appl. 20190254950.
**330** Bertrand, P., Debert, D., and Bodelin-Lecomte, S. (2001). Mascara comprising a mixture of hard waxes and of film-forming polymer. US Patent 6274131.

**331** Hsieh, I.-F., MacNeill, C.M., Zhou, X., and Hariharan, R. (2019). Nail compositions containing latex and sulfopolyester compound. US Patent Appl. 20190167556.

**332** Myers, G.L., Hiller, J.J., Jenkins, W.L. et al. (1993). Cosmetic forming compositions. US Patent 5266322.

**333** Dobbs, S.W. (1995). Cosmetic film-forming compositions which are freeze-thaw stable. US Patent 5380520.

**334** Hapern-Chirch, S., Shah, A., and Bernard, A.-L.S. (2018). Excellent water resistance with an association of two film forming polymers. US Patent Appl. 20180116944.

**335** Pays, K., Barba, C., and Samain, H. (2015). Device for the application of fibers to human keratinous fibers. US Patent 9,125,468.

**336** Bebot, C. and Pasquet, D. (2010). Cosmetic composition comprising branched sulphonic polyester and surfactant and uses in hair styling. US Patent Appl. 20100158848.

**337** Bhatt, D. and Galleguillos, R. (1992). Aqueous hair styling aid. US Patent 5164177.

**338** Pays, K., Bichon, Y., and Oliver-Mabilais, S. (2012). Easily removable water proof cosmetic care and/or makeup composition comprising at least one latex or pseudolatex. US Patent 8,211.415.

**339** Garcia, M.L., Krutak, J.J., Brown, K.C., and Rutstrom, D.J. (1999). Compositions for coloring keratinous fibers comprising sulfo-containing, water dispersible colored polymers. US Patent 5876463.

**340** Oldfield, T.A., Dobbs, S.W., George, S.E. et al. (2011). Sulfopolyesters having improved clarity in water-dispersible formulations. US Patent 7.923,526.

**341** Chen, R.G.-S., Cook, P.M., and Munjal, R.C. (2003). Hair care compositions. US Patent 6,638,992.

**342** George, S.E., Underwood, V.L., and Garber, D.M. (1997). Clear pump hair spray formulations containing a linear sulfopolyester in a hydroalcoholic liquid vehicle. US Patent 5662893.

**343** Myers, G.L., Hiller, J.J., Minga, R.L. et al. (1993). Aerosol hair spray formulations with sulfonate-containing polyesters. US Patent 5266303.

**344** McEntire, E.E., Stockl, R.R., Munjal, R.C., and Christian, V.L. (2008). Wrinkle masking film composition comprising polymer. US Patent Appl. 2008/0057090.

**345** Spina, T.N. and Maletshky, A. (2009). Water soluble depilatory comprising hot melt adhesive blend coated onto substrate, having extended shelf life. US Patent 7,507,400.

**346** Catalan, K.V., Spitzmueller, C.A., Gartstein, V.W. et al. (2016). Durable hydrophilic coating compositions for medical absorbents. US Patent 9,314,544.

**347** McEntire, E.E., Stockl, R.R., Munjal, R.C. et al. (2007). Water dispersible films for delivery of active agents to the epidermis. US Patent Appl. 20070258935.

**348** McEntire, E.E., Stockl, R.R., and Chand, R. (2007). Water-dispersible patch comprising a polymer matrix and containing an active agent for dermal delivery. US Patent Appl. 20070259029.

**349** McEntire, E.E. and McCaulley, J.A. (2007). Water resistant sunscreen formulas with sulfopolyesters and phosphate ester surfactants. US Patent Appl. 20070264216.

**350** Nagarjan, V. (2006). Method for accelerated biodegradation of aliphatic-aromatic polyesters by enzymatic treatment. US Patent 7,053,130B2.

**351** Hahn, C.W. (2006). Polyester-polyoxyalkylenes with anionic functionality as demulsifiers for petroleum dewatering. US Patent 7,041,707.

**352** Campbell, R.W. (1984). Copolymeric nucleating agent for polyester. US Patent 4,451,606.

**353** Ghosh, S.K., Chattterjee, D.K., and Majumdar, D. (1999). Method of manufacturing a ceramic article. US Patent 5,958,311.

# 5

# The Rheo-Optical Properties of Polyester Films

*Baris Yalcin*[1] *and Miko Cakmak*[2]

[1] *Global Packaging R&D, The Coca-Cola Company, Atlanta, GA, USA*
[2] *Schools of Materials and Mechanical Engineering, Purdue University, West Lafayette, IN, USA*

## 5.1 Birefringence and Its Measurement

Polyesters are an important polymer class with a wide range of applications in various markets such as packaging, textiles, automotive, construction, and paints. These polymers are characterized by the presence of an ester group along the polymer backbone. Depending on the chemical structure, the polyester can be a thermoplastic [1,2] or a thermoset [3,4]. Thermoplastic polyesters can be heated to a soft rubbery or fully molten state at high temperatures for various shaping processes. Sheet extrusion through a die and 3D part injection/compression molding processes require fully molten state, whereas fiber or film stretching and bottle or film blowing processes for polyesters typically take place in the rubbery state between the glass transition and melt temperature of the polyester. In this chapter, the scope is limited to thermoplastic polyesters and their rubbery state deformations such as experienced in flexible film operations (tenter-frame stretching, film blowing) and rigid packaging industry (blow molding) with focus on real-time monitoring of rheo-optical properties.

Real-time monitoring of rheo-optical properties such as birefringence and stress/strain fields provide an invaluable insight into the instant phase changes and molecular ordering in the material. The polyester material can go through different stages of molecular ordering and structural phases during its rubbery state deformation. These stages are highly dependent on process parameters

*Polyester Films: Materials, Processes and Applications*, First Edition.
Edited by Miko Cakmak and Jehuda Greener.

such as rate profiles (modes and sequences of deformation) and processing temperature as well as material parameters such as molecular weight and chain architecture of the polyester resin. In addition to active deformation mode (e.g. tensile, shear, and compression) profiles, the material can experience various relaxation modes when the deformation is halted during the process before assuming its frozen-in structure. The relaxation stage is as important, if not more important than the deformation stage since it provides the material with the right conditions for additional structural phase transitions to occur (e.g. crystallization) and molecular re-ordering to take place. The final product properties (mechanical, dimensional, barrier, and electrical) depend strongly on the final frozen-in structural state, e.g. degree of crystallinity, molecular orientation, and type of crystalline morphology after the process is brought to a halt. For example, a PET film that has been frozen-in (quenched) too quickly without allowing it to relax will assume completely different properties than the one that has gone through relaxation with additional crystallization and ordering processes before locking in the structure. Various examples will be presented in the ensuing sections.

In processed semi-crystalline polymers, the birefringence carries some information about the orientation, crystallinity, and to a lesser extent the residual stress [5]. In the case of single-phase polymers such as amorphous polymers, it directly carries information on the orientation and indirectly the level of residual stresses frozen in the material [6, 7]. Rapid determination of these parameters together with the "true" mechanical parameters, "true stress" and "true strain" allows an integrated monitoring of mechanical stimulus (strain) and response (stress) together with structural (birefringence) parameters. The quantitative relationships between the process dynamics such as stress and strain and the structural parameter, birefringence, particularly at large deformation levels, are needed to build constitutive relationships that are essential for predictive modeling of material behavior in shaping/forming processes.

Birefringence is a measure of the dependence of refractive index on the direction of polarized light in traveling through an anisotropic medium. It is an optical property that manifests itself by double refraction where light experiences two indices of refraction due to the optical anisotropy along its path. Birefringence is defined as the difference between the refractive indices in two orthogonal directions.

$$\Delta n = n_1 - n_2 \tag{5.1}$$

When the two directions are on the same plane, it is referred to as in-plane birefringence, and when the directions are on different planes such as $n_1$ and $n_3$, it represents out-of-plane birefringence. In this case, $n_3$ points in the $z$-direction, typically the thickness direction in films.

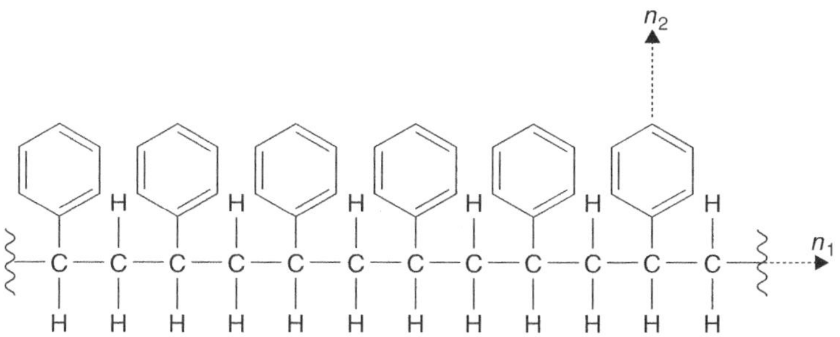


**Figure 5.1** Directions of two refractive indices $n_1$ and $n_2$ in a polymer (polystyrene) chain.

Molecular chains of most polymers exhibit birefringence owing to the asymmetry of their chemical structures. For most polymers, the refractive index parallel to the main chain encounters primarily C—C bonds, which have low polarizability and hence a refractive index close to 1. In a direction perpendicular to the molecular backbone, many polymers have side groups that retard the light to a greater extent giving rise to a higher refractive index. For example, if the polarized light were to interact with a single polymer molecule shown in Figure 5.1, with plane-polarized electric vector vibration along C–C bonds along the main chain direction (direction 1) and along phenyl groups in the perpendicular direction (direction 2), the polarized light vibrating along the main chain would be retarded less compared to that in the perpendicular direction due to the higher refractive index $n_2$ in this direction. Since retardation of light in a material is related to its ability to oscillate the electrons of the material, it is a function of the polarizability of the bonds in the material. In the example shown below, main chain C–C bonds do not have significant polarizability, while double bonds in the phenyl group are highly polarizable and exhibit higher refractive index.

Unlike single molecular chains, bulk polymer material, in the absence of deformation, is macroscopically isotropic and exhibits no birefringence because 3D spatial arrangement of the chains is random in all directions and polarizability of each repeating unit cancel the intrinsic optical anisotropies of each other Figure 5.2 (left). In contrast, when polymer films or fibers are oriented, bulk polymer starts to develop birefringence due to the preferential orientation of intrinsically birefringent molecular chains in the same direction; in a state where they can no longer cancel out each other's individual polarizability Figure 5.2 (right).

In the case of uniaxially drawn polymer films, the orientational birefringence $\Delta n$ is defined as $\Delta n = n_{//} - n_{\perp}$, where $n_{//}$ and $n_{\perp}$ are the refractive indices for light polarized in directions parallel and perpendicular to the chain axis, respectively. The difference is defined as the intrinsic birefringence of the polymer chain.

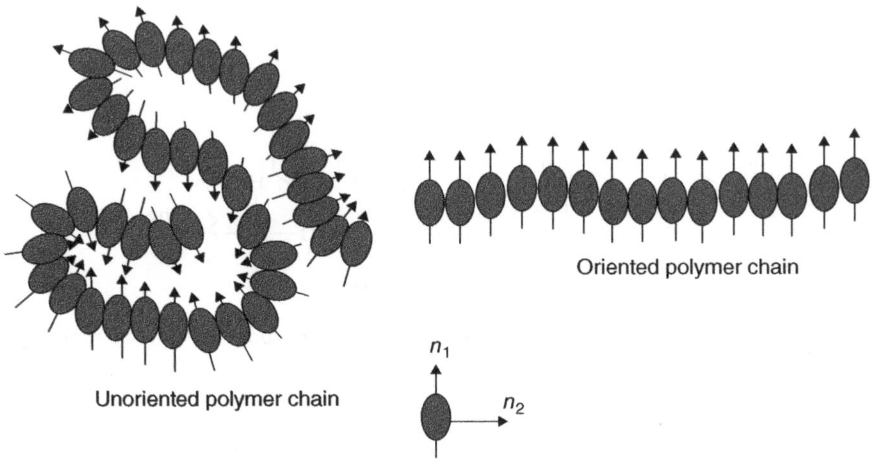


**Figure 5.2** Unoriented and oriented polymer chain-arrow pointing in the direction of optical axis of each repeating unit and representing anisotropy.

Polymers are positively birefringent if the orientational birefringence values during uniaxial stretching are positive $\Delta n > 0$ and negatively birefringent if orientational birefringence develops negative values $\Delta n < 0$ as illustrated in Figure 5.1. The sign of the birefringence is an inherent property of the polymer chains. One can prepare polymers with zero intrinsic birefringence by combining positively and negatively birefringent monomer structures via reactive copolymerization to prevent the development of orientation birefringence even after film processing [8–10]. Such structures are useful to control degradation of optical performance of LCDs which rely on maintenance of the polarization state of the incident light [11].

When light experiences double refraction in a given material, that material is said to be birefringent. Light entering a birefringent material is modeled in terms of the light being split into two rays called the ordinary ray (o-ray) (fast) and the extraordinary ray (e-ray) (slow). The angle at which the incident light enters the material is very critical in determining how the two rays will behave, and birefringence will manifest itself. In anisotropic crystals or polymers, orientation of the crystalline axis or main chain axis with respect to the incident light angle determines the relative speeds and trajectories that the ordinary and extraordinary rays will attain during their passage through the material. If incident light enters the optical axis of an anisotropic material, light will behave as though it entered an isotropic structure and travel at a single velocity without experiencing double refraction, i.e. birefringence. However, the same light entering the same material will refract into two rays, i.e. o-ray and e-ray, and experience birefringence if it enters at a non-zero oblique angle with respect to the optical axis of the material.

Both rays will be polarized with vibration directions oriented at right angles to one another and travel at different speeds. Because the two components travel at different velocities, phase difference between the waves occurs as they recombine at the exit of the birefringent material. This phase difference is a direct measure of retardation and is dependent on the sample birefringence, thickness, and wavelength of light in vacuum as shown in Eq. 5.2. The greater the thickness or difference in refractive indices, the greater the degree of retardation between waves. On the other hand, waves with smaller wavelengths are retarded to a greater extent.

$$\Gamma = \frac{2\pi \times \Delta n \times d}{\lambda_0} \tag{5.2}$$

When the two rays exit the material with a phase difference, the polarization state of the incident light is changed, for example, from linear polarization to elliptical polarization.

Figure 5.3 depicts three different scenarios where light enters at a nonzero and zero angle with respect to the optical axis of the material. In Figure 5.3c where incident polarized light enters parallel to the optical axis of the material, i.e. zero angle, the light behaves as ordinary ray only and wave separation into o and e-ray cannot be distinguished. However, in Figure 5.3a and b where light enters at a non-zero angle with respect to the optical axis, it splits in both cases into ordinary and extraordinary waves with different trajectories. Only in the case of light entering the material at 90° angle with respect to the optical axis, the pathways of the ordinary and extraordinary waves coincide, i.e. follow the same trajectory, with the only difference being their velocity and the corresponding phase shift, i.e. retardation. In the case of Figure 5.3a, not only the relative velocities but also the

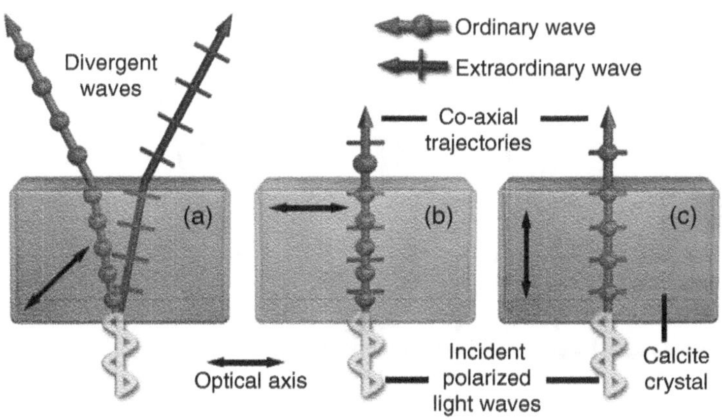


**Figure 5.3** Light entering at an (a) oblique, (b) perpendicular, and (c) parallel angle to the optical axis of the crystal. https://www.olympus-lifescience.com/en/microscope-resource/primer/java/polarizedlight/crystalwavefronts/

trajectories of the two rays are different. These are important points to consider when aligning samples and incident light directions in the optical path of the measuring setup.

We now briefly review the behavior of birefringent materials in between crossed polarizers because this is the underlying setup used when measuring birefringence via the compensator technique or the intensity techniques.

In the set-up shown in Figure 5.4, white light is plane polarized in direction P and travels in a direction perpendicular to the optical axis of the anisotropic crystal (configuration shown in Figure 5.3b). Upon entering the material, polarized light will split into two separate components (o and e-ray) since it is traveling at a non-zero angle with respect to the optical axis of the crystal and will be polarized vibrating parallel to the crystallographic axis (1 and 2) and at directions perpendicular to each other. The light rays will then travel through the analyzer, which allows only those components of the light waves that are parallel to the analyzer transmission azimuth to pass. The component of these waves will be vibrating in the same plane and retarded with respect to one another. This causes either constructive or destructive wave interference to form as they pass through the analyzer creating a spectrum of color when observed in white light through crossed polarizers. When an optical detector is used behind the analyzer, the spectral light intensity is computed from

$$I = Io\sin^2(2\theta)\sin^2\left(\frac{\Gamma}{2}\right) \tag{5.3}$$

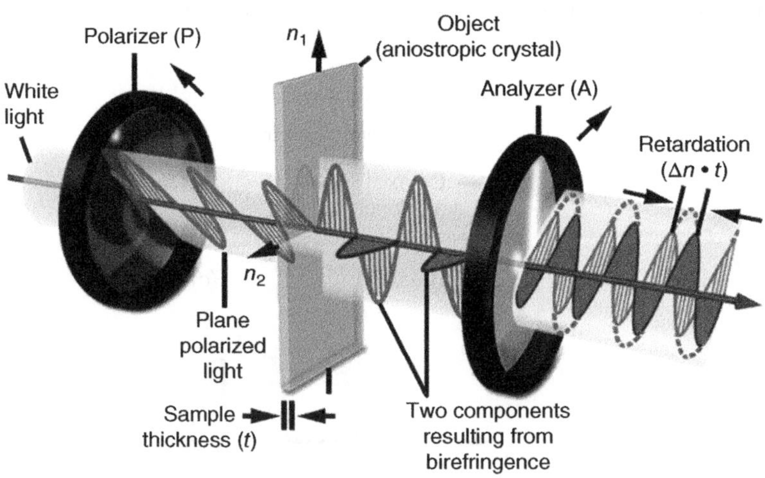


**Figure 5.4** Birefringent materials in between crossed polarizers. https://www.olympus-lifescience.com/en/microscope-resource/primer/java/polarizedlight/crystalwavefronts/

where $Io$ is the intensity of light emerging from the polarizer and $\theta$ is the angle between the polarizer and the sample principal axis. As the equation indicates, the measured intensity is zero for $\theta = n\pi$ ($n = 0, 1, 2...$) and for $\Gamma = 2n\pi$ ($n = 0, 1, 2...$).

Historical sequence of the polarized light technique for birefringence measurements can be listed as follows:

- Single wavelength methods;
- Dual wavelength methods;
- Online dual wavelength methods;
- Multi-wavelength method;
- Discrete multi-wavelength method; and
- Continuous wavelength method (spectral birefringence).

Stein and his co-workers [12, 13] measured time-dependent birefringence, i.e. dynamic birefringence, of high polymers in 1960s using single wavelength techniques. They used monochromatic light from a filtered parallel mercury vapor light which was polarized 45° to the stretching direction of the sample. Polarized monochromatic light was simply passed through the sample, which was clamped in vertical direction and through the analyzer oriented perpendicular to the polarizer and finally detected by a photomultiplier as shown in Figure 5.5.

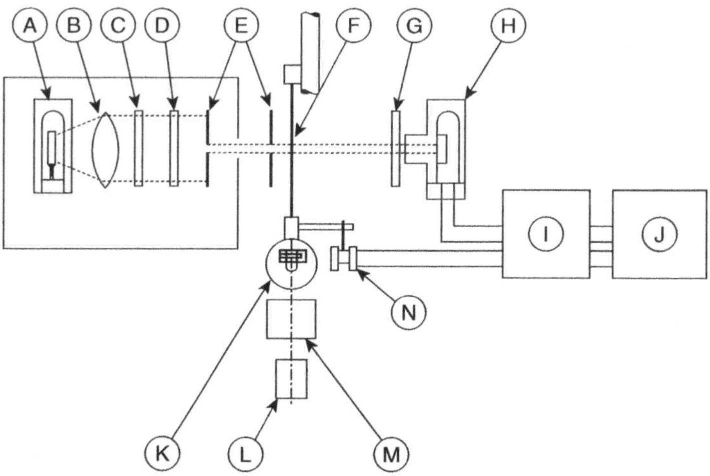


**Figure 5.5** The dynamic birefringence apparatus with single wavelength method: (a) mercury light source, (b) condensing lens, (c) monochromatizing filter, (d) polarizer, (e) slits, (f) sample, (g) analyzer, (h) photomultiplier, (i) two-channel amplifier, (j) two-channel recorder, (k) eccentric drive, (l) motor, (m) variable-speed transmission, and (n) linear variable differential transformer. *Source:* Adapted from [14].

The output was amplified and recorded whereas the reading which is a function of birefringence was adjusted against using a Babinet compensator. Later, Asada et al. [15] used a similar experimental design but with sodium light as the light source in a cone-and-plate rheometer with a transparent cone and plate made of quartz with an optical system. The latter technique was significantly successful in measurement of rheological and optical properties of a sheared polymeric system. However, all these single-wavelength methods employed nonautomated calibration procedures and were limited to measure retardations smaller than one wavelength and incapable of detecting inversion of refractive index trends. Galay and Cakmak [16,17] solved this problem by studying the inversion point of the birefringence using two different wavelengths instead of one. They used a two-color laser optical measurement system capable of analyzing fast structural changes taking place in uniaxially oriented PEN as well as PEN/PEI blends through correlation with retardation data. Following two color techniques, Cakmak and coworkers utilized the developments by Yang, Harding [18,19], Burghardt [19–22], and Posthuma and developed an extensive analysis system utilizing the principles of spectral birefringence technique based on the entire visible spectrum and made it fully automated as demonstrated in publications by Serhatkulu and Cakmak [23–27]. This enabled real-time measurement of very rapid retardation (and thus birefringence measurement) changes in as little as 0.2 s that take place during annealing of oriented films under constraint indefinitely.

## 5.2 Polyester Films

Cakmak and coworkers studied real-time birefringence development during rubbery state film processing of various polyester films and their blends along with stress and strain fields in an attempt to shed light into the structural changes taking place during orientation, relaxation, and heat-setting (annealing) of this important class of polymers. Polyesters, as mentioned earlier, are a large class of polymers formed by condensation polymerization of difunctional monomers of dicarboxylic acid or diesters with aliphatic diols. These polyesters can be aliphatic or aromatic depending on the type of dicarboxylic acids used. Examples of aliphatic polyesters are poly glycolic acid, polycaprolactone, poly lactic acid, polyhydroxyalkanoate, polyhydroxybutyrate, polyethylene adipate, polybutylene succinate and poly(3-hydroxybutyrate-co-3-hydroxyvalerate). Commonly used aromatic polyesters are polyethylene terephthalate (PET), polyethylene naphthalate (PEN), polyether ether ketone (PEEK), and polybutylene terephthalate (PBT). Aromatic polyesters have superior thermal properties compared to aliphatic polyesters due to their large aromatic ring structures with alternating double-single bonds. Among the aromatic copolyesters which are processed in the rubbery state for film, sheet and blown rigid container applications, PET, PEN, PEEK, and their blends with each other or other polymers and copolymers, are the most common. PBT, which is very similar in structure to PET, is not suited for film processing because it crystallizes too fast due to its longer aliphatic diol (1,4 butane diole), which imparts added flexibility to the chain backbone. As a result, it is typically used in processes where faster crystallization is desired such as in injection molding. In contrast, PET, PEN, and PEEK are classified as slow crystallizing polymers, which makes them suitable for rubbery state processing. Slow crystallizing polymers can possess various degrees of crystallinity depending on the cooling rate from the melt. When quenched rapidly, their stiff backbones do not have time to rearrange and therefore assume a completely amorphous state. They can then be heated above their $T_g$ but below their cold crystallization temperature to soften them without crystallizing. In this temperature range, these polymers can be oriented to impart stress-induced crystallization and further heat set (annealed under lateral constraint) to form thermal crystallites. More details are given below.

A typical filmmaking process [28] involves three main stages, i.e. (i) melt extrusion and casting, (ii) tenter-frame sequential or simultaneous biaxial stretching, and (iii) heat-setting stages as shown in Figure 5.6. Melt casting is carried out through a sheet casting die onto a chilled roll and the film contact is typically promoted by electrostatic charge bar to ensure the molten sheet makes a firm contact with the chill roll surface whereby the polymer rapidly quenches into amorphous state. As the differential scanning calorimetry (DSC) diagram indicates in Figure 5.6 inset, the film can be crystallized if the temperature on the chilled roll is raised to $T_{cc}$ (cold crystallization) temperature region that is situated roughly between the $T_g$ and $T_m$.

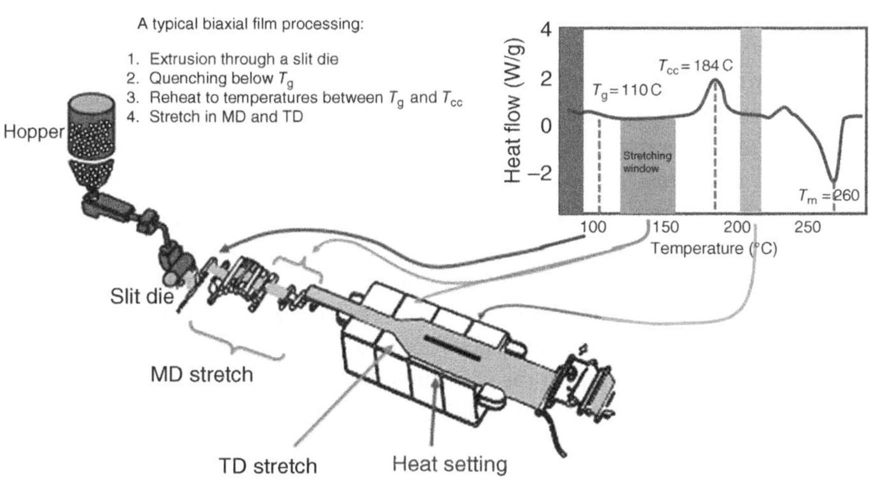


**Figure 5.6** Tenter-frame film process.

Stretching temperatures are typically chosen within the $T_g$-$T_{cc}$ window that is sufficiently away from the $T_{cc}$ (see stretching window rectangle in the inset of Figure 5.6) in order not to initiate thermally induced crystallization (nucleation and growth through chain diffusion to the growing front) prior to stretching. Once this crystallization is initiated, these materials lose the ability to be uniformly stretched easily unless they are heated to temperatures where some of these crystals are partially molten. This is usually avoided for a variety of reasons. These films rapidly lose their strain-hardening ability when partially molten as these polymers – e.g. polycondensates – do not generally have high molecular weights like polyolefins, and hence possess lower entanglement densities and lesser ability to promote strain hardening during stretching. The activation and promotion of strain hardening mechanism is critical to obtaining uniform thickness films at the end of the stretching steps. In a sequential biaxial mode, stretching is accomplished in sequence by first stretching in the machine direction (MD) and then in transverse direction (TD) as shown in Figure 5.6. In the MDO (machine direction orientation) roller systems, the film is increasingly accelerated in the rubbery state ($T_g$-$T_{ll}$ window) between the hot rolls inducing very high preferential chain orientation to the film. Beyond a critical strain level, the material undergoes orientation-induced crystallization when the polymer chains become parallel to each other and register in crystalline state. It is important to distinguish the mechanism of orientation (also called strain) induced crystallization from thermal crystallization. In both cases the crystallization occurs when the polymer chains become aligned parallel to each other. They differ as to how this parallel orientation occurs. In orientation-induced crystallization chains are mechanically oriented to become parallel to each other by external action. (shear, elongation, planar, biaxial flow) hence polymers do not necessarily need to diffuse to crystal growing front for crystallization to proceed. Whereas, thermally activated crystallization exclusively rely on the growth of crystal phase by the diffusion of polymer chains to the growing front and become parallel to the chains previously deposited at this front. Orientation induced crystals are typically very small and do not affect transparency of the films, while crystals that are formed by thermal crystallization could be large enough to scatter light and cause white (hazy) appearance. After MDO rollers, the material in this next stage is subjected to transverse stretching (TD stretching) using the tenter-frame stretching process and this step in the process balances properties in the plane. In both previous stretching stages, the level of crystallinity does not typically exceed 25% and the films contain a great deal of oriented amorphous chains. Films with such structures tend to be unstable above their $T_g$ and in order to thermally stabilize them, they are subjected to constrained annealing – sometimes called heat-setting – at temperatures where the thermal crystallization rates are maximum (e.g. for PET ~180 °C) that would boost the crystallinity while relaxing some of the oriented amorphous chains, that are

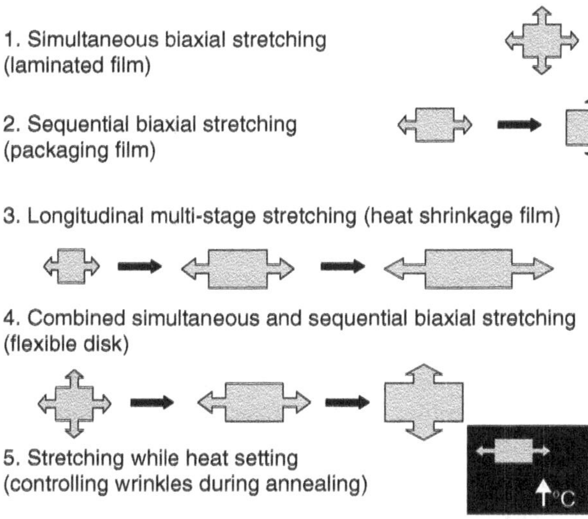


**Figure 5.7** Most common stretching protocols practiced in the film industry. *Source:* Cakmak et al. [29]/with permission from AIP Publishing.

responsible for shrinkage that may occur at elevated temperatures. Annealed films remain dimensionally stable nearly up to their crystalline melting range.

The biaxial stretching (or tenter-frame) process is used widely for manufacturing polyester films. When producing biaxially oriented polyester films, processing conditions such as temperature, stretching rate, and sequence of stretching are altered to obtain desired effects on the properties of the films to suit the given application. These properties include optical transparency, optical anisotropy, thermal stability, particularly shrinkage, and balance of in-plane mechanical properties. Typical industrial processes that use multistage, multiaxial film processing is graphically summarized in Figure 5.7.

## 5.3 Online Optical Measurements

In order to simulate different stretching protocols, Cakmak et al. have developed laboratory scale uniaxial [14, 30–32] biaxial stretching [29, 33–35] and heat-setting [36, 37] machines that are equipped with measurement systems that enable real-time measurement of true stress, true strain, birefringence and IR dichroism [38] as they are subjected to deformation fields at industrially realistic speeds at elevated temperatures where they are partially molten or in a rubbery state. These machines simulate deformation modes that are typically encountered in industrial processes including, uniaxial free width, uniaxial constant width,

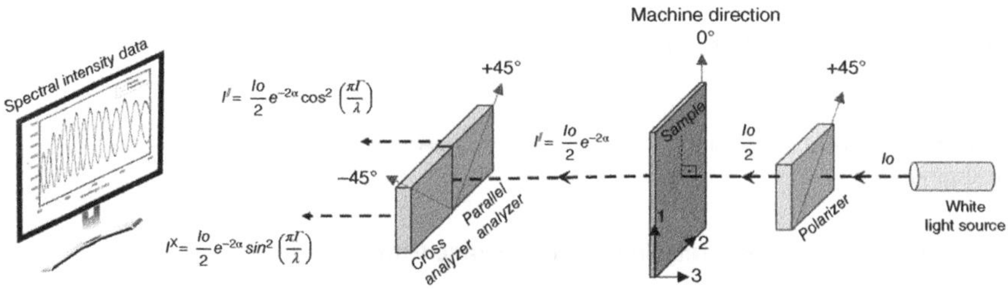


**Figure 5.8** Optical train set-up in lab scale uniaxial, biaxial, and heat-setting machines by Cakmak et al. for retardation in normal incidence.

unequal simultaneous biaxial, equal simultaneous biaxial, and sequential biaxial. They also apply programmed motion profiles, such as linear motion on both axes as well as user programmable motions that includes typical temporal position profiles such as logarithmic, exponential, sinusoidal, and saw tooth. The machine is also capable of receiving deformation temporal profiles from industrial machines and replicating them under well-controlled temperatures. Next, we describe real-time spectral birefringence, stress and strain measurement techniques employed in these machines. Optical train used to determine real-time retardation of films in these machines consisted of white light source, one polarizer (at +45°), sample at 0° and two analyzers (one at −45° and one at +45°) to obtain parallel and cross-polarized light as shown in Figure 5.8.

Governing equations obtained from Mueller matrix/Stokes vector analysis of the optical trains [39–41] are as follows:

The parallel light intensity is

$$I^{//} = \frac{Io}{2} e^{-2\propto} \cos^2\left(\frac{\pi\Gamma}{\lambda}\right) \tag{5.4}$$

The cross-light intensity is

$$I^{\times} = \frac{Io}{2} e^{-2\propto} \sin^2\left(\frac{\pi\Gamma}{\lambda}\right) \tag{5.5}$$

where $\Gamma = \Delta nd$

A typical data is given in Figure 5.9 for parallel and cross-polarized light intensity

In order to eliminate the absorption (exponential term), the two equations are normalized (Eq. 5.6).

$$N^{\perp} = \frac{I^x}{I^x + I^{//}} = \frac{\frac{Io}{2} e^{-2\propto} \sin^2\left(\frac{\pi\Gamma}{\lambda}\right)}{\frac{Io}{2} e^{-2\propto}\left\{\cos^2\left(\frac{\pi\Gamma}{\lambda}\right) + \sin^2\left(\frac{\pi\Gamma}{\lambda}\right)\right\}} = \sin^2\left(\frac{\pi\Gamma}{\lambda}\right) \tag{5.6}$$

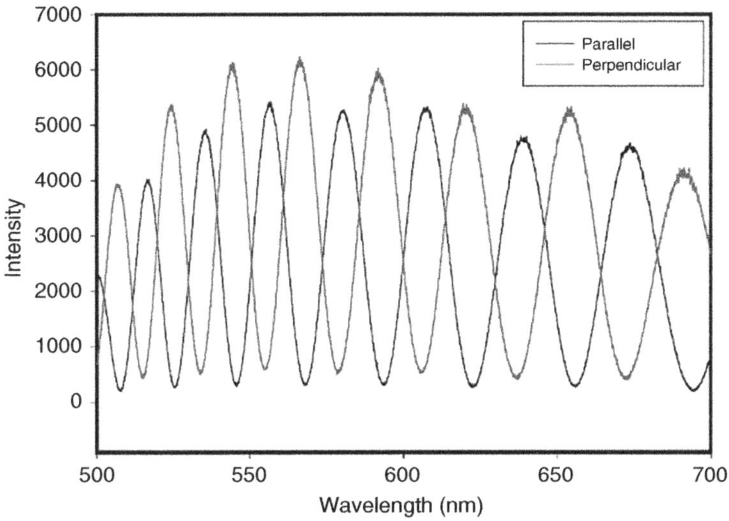


**Figure 5.9** Cross and parallel polarized light spectra.

Retardation is then calculated using the equation below in Figure 5.10 by graphical determination of peak and zero crossing values.

The resulting retardation vs. wavelength data is fitted to Cauchy's equation, and retardation at the desired wavelength is read from the chart. Figure 5.11 illustrates retardation determination with wavelength chosen at 546 nm.

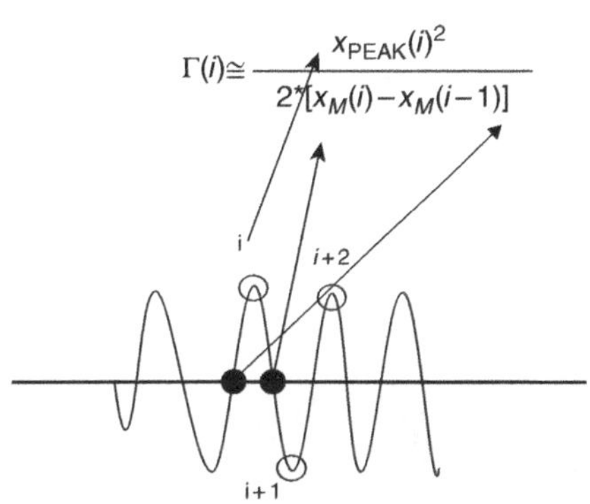


**Figure 5.10** Determining retardation from peak points on the normalized intensities curve.

In-plane birefringence $\Delta n_{12}$ (Eq. 5.7) is found by dividing retardation at normal incidence $\Gamma_0$ (set up shown in Figure 5.8) by the thickness of the film $d_0$.

$$\Delta n_{12} = \frac{\Gamma_0}{d_0} \tag{5.7}$$

To measure the birefringence, the thickness must be determined at the same time and at the same location where the birefringence is measured. In the uniaxial stretching tests, this is accomplished by moving the two cross-heads of the

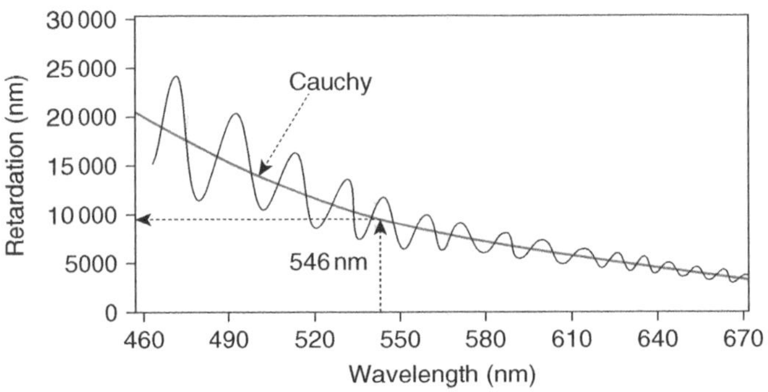


**Figure 5.11** Fitting a curve on the actual data and obtaining the retardation.

stretching frame in opposing directions allowing the narrowest symmetry plane of the sample to remain spatially stationary. A laser micrometer mounted at 45° to the horizontal plane is focused at this stationary symmetry plane such that it continuously monitors the width of the sample during stretching and holding stages as shown in Figure 5.12.

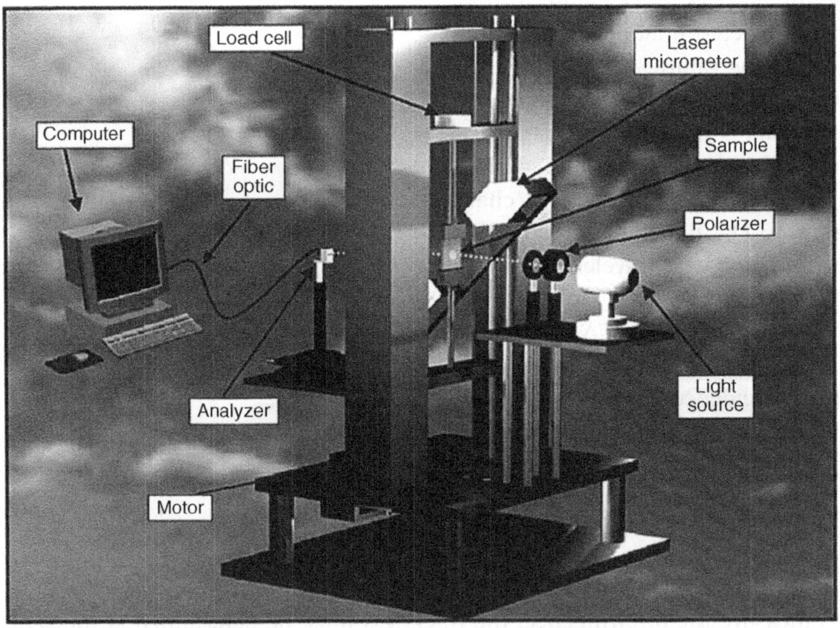


**Figure 5.12** Uniaxial stretching machine equipped with spectral birefringence and width measurement equipment. Adapted from Cakmak et al. [32].

The temporal development of thickness and thus cross-sectional areas are then determined utilizing uniaxial symmetry Eq. 5.8 with the knowledge of initial thickness and widths.

$$\frac{W_t}{W_0} = \frac{D_t}{D_0} \tag{5.8}$$

where

$W_t$: real-time width of the film
$W_0$: initial width of the film
$D_0$: initial film thickness
$D_t$: real-time film thickness

In order to couple the birefringence information with stress- strain information during uniaxial stretching, the following equations are used.

For elongation, three different parameters were typically studied, i.e. engineering strain, true strain, and Hencky strain. Engineering strain is defined as the ratio of elongation (difference between the final length and the initial length) to the initial length of the sample. As for true strain which applies nominal (fully lagrangian) definition, the elongation was related to the reduction in the local width of the sample using transverse isotropy (Eq. 5.8) assumption and incompressibility (Eq. 5.9) assumption to formulate width-based elongation (Eq. 5.10). This is the same local region where the retardation measurements are made.

$$D_0 W_0 L_0 = D_t W_t L_t \tag{5.9}$$

$L_0$: initial length of the film
$L_t$: real-time length of the film

$$\text{True strain} = \frac{\text{Elongation}}{\text{Initial length}} = \frac{L_t - L_0}{L_0} = \frac{\Delta L}{L_0} = \left(\frac{W_0}{W_t}\right)^2 - 1 \tag{5.10}$$

For Hencky strain, logarithmic definition of width-based strain measurements is used (Eq. 5.11).

$$\text{Hencky strain} = \text{Ln}\left(\frac{L_t}{L_0}\right) = \text{Ln}\left(\frac{W_0}{W_t}\right)^2 \tag{5.11}$$

Finally, true stress is calculated in the uniaxial stretcher by

$$\text{True stress} = \frac{\text{Force}}{\text{Cross sectional area}} = \frac{F_t}{W_t D_t} = \frac{F_t}{\left\{\left(W_t^2 / W_0\right) D_0\right\}} \tag{5.12}$$

Where $F_t$ = force

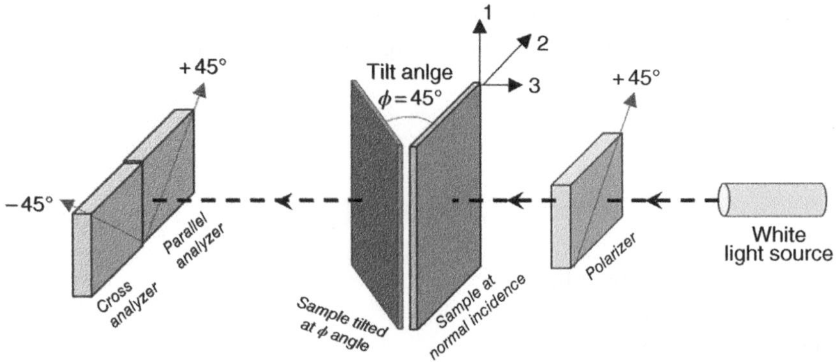


**Figure 5.13** Sample tilting for out-of-plane birefringence

For uniaxial films, in plane birefringence $\Delta n_{12}$ data is enough to track orientation as the film is isotropic in other planes. For biaxially oriented films, on the other hand, both the in-plane and out-of-plane birefringence are tracked simultaneously. In-plane birefringence is calculated using normal incidence retardation as shown in Eq. 5.7. In order to determine out-of-plane birefringence $\Delta n_{23}$, sample tilting analysis is conducted as shown in Figure 5.13 and Eq. 5.13 is used.

$$\Delta n_{23} = -\frac{1}{d_0}\left[\frac{\Gamma_0 - \Gamma_\phi\left(1 - \frac{\sin^2\phi}{\bar{n}^2}\right)^{1/2}}{\frac{\sin^2\phi}{\bar{n}^2}}\right] \tag{5.13}$$

Where

$d_0$ = thickness of the film
$\Gamma_0$ = retardation of the film for normal incidence
$\Gamma_\phi$ = retardation of the film tilted through the angle $\phi$
$\bar{n}$ = average refractive index of the film which is assumed to be a constant value

$$\bar{n} = \frac{n_1 + n_2 + n_3}{3} \tag{5.14}$$

The third birefringence $\Delta n_{31}$ is automatically defined through

$$\Delta n_{12} + \Delta n_{23} + \Delta n_{31} = 0 \tag{5.15}$$

For practical reasons, instead of sample tilting, two optical trains are used for gathering simultaneous in-plane and out-of-plane birefringence data, i.e. one optical train for normal incidence and another optical train tilted at an angle with respect to the film as shown in Figure 5.14 in the biaxial stretching platform equipped with spectral birefringence set-up used by Cakmak and coworkers.

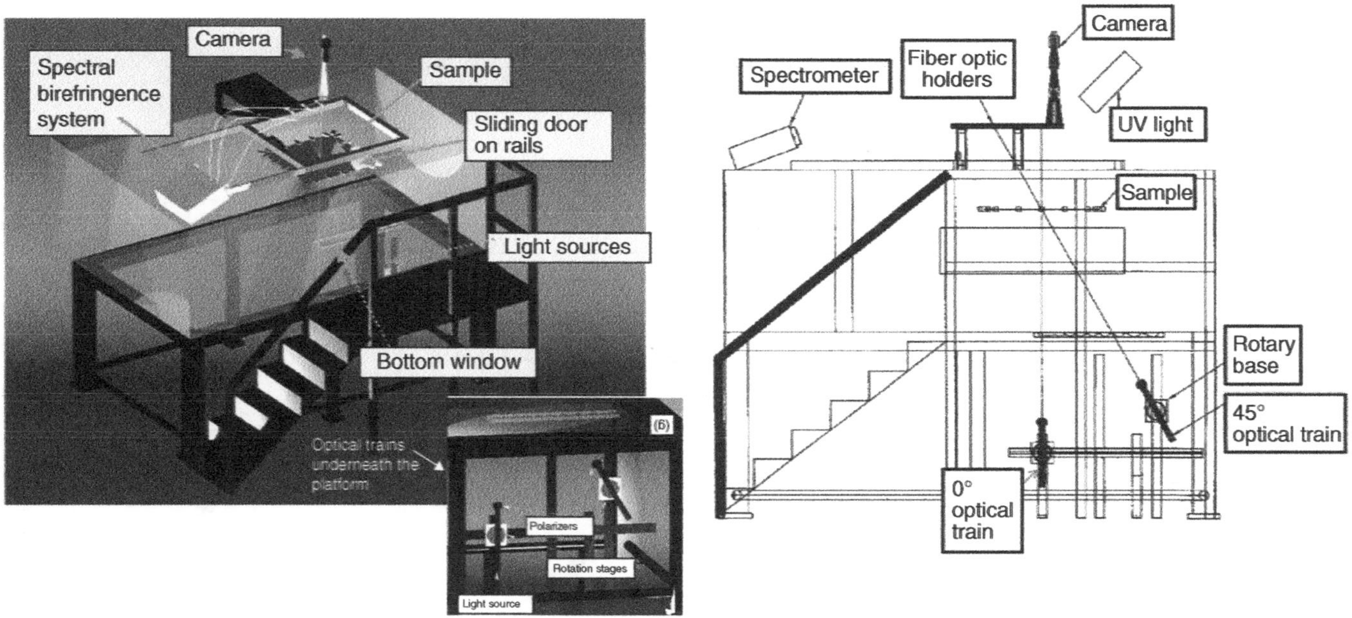


**Figure 5.14** Spectral birefringence- Biaxial stretching platform. Adapted from Cakmak et al. [29].

On the detector side, a high-speed CCD camera is mounted and directed toward the sample as shown in Figure 5.14. This camera is used to capture the pre-painted 24-dot matrix pattern shown in Figure 5.15. These are used to determine and temporal evolution of strains in $X$ and $Y$ directions. These data were then used to calculate the real-time thickness at the geometric center of the sample. The image processing software determines the spatial location of all 24 dots and from time sequence of images determines strains in both $X$ and $Y$ directions. The real-time thickness is calculated assuming material incompressibility:

The real time volume, $V(t)$ = The original volume, $V(0)$,

$$X(t)\cdot Y(t)\cdot T(t) = X(0)\cdot Y(0)\cdot T(0).$$

From initial measurements of the sample $X(0)$, $Y(0)$, $T(0)$, and calculation of $X(t)$ and $Y(t)$ from the movements of the dot pattern centers on the film surface,

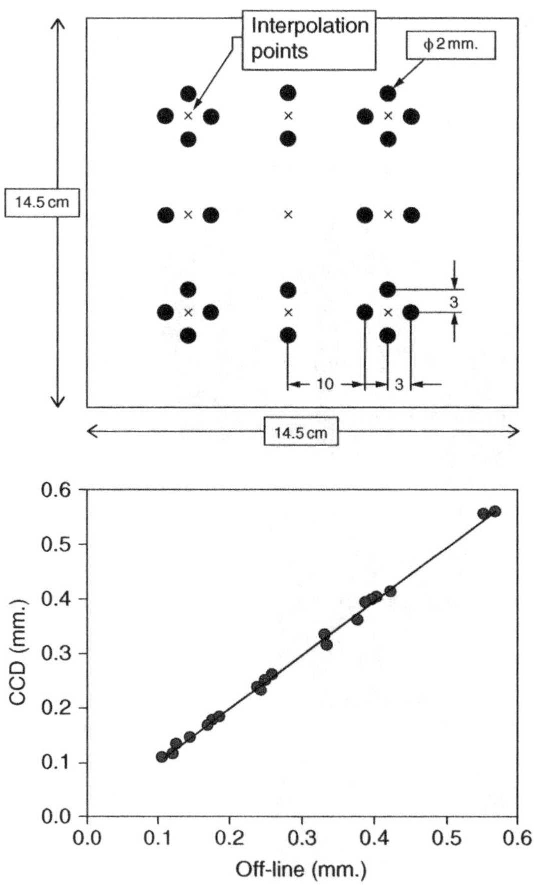


**Figure 5.15** Dot pattern for tracking the and X and Y deformations and determining thickness. *Source:* Cakmak et al. [29]/with permission from AIP Publishing.

thickness $T(t)$ at the point where all of the retardation measuring beams converge is calculated. From these data true stress, true strain, in-plane birefringence, and out-of-plane birefringence are all calculated and tabulated in an output file. Validation of measurements for force, thickness, birefringence, and strain measurements are reported in detail elsewhere [29]. Thickness validation is shown in Figure 5.15 (right), where a series of samples were stretched at different stretch ratios and allowed to cool down below the glass transition temperature to insure that there is no more deformation in the film, while monitoring the thickness change with the CCD camera. The final value of thickness measured by CCD is compared with the offline measured values by a precision micrometer. The comparison shows a very good agreement between online and offline measurements.

The next critical film processing set-up after biaxial stretching is the heat-setting spectral birefringence set up shown in Figure 5.16. After the film is biaxially stretched, it is cooled rapidly using both the cooling system of the stretching machine and an external fan. It is then affixed between especially designed circular clamps and inserted into the heat-setting spectral birefringence set up shown in Figure 5.16. Similar to the biaxial stretching set-up, the heat-setting unit is equipped with multi-angle retardation measurement optics in order to determine both the in-plane and out-of-plane birefringence components [37]. The oven has guide rails to guide the sample holder during inserting and removing the sample from the thermal chamber. The sample is mounted on a circular holder that constrains the film during annealing. The holder is inserted into the chamber through a spring-loaded trap door at the top of the chamber. The point of measurement on

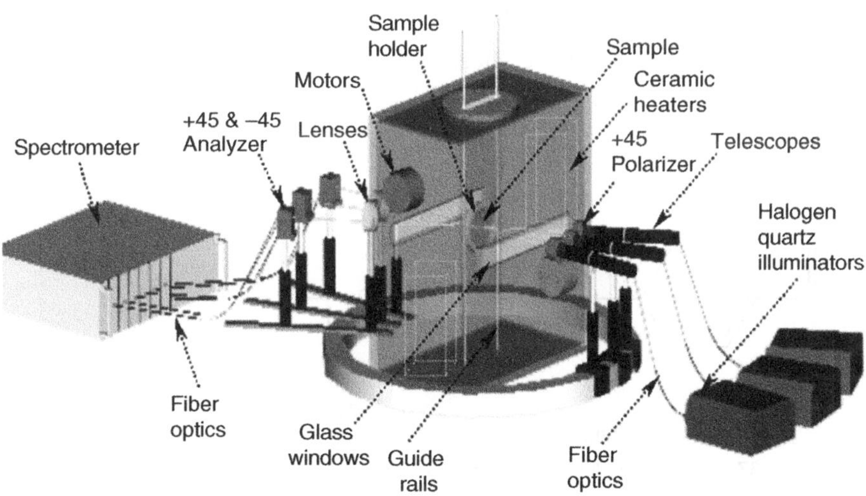


**Figure 5.16** Heat-setting set up with in-plane and out-of-plane multi-angle spectral birefringence set up. *Source:* Hassan and Cakmak [37]/with permission from John Wiley & Sons.

the sample is predetermined, and then the sample holder height is adjusted for that point using a removable latch. After the heat-setting step is complete the sample is removed by opening the latch. The sample with the holder is pushed down the rails and exits through a spring-loaded trap door at the bottom of the chamber to be quenched by immersing in a water bath. The loading and unloading of the sample are done very quickly (using this mechanism in order not to alter the thermal equilibrium inside the heated chamber. The light used for the retardation measurements passes through windows on the front and back of the oven. Windows are made from a special optical flat tempered glass that is optically isotropic such that all the measured birefringence values are intrinsic to the test sample.

Using the three birefringence set-ups discussed above, Cakmak and coworkers investigated the rheo-optical properties of several polyester films in uniaxial, biaxial, and heat-setting modes. In the remainder of the chapter, we will review the results for several polyester film types.

## 5.4 Polyethylene Terephthalate

Polyethylene terephthalate is perhaps the most widely studied polyester due to its extensive use in fiber, film, and rigid container applications [42]. Its chemical structure is shown in Figure 5.17. PET exhibits a glass transition temperature of 67–81 °C depending on the level of crystallinity and a melting temperature of about 255 °C [43].

### 5.4.1 Simultaneous Stretching

Several studies investigated the effect of processing conditions and modes of deformation on the final properties of PET [44–49]. Starting with an amorphous phase, PET undergoes a series of structural phases/regimes during orientation in

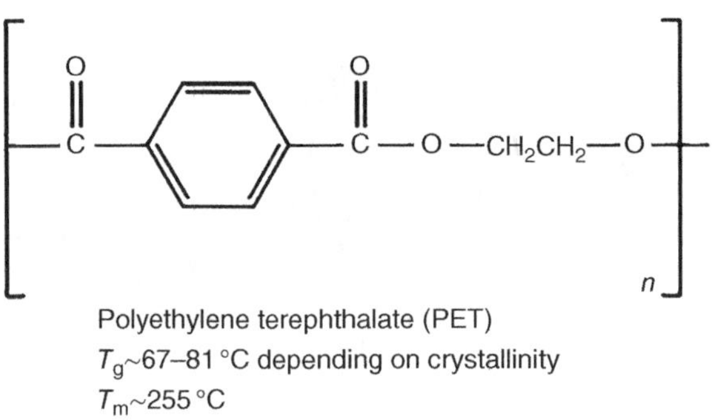


**Figure 5.17** Polyethylene terephthalate structure.

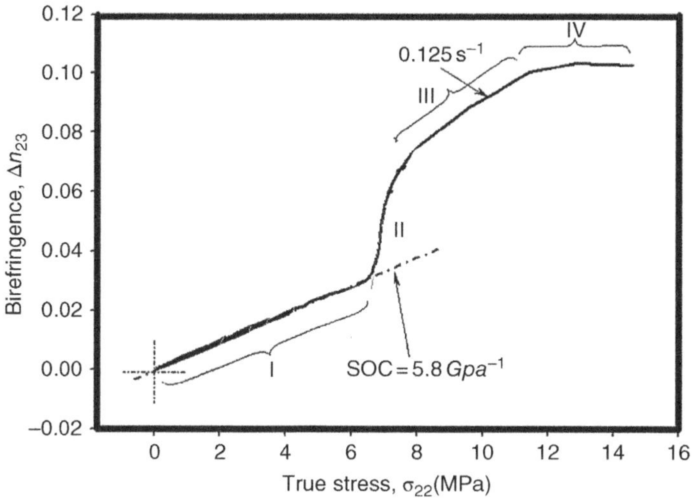


**Figure 5.18** Multi-regime behavior of PET during simultaneous biaxial stretching at 95 °C as shown by birefringence vs. true stress. *Source:* Image modified from reference [35].

the rubbery state and subsequent relaxation and/or heat-setting stages (if employed). These structural regimes and transitions become apparent when mechano-optical data are collected during processing. For instance, multi regime behavior of PET film during simultaneous biaxial stretching at a strain rate of $0.125\,s^{-1}$ is shown in an out-of-plane birefringence vs. true stress graph in Figure 5.18. The initial linear birefringence true stress regime (I) is followed by rapid increase (II) that leads to plateauing (III) and (IV) stages.

In order to elucidate the structural changes that the amorphous cast PET film undergoes in each stage, Hassan et al. [35] conducted systematic stretching experiments at different strain rates ranging from 0.00125 to $0.125\,s^{-1}$ at 95 °C, and stopped the process at strategic strain levels to perform offline DSC, Wide angle X-ray (WAXD) and Raman Spectroscopy. These measurements along with real-time birefringence development shed light into the development of amorphous ordering, gauche-trans conformational transitions (Raman), and accompanying crystallization behavior. Figure 5.19 shows the combined information as a function of stress when stretching at a strain rate of $0.125\,s^{-1}$. The underlying changes during stretching corresponding to each regime are discussed below.

The as-cast film, which was significantly amorphous (~96–97%), had a $T_g$ of ca. 79 °C, a cold crystallization temperature, $T_{cc}$ of ca. 142 °C and a melting temperature, $T_m$ of ca. 254 °C.

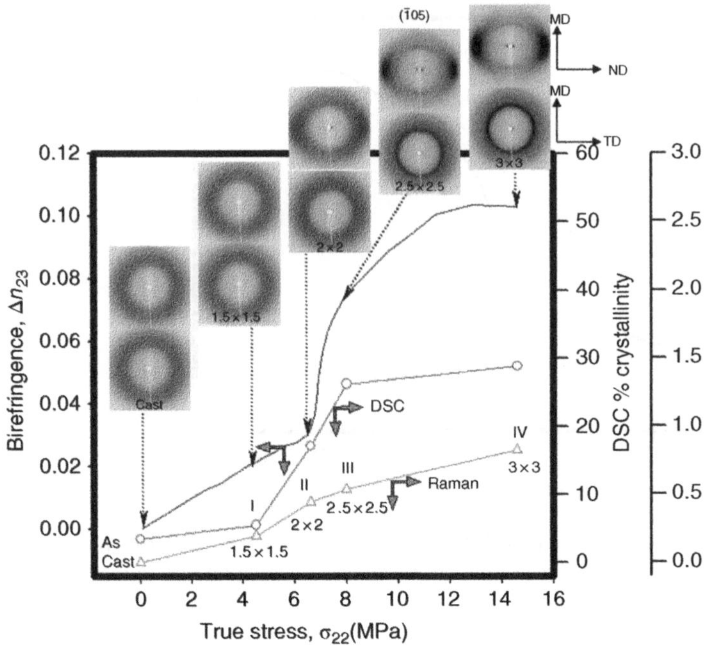


**Figure 5.19** Evolution of the birefringence, trans isomer, crystallinity, and X-ray patterns (thru TD "top" and ND "bottom" WAXD patterns) with the stress level at the rate of $0.125\,s^{-1}$. MD: machine stretching direction; TD: transverse (width); ND: normal thickness direction.

#### 5.4.1.1 Regime I

In this regime, the relationship between the stress and birefringence was found to be linear as shown in Figure 5.18 with a stress optical coefficient (SOC) of $5.8\,GPa^{-1}$. The as-cast film maintains its amorphous state to a great extent at low stretch ratios (e.g. 1.5×1.5×) and especially at low rates (not shown) but the decrease in the $T_{cc}$ and the area under the cold crystallization peak implies that the amorphous chains are developing orientation (compare $T_{cc}$ peak as cast and 1.5×1.5 in left DSC thermogram of Figure 5.20). From a theoretical point of view, the validity of the SOR in the low stretch ratio region (Regime I) indicates that, the molecular origin of the stress as well as the birefringence is the orientation of the amorphous chains [50, 51]. At this regime, the system behaves like an ordinary polymer melt. The stretching process leads to preferential chain orientation, and this change causes the polymer segments to move from the high-energy gauche conformation into the lower-energy trans conformation. The result is a slight increase in the transformation from the gauche into the trans conformation with increasing the stretch ratio and rate indicating a small increase in crystallinity.

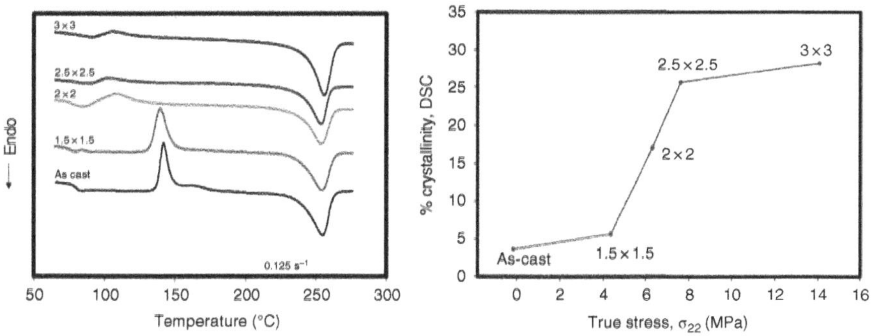


**Figure 5.20** DSC Thermograms (left) and crystallinity vs. true stress (right) -biaxially stretched films at various stretch ratios.

As is well known, only trans conformers exist in the crystalline domains of PET. DSC also supports few percent increase in crystallinity when comparing as cast and 1.5×1.5 stretched film in Figure 5.20 (right). The WAXD patterns also support the DSC data in Figure 5.20 showing significantly remaining amorphous structure in the low deformation levels in regime I. With increasing ratios, an oriented amorphous structure is observed [52].

As the stretch rate increases, linear Regime I is extended as shown in Figure 5.21, retarding the onset of Regime II which coincides with stress-induced crystallization

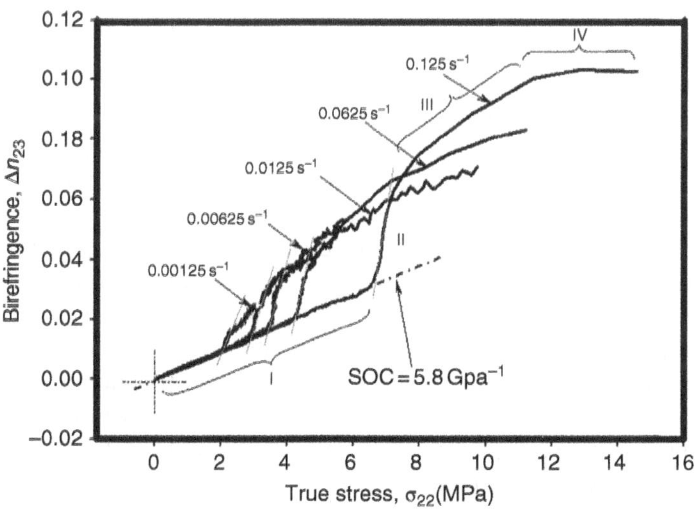


**Figure 5.21** Retardation of Regime 1 at increased strain rates.

#### 5.4.1.2 Regime II

End of Regime I and start of Regime II is marked by a rapid increase of birefringence, which also coincides with a parallel increase in the trans content and crystallinity. Crystallinity increases instantly from ~5 to 20% at the beginning of Regime II and is steeper than trans content development in these films. The rapidity of increase in birefringence, trans content and crystallinity all point to the formation of long range inter-connected network at this I to II transition. This leads to a higher efficiency in stress delivery to local chain segments and increase in orientation corresponding to a gauche to trans transformation. Consequently, crystalline domains are developed as all the chains in the crystalline lattice are in trans conformation. Throughout the course of Regime II, birefringence increases linearly while degree of crystallinity grows from 20 to ~25%. As observed in WAXD patterns taken in two directions, TD and ND (top and bottom patterns for each condition in Figure 5.19), these films exhibited in-plane isotropy and crystalline regions exhibit rather high lattice distortions evidenced by diffuse yet discrete diffraction spots $(0\bar{1}1)$ and (010).

#### 5.4.1.3 Regime III

End of Regime II and start of Regime III is marked by a slowing down of birefringence development with stress. In the beginning of this regime, the chains are already highly stretched with a considerable degree of crystallinity (~25%). The highly oriented and extended chains resist the deformation, and this causes the stress originating from the chain orientation to increase rapidly. At higher draw ratios, the stress increases faster than the birefringence. This is attributed to the non-Gaussian nature of the highly extended chains [53] as the chains are approaching their finite extensibilities. Crystallinity increases further from 25 to 27% in this regime. Due to high degrees of chain extension, mobilities in the long-range connected network decrease and thus decreasing the chances of crystallization of the chains by aligning parallel to each other.

With increasing deformation going into Regime III, more phenyl rings orient parallel to the surface of the film and the plane normals point perpendicular toward the film surface as started to be seen by the appearance of the $(\bar{1}05)$ peak in the WAXD. Since the chain axes are randomly distributed in the plane of the film, the crystallites orient equally in all directions in the plane of the film surface. In reheat blow molded PET applications, for example, starting from an amorphous PET preform, it is very important to design the preform stretch ratio such as that when the preform touches the mold surface, material orientation is in somewhere in Regime III. This is because Regime III marks the onset of strain hardening allowing thickness variations in the bottle to disappear. Regime III is the design window for the blow molding process. Further into the next regime, bottles may attain a pearlescent color indicating that the crystallites are over stretched, and breaking the crystallites causing stress whitening.

#### 5.4.1.4 Regime IV

At this stage, the chains cannot extend further, and crystallinity does not increase anymore. The result is leveling off (saturation) of birefringence while the stress is still increasing. This "muscle bound" structure can only be attained at very high deformation rates ($0.125\,s^{-1}$), where the orientation relaxations are highly suppressed at high stretch ratios.

As evidenced in gauche- trans Raman signal in Figure 5.19, even though the overall structure of the material had become highly immobilized "muscle-bound" reducing the further increases in crystallinity, the trans conformation content continues to increase noticeably as the chains in the amorphous regions continue to experience stress.

The four-regime behavior of simultaneous biaxially stretched PET is depicted in Figure 5.22. In Regime I the material remains amorphous. As the polymer chains become oriented in the plane, they start registering side-by-side with one another in parallel orientation that is needed for crystallization. If enough them form this side-by-side alignment, they start influencing the birefringence and stress development as the network composed of entanglements is enhanced by the formation of these crystalline "nodes" causing rapid rise in efficiency of orienting polymer chains while suppressing the relaxation processes. This manifest itself as the rapid increase in birefringence (Regime II), and beyond this stage the stress optical behavior becomes nonlinear (Regime III) and tend to reach a plateau as the limit of extensibility (Regime IV) is reached.

### 5.4.2 Sequential Stretching

Sequential biaxial stretching is more advantageous to use than simultaneous biaxial mode for slow crystallizing polymers such as PET, PEN, and PEEK. This is because, the uniaxial stretching applied in the first stage accelerates the stress-induced crystallization as the polymer chains preferentially orient in the stretching direction – their ideal arrangement to promote crystallization- thereby leading

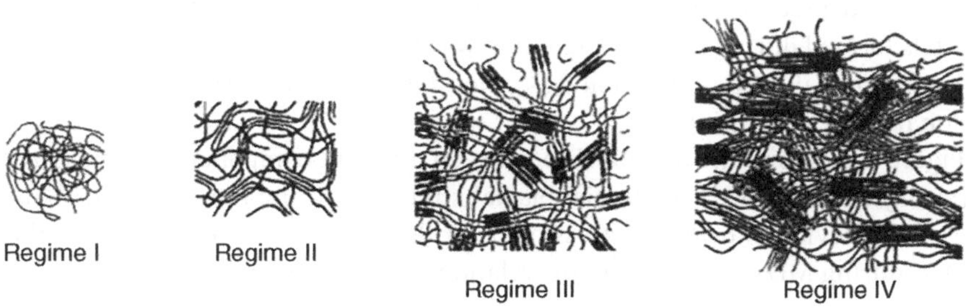


**Figure 5.22** Structural model for simultaneous biaxially stretched films.

to self-leveling deformation and uniform thickness products at smaller areal expansion ratios ($\lambda_{MD} \times \lambda_{TD}$). At equal MD and TD draw ratios, one obtains unbalanced biaxial orientation in the polyester films. This contrasts with the equal biaxially stretched films in simultaneous biaxial mode that exhibit balanced orientation. For fast crystallizing polymers (e.g. polypropylene and Nylon) with large super structural features (e.g. crystallites), reports in the literature suggest that simultaneous biaxial stretching is advantageous [54]. In such systems, in-plane balanced properties are easier to achieve in simultaneous mode [55] that is also reported to provide higher mechanical properties.

#### 5.4.2.1 Regime I

Hassan et al. studied sequential biaxial stretching of PET films and identified multi-regime rheo-optical behavior [34]. They stretched the films in two stages as per the sequential stretching protocol; first in the MD at constant width and then in the TD at constant length. They obtained films with the following stretch ratios ($\lambda_{MD} \times \lambda_{TD}$): $1.5\times1, 1.5\times1.5, 1.75\times1.75, 2\times1, 2\times1.5, 2\times2, 2.5\times1, 2.5\times1.5, 2.5\times2, 3\times1$ and $3\times1.25$ at three different rates of $0.00125\,s^{-1}$ ($0.15\,mm\ s^{-1}$), $0.0125\,s^{-1}$ ($1.5\,mm\ s^{-1}$) and $0.125\,s^{-1}$ ($15\,mm\ s^{-1}$). They observed through WAXD patterns that the films, in the first stage of sequential biaxial stretching, display chains that are planar in the MD-TD plane with more of the chains oriented in the MD direction. An increase of the transverse stretching causes the chains to be more randomly planar in the film plane with uniform distribution between the MD and TD directions. The phenyl rings quickly align parallel to the film plane during the TD stretch. Similar to the earlier studies, they observed that a balanced biaxial orientation distribution could be achieved at a TD draw ratio below that of the MD draw ratio. This means that the only way to achieve a uniform orientation distribution is through the unbalanced stretch ratios. There was evidence in their real-time biaxial measurement system that, during the MD stretch, there was already indirect stress in the TD direction working on orienting the chains in the TD direction. Consequently, the degree of stretching required to obtain the balanced orientation in TD direction was less than the MD stretching as the chains already had some degree of orientation in TD direction. In addition, they noted that crystals formed during the MD stretching was in the TD direction. Therefore, the balanced crystalline reorientation was faster than the balanced amorphous orientation in the TD direction.

Figure 5.23 shows the relationship between different components of stress and birefringence for $2.5\times2$ sequential biaxial stretching at the rate of $0.125\,s^{-1}$ and 95 °C. The deviation from the stress optical rule starts around a true strain value of 0.65 during MD stretch, where the relationships start to be nonlinear.

Hassan et al. described the behavior of PET during sequential stretching to cover three regimes. Regime 1 was the low-stress region (below a true strain of 0.65)

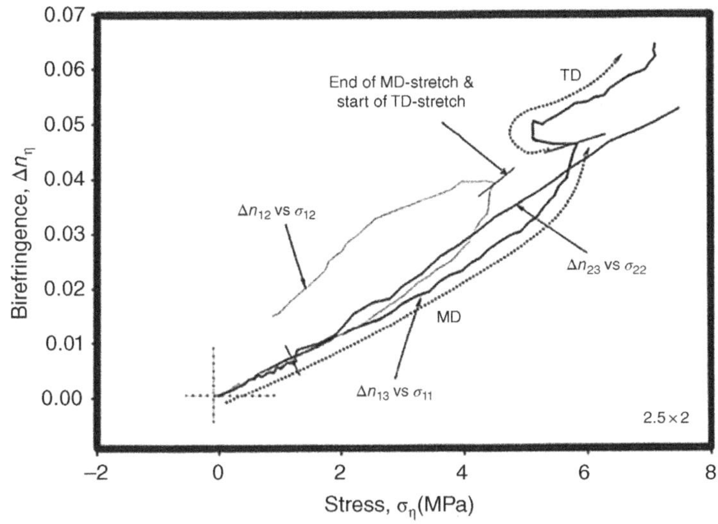


**Figure 5.23** Relationship between different components of stress and birefringence for 2.5 × 2 sequential biaxial stretching at the rate of 0.125 $s^{-1}$ and 95 °C.

in which the stress increased slowly with the draw ratio. During the MD-stretch, the chains are oriented in the MD causing an increase in the refractive index in the MD direction. The result is increasing $\Delta n_{12}$ with the stretch ratio. Even though the main orientation direction during the MD-stretch is the MD, there is still an indirect stress in TD direction as described above. This indirect stress causes the chains to start aligning in the TD direction. Stretching up to a true strain of 0.65 and during the MD-stretch, the relationships between $\sigma_{11}$ vs. $\Delta n_{13}$, $\sigma_{22}$ vs. $\Delta n_{23}$, and $\sigma_{12}$ vs. $\Delta n_{12}$ are all linear and laying on the same line. This stage is a uniaxial constant width stretching, at which the stress-optical-rule is holding with a stress-optical constant of 5.8 $GPa^{-1}$ (5800 Brewsters) as was found in the simultaneous biaxial stretching [35]. Up to a true strain of 0.65 for all stretching rates, the Raman measurements showed the lowest degree of the trans isomer content, as a sign of a lowest degree of crystallinity and order in the films compared with the higher stretch ratios.

#### 5.4.2.2 Regime II

Regime II takes place above a true strain of 0.65 when stretching the film in MD direction. The network formation and crystallization occur when the TD stretch starts above this true strain. Once crystalline regions are established, segmental relaxation slows down since the crystallites act as physical crosslinking nodes acting similar to entanglements [56]. The result is a higher degree of orientation causing a rapid increase in birefringence. The rate of the birefringence increase at this stage is higher than the stress increase, which corresponds to deviation from the SOR in a nonlinear manner. After the strain-induced crystallization, TD stretching causes increase in out-of-plane birefringence in the rates of 0.0125 and $0.125\,s^{-1}$ (Figure 5.24 (left)) as there is already now a network, which is holding the state of orientation and preventing the chains from retracting back. In the case of the $0.00125\,s^{-1}$ rate, TD causes a decrease in out-of-plane birefringence for all ratios as no strain-induced crystallization was observed and hence the chains could retract freely during the TD stretch and the material stays in Regime I throughout Figure 5.24 (right).

After the onset of the strain-induced crystallization, an increase in $\lambda_{TD}$ first decreases the crystallinity and after reaching a minimum, the crystallinity starts to increase again (Figure 5.25).

#### 5.4.2.3 Regime III

Regime III was observed above the stretch ratio of $2.5\times1.5$ at the rate of $0.125\,s^{-1}$. Further increase of the stretch ratio above this level caused the stress to increase more rapidly than the birefringence due to the higher extensibilities of the chains, combined with an increase in crystallinity ratio).

### 5.4.3 Relaxation

Hassan et al. supplemented the mechano-optical study of simultaneous and sequentially stretched PET with the study on relaxation of stretched films in constrained state [33]. During stretching processes such as in the tenter-frame, oriented films are transported from station to station while constrained. During these transition stages, the material undergoes stress relaxation and other related physical changes. The level of orientation obtained at the end of the stretching is modified by the relaxation, especially if high stretching temperatures and low deformation rates are applied. The recovery that occurs during relaxation above $T_g$ may cause pure recoiling of the oriented amorphous chains or induce/enhance crystallization. The latter typically occurs at high stretch ratios when stretching creates well-oriented taut chains but with unfavorable orientations adjacent to each other. During relaxation, these chains slip into the correct conformations for the crystallization to occur or enhance. The relaxation of polymeric materials can be categorized into two classes: the orientation relaxation of a coiled chain as a whole and that of local segment of molecular chains. According to Qian [57], the segmental relaxation time of amorphous PET is on the order of seconds or even less depending on how long the local segment is involved in the molecular motion. It is relatively easy for local segments to relax because of their mobility. On the

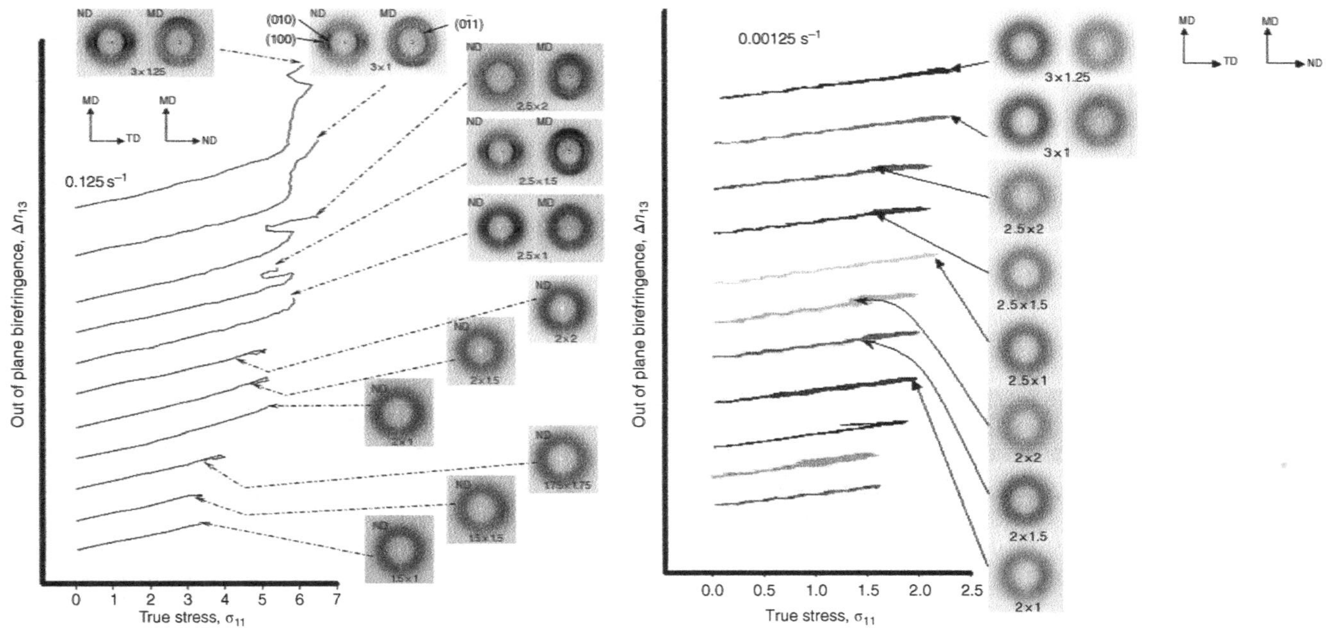


**Figure 5.24** Out of plane birefringence $\mathbf{\Delta}n_{13}$ and true stress $\boldsymbol{\sigma}_{11}$ at 0.125 s$^{-1}$ (left) and 0.001 25 s$^{-1}$ (right) along with WAXD patterns.

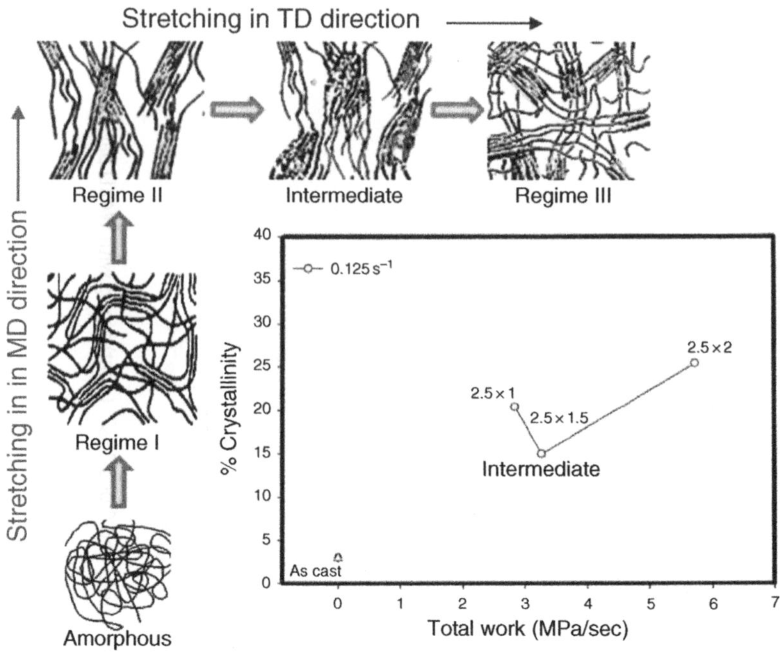


**Figure 5.25** Three-regime behavior of the sequentially biaxially stretched PET.

other hand, the whole chain relaxation time is on the order of minutes or longer at the drawing temperature above $T_g$. Buckley and Jones [58] in explaining the slippage of entanglements (reptation) at high temperatures and long times, they argue that, conformation stresses can relax by a process of axial diffusion of entire molecules, whereby they disengage from the topological constraint of other molecules. Using stress-birefringence measurements, Ryu [59] studied relaxation of simultaneously biaxially stretched PET samples from various deformation levels and found out that at low stretch ratios where the stress-optical law holds, birefringence and stress decreased along the same straight line as in stretching. At high stretch ratios $\lambda = 3.5$, on the other hand, the birefringence increased while the stress decreased during relaxation. Interestingly, when the relaxation took place just above the point at which the deviation from the stress-optical law was detected, the birefringence first decreased but eventually increased with time while stress decreased at all times. In other words, the deviation point represented the starting of spontaneous molecular orientation, which was associated with the crystallization process. Boyce [53] reached the same conclusion that, after the strain-induced crystallization, there was an arrest in viscous flow and this arrest correlated with network stretching.

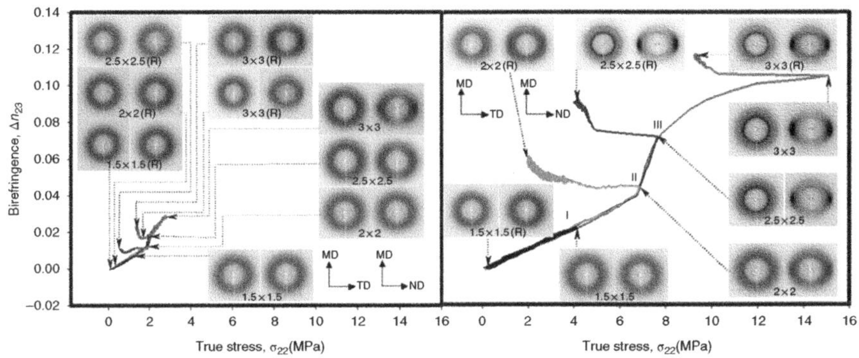


**Figure 5.26** WAXD patterns for the as stretched and the relaxed simultaneous biaxially stretched samples at the rate of 0.001 25 $s^{-1}$ (left) 0.125 $s^{-1}$ (right) and different stretch ratios.

Hassan et al. made similar observations when they stretched the PET films simultaneously and sequentially and held at the stretching temperature at fixed dimensions while measuring true stress, true strain, in and out of plane birefringence values. The mechano-optical behavior was found to be dependent on the process prehistory including extent and rate of deformation. Figure 5.26 shows the WAXD patterns for both the as-stretched and the relaxed stages at selected conditions on the birefringence- true stress curve to illustrate the key stages of molecular alignment. For all stretching rates, relaxation in Regime 1, where the linear stress optical behavior was observed, caused the birefringence and stress to relax on the same path as during stretching and the WAXD patterns did not show any sign of crystallization either during stretching or relaxation. Relaxing in Regimes II and III, on the other hand, showed retention of as stretched birefringence levels with subsequent increase in birefringence at later times during relaxation (see Figure 5.26 – Right panel simultaneous stretched films, $2\times2, 2.5\times2.5, 3\times3$ at a rate of $0.125\,s^{-1}$). When comparing the as stretched and the relaxed patterns, intensification and sharpness in the crystalline peaks in relaxed samples are apparent indicating better ordering for the crystalline entities in regime II and growth observed in regime III, where the distinct crystalline peaks became more pronounced. In all cases, the patterns taken with the X-ray in normal direction (ND) indicate that, these films exhibit in-plane isotropy.

For the rate of $0.00125\,s^{-1}$ there was not much change in the WAXD patterns when comparing the as stretched and the relaxed patterns except for the $3\times3$ ratio, which started to show very weak crystalline diffractions.

WAXD diffraction patterns for the as stretched and the relaxed sequentially biaxially stretched samples at different stretch ratios are shown in Figure 5.27. The main observation was the appearance of crystalline diffractions after

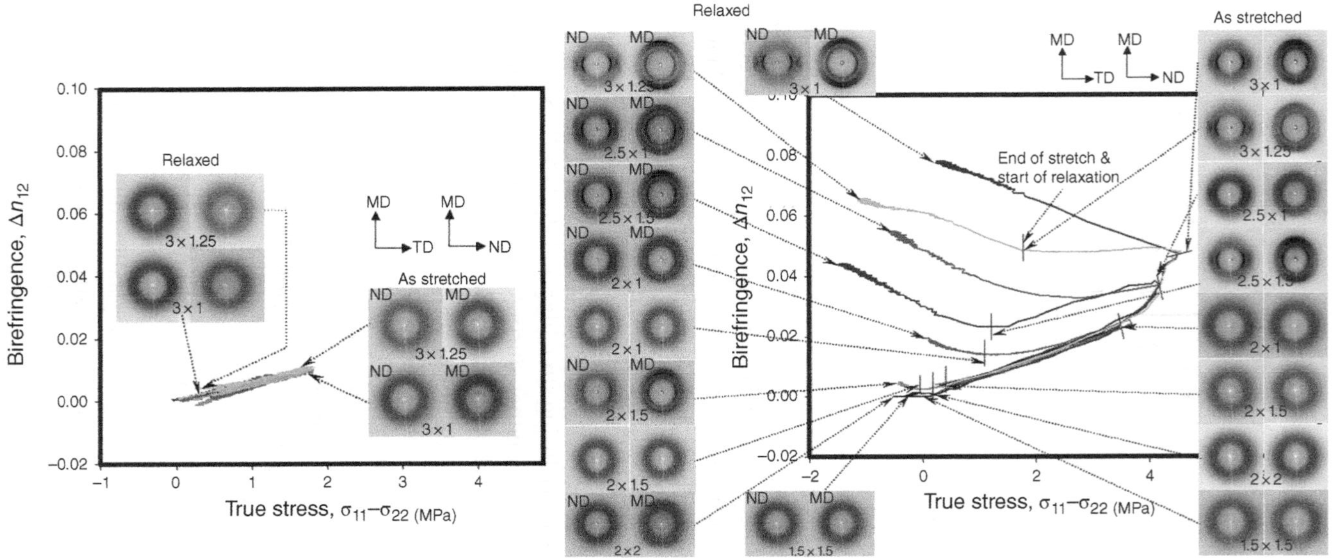


**Figure 5.27** WAXD patterns for the as stretched and the relaxed sequentially biaxially stretched samples at the rate of $0.001\,25\,s^{-1}$ (left) $0.125\,s^{-1}$ (right) and different stretch ratios.

relaxation which were absent after sequential stretch ratio series of 2 (2×1, 2×1.5, and 2×2) indicating that crystallization occurred during relaxation. The X-ray patterns taken at the intermediate stages of relaxation for the samples that experienced a minima in the birefringence as in the case of the 2×1 and 2×1.5 samples at the rate of $0.125\,s^{-1}$, showed no distinct crystalline peaks at those mid stages of relaxation, while on further relaxation the samples showed crystalline diffractions with increasing birefringence. The samples that showed this behavior needed a longer time for the crystallization from the oriented state. The time and relaxation temperature enabled the chains to work cooperatively in registering into crystalline domains. In all films with the lowest rate used $0.00125\,s^{-1}$, crystallization did not take place even after relaxation from all ratios (Figure 5.27- Left). In these samples, relaxation dominates the orientation process.

The authors concluded that the relaxation behavior of oriented films was significantly influenced by stretch ratio and rate. In the first regime where the linear stress-optical rule applies, the material remains amorphous during stretching and relaxation. At intermediate deformation levels, the polymer chains may remain amorphous initially. But during the relaxation stage they begin to exhibit strain-induced crystallization as the adjacent oriented amorphous chains become locally parallel to each other and the matrix crystallizes. Once this crystallization occurs, previously decreasing birefringence begins to increase while the stress continues to decrease. There appears to be an overall critical minimum degree of orientation required for the chains to start crystallizing during relaxation. When the relaxation starts from a point where the material has developed a well-connected network during deformation, the birefringence does not experience any decrease and starts to increase immediately upon the start of relaxation, forming higher orientation and/or higher crystalline order.

### 5.4.4 Heat-setting

Constrained annealing (heat-setting) step is applied after stretching when thermal stability of the films is desired. Heat-setting stage is omitted in applications such as heat shrink film or blow-molded PET bottles for carbonated beverages where high creep resistance is needed. This is because heat-setting relieves the required amorphous orientation between crystalline nodules. During heat-setting, constrained films are exposed to elevated temperatures where the polymer exhibits its highest rate of crystallization usually about halfway between $T_{cc}$ and $T_m$ (peak). Depending on the temperature chosen, the heat-setting process may result in relaxation, partial melting, and/or crystalline growth [60, 61]. Substantial amount of "after the fact or off-line" studies were carried out to understand the structural changes that take place during the heat-setting process of the uniaxially oriented PET fibers and films [62–77] and the biaxially oriented PET films [78–86].

The structures of films obtained immediately after the stretching process, either uniaxially or biaxially, are metastable, i.e. far from thermodynamic equilibrium. As they contain large fraction of oriented amorphous chains, these structures are converted to the thermodynamically more stable structures via heat-setting. Final dimensional stability and other properties of the films and fibers are set at this stage. Because these oriented films contain very highly oriented and amorphous chains, the sudden exposure to high temperatures causes the structure to rearrange very fast. Although studying this phenomenon by offline methods is useful, they do not provide kinetics information. Because the changes that occur during the heat-setting process are very fast, the monitoring technique should also acquire the data at comparable speeds. Infra-red absorption, ultraviolet light transmission, and scattering patterns obtained from synchrotron radiation, were used to track the structural changes in heat-setting films. With uniaxial deformation the PET chains preferentially orient in the MD, while the simultaneous biaxial stretching leads to in-plane isotropy as the chains are randomly biaxially oriented in the plane while the phenyl rings were observed to orient in the plane of the film [44, 78]. Annealing further increases the in-plane chain and phenyl in-plane orientation. In case of the sequential biaxially stretched films, Salem [68] observed that, after the annealing there is some redistribution of the amorphous orientation in the high stretch ratio samples, such that the proportion of the TD-oriented chains diminishes. In the case of the annealed films a balanced orientation distribution is achieved at a TD draw ratio lower than that of the unannealed films. Annealing the sequentially biaxially stretched films causes a larger proportion of chains to relax along the TD than along the MD at low TD draw ratios, due to higher molecular constraints in the MD direction, where, the crystalline domains are better developed.

Matsumoto observed in the case of sequentially biaxially stretched films, a lower population of TD oriented crystals, which immediately increase with annealing of the films [87]. In this region there is an increase in molecular orientation, but at the cost of loss of some crystallinity and corresponding decrease in physical cross-links.

Hassan et al. [37] studied the structural changes that take place during the heat-setting of simultaneously and sequentially biaxially stretched films of PET by real-time birefringence, offline Raman spectroscopy, DSC thermal analysis, and X-ray techniques. They reported three different behaviors of the stretched films during the annealing process depending on the orientation and percent crystallinity of the pre-stretched films. These behaviors (categories) are discussed next.

#### 5.4.4.1 Category I

This category is characterized by low orientation and crystallinity levels of the pre-stretched films and shown for sample $1.5 \times 1.5$ in Figure 5.28. In this category, the samples experience a little increase in birefringence for a very short time after

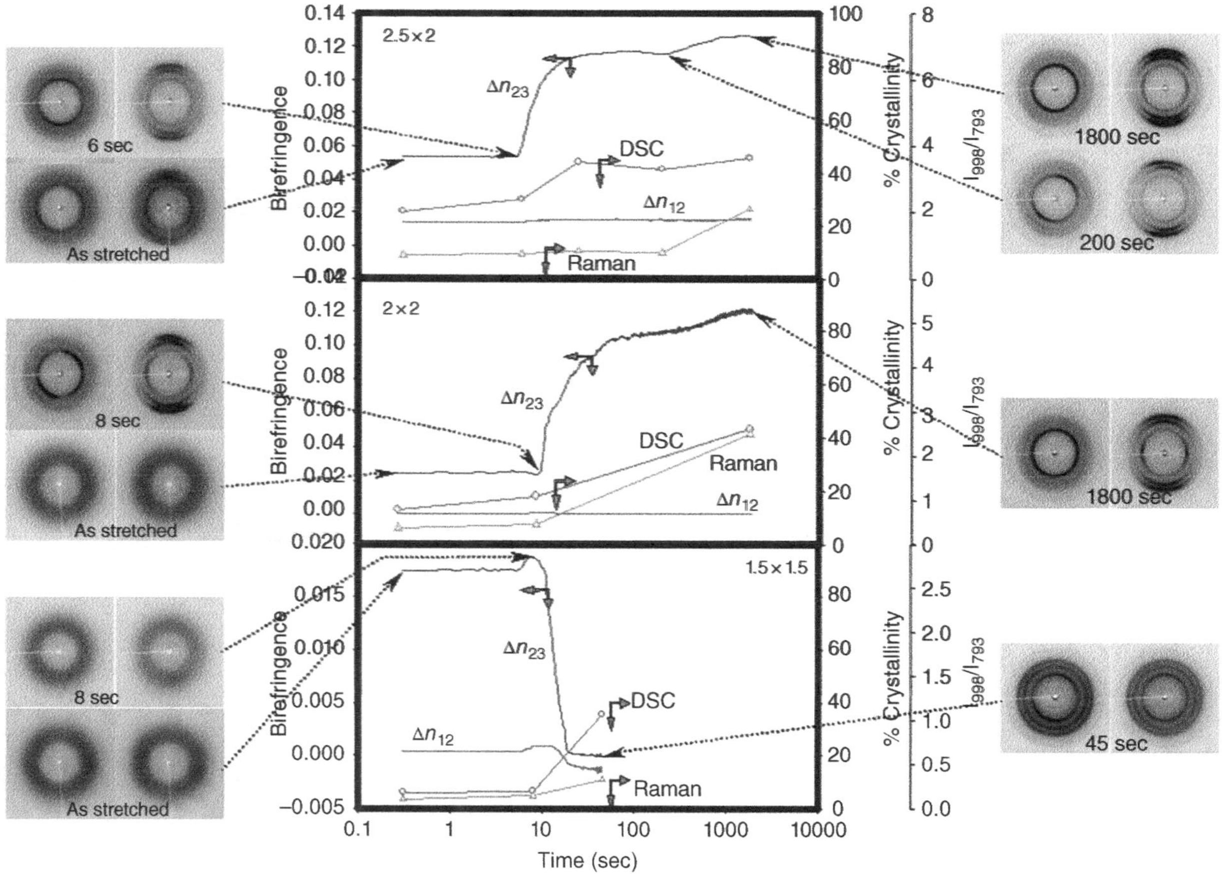


**Figure 5.28** Temporal evolution studies for different sequentially biaxially stretched and annealed samples. *Source:* Hassan et al. [37], 2016/John Wiley & Sons.

the induction period then a relaxation during the rest of the annealing process. The high thermal energy transferred to the films works on orienting the chains after the induction period, but as there is no physical network to hold that state of orientation, the chains relax very quickly. This behavior is apparent in the low deformation levels prior to the onset of the strain-induced crystallization. The high annealing temperature causes the oriented chains to relax upon heating, as there are no physical junctions or crystalline network to prevent the relaxation of the chains. The low orientation and crystallinity levels results in a loose structure, which, upon heating, the high energy put into the material causes the oriented chains to recoil, resulting in a relaxation process, and if the temperature is high enough, thermal crystallization through nucleation and growth process typically ensues.

#### 5.4.4.2 Category II

This category is observed in the intermediate deformation levels (e.g. 2×2 sample in Figure 5.28. This regime starts with a small decrease in birefringence after the induction period which then starts to increase. The small decline of birefringence is a result of reduction of orientation as some of the oriented chains relax at the annealing temperature. The subtle relaxation of the chains helps cooperative organization that lead to parallel registration of the chains and formation of crystalline domains. The high increase in birefringence, crystallinity, and crystallization kinetics in this category is due to the ability of the samples to attain higher crystallinities after substantial amount of oriented amorphous chains were developed during the stretching stage that transform upon annealing.

#### 5.4.4.3 Category III

This category includes the high deformation samples (e.g. 2.5×2 in Figure 5.28). These films experience little to no relaxation in the amorphous domains following the induction period followed by the rapid increase in birefringence. Also, these samples exhibit a small decline in birefringence before the upturn. The small decline is associated with a decrease in crystallinity as a result of partial melting and reformation of the crystallites. These samples were stretched above the onset of the strain-induced crystallization and exhibit well-formed long-range connected physical networks. In Category III, the birefringence and crystallinity increase are lower than in Category II because the previously formed crystalline domains during the stretching process may lock in a large amount of oriented amorphous chains in unfavorable orientations with respect to neighboring chains that prevents the amorphous chains from transforming into crystalline domains, thereby reducing the crystallizability. Another reason for the decrease in crystallizability, which is also the reason for the leveling off behavior in the crystallinity and the trans isomer content at the high annealing times, is that, at these conditions most of the crystal structure has already been formed either by the stretching or annealing processes, reducing the amount of amorphous chains that could convert during further annealing.

The authors also reported that the sequential mode of stretching provides better environment for crystallinity development and associated trans conformational transformation than the simultaneous mode of biaxial stretching.

The studies mentioned above utilized linear stretching profiles where the rate of stretching was constant. Nonlinear stretching profiles of PET result in the development of crystalline and orientation structures that are different from those that are induced by linear stretching profiles. Martins et al. [88] showed this effect through online rheo-optical studies by evaluating four stretching rate profiles on PET films with different molecular weights (intrinsic viscosities), i.e. linear, sigmoidal, logarithmic and exponential, in uniaxial constrained width mode (Figure 5.29).

All the stages, I, II, II, and IV, described in Figure 5.18 were still observable in logarithmic and exponential stretching modes. Sigmoidal deformation showed only the first two regimes even though the film was stretched to the same total engineering strain as applied to all profiles.

Figure 5.30 shows mechano-optical behavior of low molecular weight PET (Mn 22 800 with intrinsic viscosity of 0.69 dl $g^{-1}$) for each type of the deformation profiles studied. The use of logarithmic profile caused the crystallinity development to start earlier which in turn, translated into early development of strain hardening that is directly responsible for the achievement of thickness uniformity at lower strains. In logarithmic mode, the use of slower speeds at high strains also helped improve the organization of the crystalline domains by

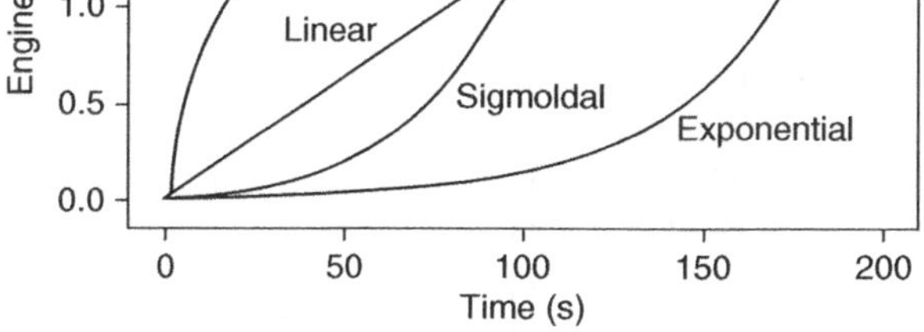


**Figure 5.29** Types of deformation applied to PET film stretching: engineering strain vs. time (global deformation history). *Source:* Adapted from [87].

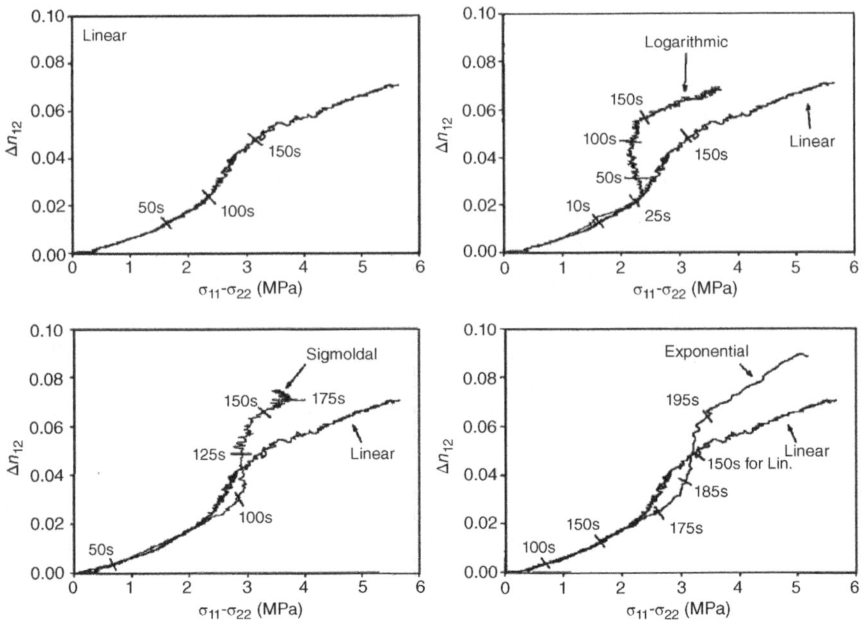


**Figure 5.30** Temporal evolution of mechano-optical behavior of LPET (lower MW PET) for each type of deformation profile. *Source:* Adapted from [87].

facilitating the partial relaxation of polymer chains that in turn allowed them to register with oriented crystalline domains, facilitating their growth as well as their perfection.

In the Exponential profile stretching, the early stage of deformation was dominated by the relaxation processes due to low strain rates in the beginning. The delay in strain-induced crystallization delayed the strain hardening allowing for an increased orientation of the polymers chains as they were not constrained by the long-range network until high strains were achieved. This, in turn, allowed further development of birefringence. The exponential deformation profile yielded the highest crystallinity for the same total true strain. In other words, the exponential application of strain keeps the structure "looser" while the polymer chains become oriented preventing the formation of long-range network at intermediate states, thereby facilitating higher strain-induced orientation and consequent crystallinity development at the end of the deformation period. The authors recommended that if a uniform thickness profile is desired at lower strains, the logarithmic deformation profile should be used. However, if higher crystallinity with higher orientation is desired, it is recommended to use the exponential deformation profile for the same total strain application to the films.

## 5.5 Poly(Ethylene Terephthalate) (PET) and Poly(Etherimide) (PEI) Blend

The functionality of PET films for use as flexible substrates can be improved by blending with various polymers. One useful blend is formed by compounding PEI into PET. PEI is a linear amorphous high-performance polymer with a high glass transition temperature ($T_g$) of ~215 °C. The bulky aromatic imide group provides PEI with the high rigidity, creep resistance, and high thermal stability while the flexible ether linkage allows for good melt processibility. Choi et al. reported the several benefits of incorporating PEI in PET-rich blends not only for the final utility of the films but also during film stretching and heat-setting processes during their manufacture [36, 89]. The increase in $T_g$ of the films due to the presence of PEI (Figure 5.31-left a) is desirable since it widens the application window of the films by increasing their use temperature. Besides the increase in $T_g$, cold crystallization temperature, $T_{cc}$, also increases as shown in Figure 5.31-left c). Although the increased $T_{cc}$ is accompanied by the increased $T_g$ with increased PEI concentration, the magnitude of $T_{cc}$ is even larger at a given PEI concentration. Thus, the net result is a wider processing window, defined by the temperature difference between $T_g$ and $T_{cc}$ (Figure 5.31-left d), in which the process (i.e. deformation, followed by heat-setting) can be operated in a more favorable environment. The crystallization rate of PET in the blend films was dominated by two independently counteractive factors: PEI concentration and preferred orientation. PEI lowers the crystallization rate due mainly to the dilution effect that spatially prevents the crystallizable polymer chains from coming together, while preferred orientation boosts it by reducing the activation energy for transport for crystallization. Although the order of crystalline structure decreases on stretching as the PEI concentration increases, the stress level becomes higher, giving rise to strain hardening at smaller stretch ratios. This deformation behavior has a benefit of being able to strain harden the films with a higher degree of amorphous orientation without locking in the structure by crystallization, which may impart more flexibility to the films during deformation stage in the tenter-frame process. The authors found that the increase in PEI concentration led to increase in highly defective crystallites as the bulky PEI chains intrude into the PET crystalline regions (see WAXD patterns in Figure 5.31-right). Above 30% PEI, the blend crystallizability disappears. Three categories of distinct structure evolution was observed during stretching of PET/PEI blends with 100/0, 90/10, and 70/30. In 100/0, amorphous orientation takes place rapidly, followed by crystallization. In 90/10, amorphous PET and PEI intermingled molecules steadily orient with suppressed crystallization of PET due to the dilution effect and insufficient stress level to overcome it. At 70/30, PET/PEI composition, amorphous molecules orient significantly even at small stretch ratios due to the high-stress level; however, the crystallization is

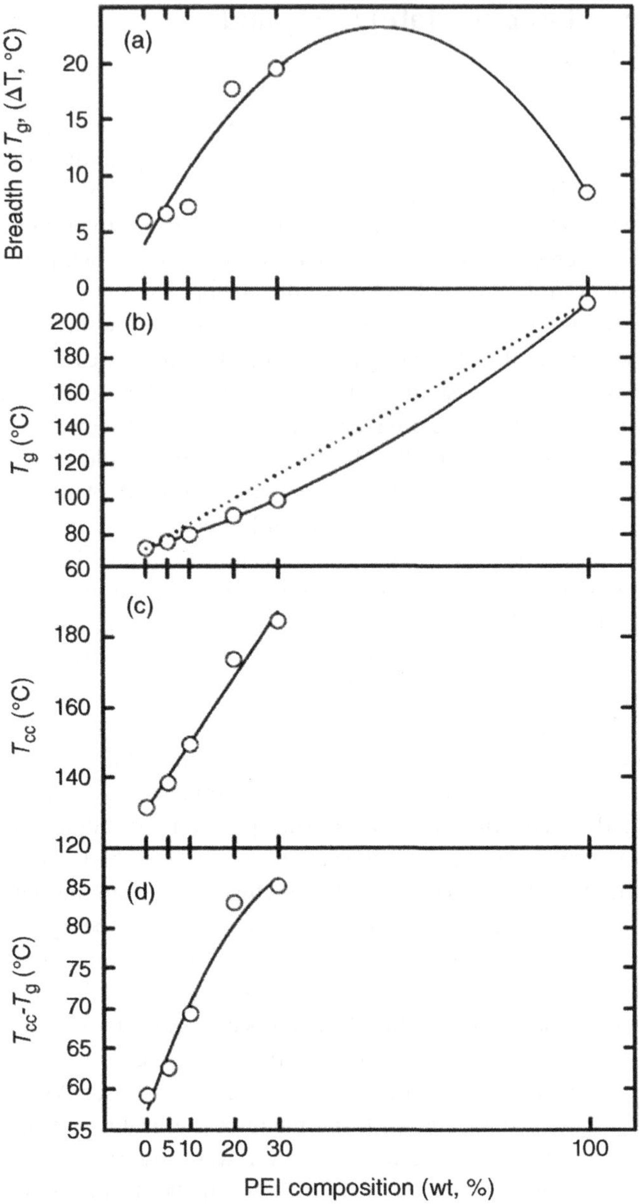


**Figure 5.31** Breadth of $T_g$, $T_g$, $T_{cc}$, and $T_{cc}$-$T_g$ for PET/PEI films as a function of PEI composition (left) true stress, degree of crystallinity, and WAXD of PET/PEI films as a function of stretch ratio: (a) 100/0, (b) 90/10, and (c) 70/30 (right).

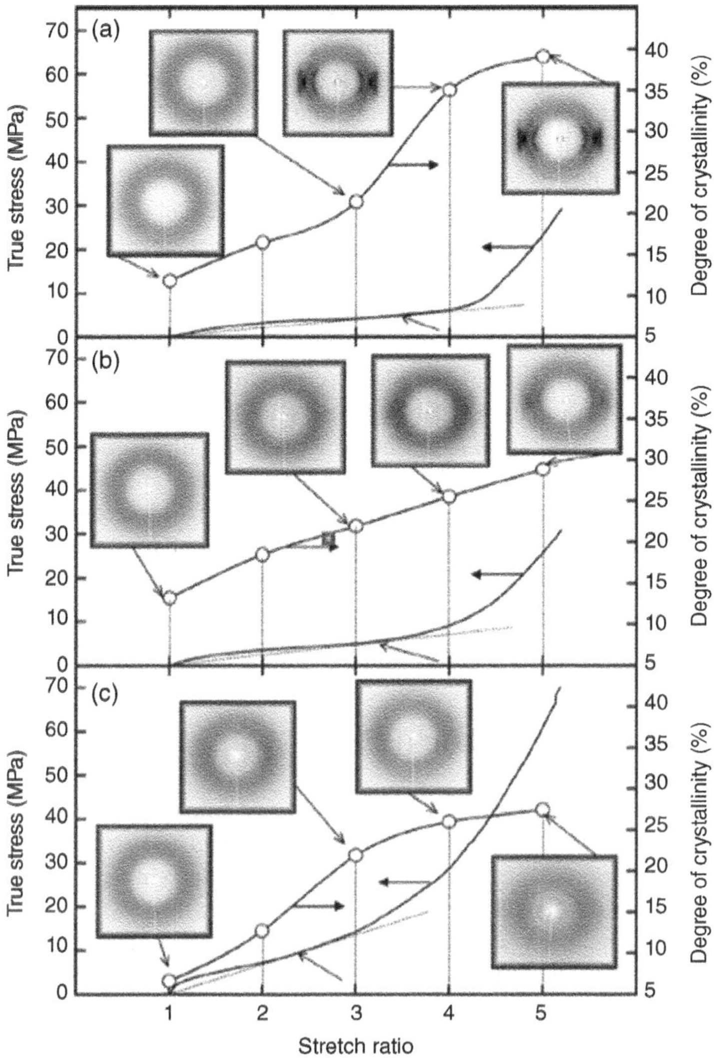


**Figure 5.31** (Continued)

greatly restricted because of the high dilution effect, forming a defective crystalline structure. The slowing down of the crystallization kinetics by the addition of PEI, also shown by the increase in $T_{cc}$ (Figure 5.31- left c) allows a prolonged deformation process that is conducive to increased molecular order without locking in the structure, and hence improved physical and gas barrier properties on which the utility of the films may be achieved.

Annealing of PET/PEI blends that are stretched to low deformation levels before strain hardening leads to complete elimination of birefringence due to relaxation in blends containing as little as 10% PEI [88]. At deformation levels (~5×) which is beyond the onset of strain hardening, annealing leads to initial relaxation nearly to optical isotropy for 70/30 blend and to some extent for the 90/10 blend, but this is followed by renewed increase at later stages of annealing indicating that oriented nuclei that are formed during deformation survive the initial relaxation phase and dominate the crystallization at longer term annealing (Figure 5.32).

Mechanical properties of PET/PEI films stretched to 5× and heat-set for (a) 1 s, (b) 5 s, and (c) 4000 s are shown in Figure 5.33. The final mechanical properties are highly dependent on the structure of the blend (crystallinity and orientation), which is influenced by deformation level and heat-setting time. The dramatic increase in toughness of 70/30 at 5 s is attributed to the relaxation process as a result of the suppressed crystallization. On the other hand, in 100/0 and 90/10, due to the dominant crystallization process over relaxation, the strength increases with increasing the heat-setting time, and hence toughness tends to decrease (Figure 5.15d). The brittleness of 70/30 (Figure 5.15c) at the end of the heat-setting may be attributed to the considerable inhomogeneity induced from the phase separation between PET and PEI in amorphous regions as well as crystallization of PET.

## 5.6 Polyethylene Naphthalate (PEN)

Rheo-optical properties of PEN during stretching and relaxation in the rubbery state are very similar to those of PET since they are closely related in their chemical structures with the exception of the bulky naphthalene rings shown in Figure 5.34. The bulky naphthalene rings renders PEN stiffer with higher Tg than PET.

Martins et al. [90] reported structural and mechano-optical behavior of amorphous PEN films stretched uniaxially in the rubbery state as a function of processing temperature, draw rate, and molecular weight ($M_w$). Similar to PET, a three-regime stress optical behavior was identified as shown in Figure 5.35. Regime I corresponds to classical stress-optical behavior of amorphous materials where PEN remains completely amorphous during stretching. Oriented crystallization and long-range connected network formation manifests itself as a positive slope change in birefringence/stress curve into regime II. In the last regime, III, the formation of "muscle bound" structure begins to level off birefringence as it approaches the finite extensibility of polymer chains.

Birefringence and true stress curve coupled with WAXD patterns in Figure 5.35 suggest that the material remains amorphous in Regime I. Right after the transition to Regime II, the amorphous chains are highly oriented but not crystalline because

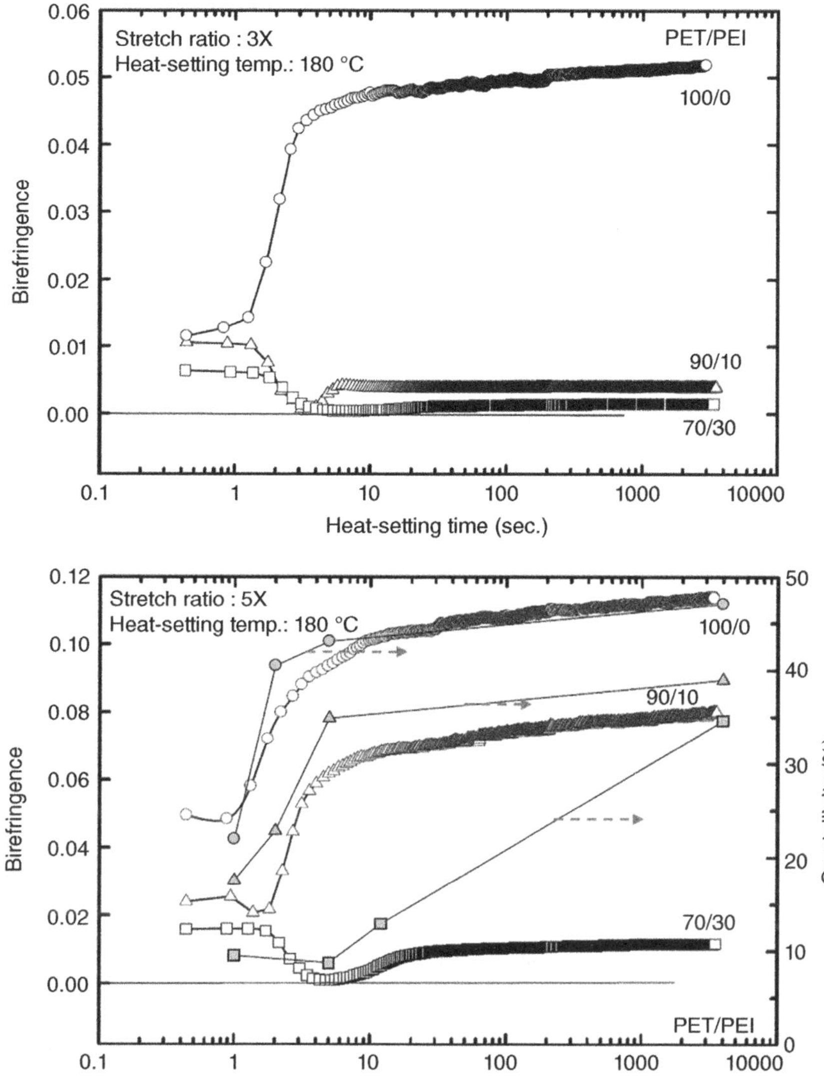


**Figure 5.32** Birefringence of PET/PEI 100/0,90/10, and 70/30, stretched by 3× (left) and 5× (right) at the rubbery state temperatures. Heat-setting temperature is 180 °C.

they possess low translational order, which prevents them from crystallizing. With further stretching, the structure evolves very rapidly to a crystalline structure, with approximately 40% crystallinity that is maintained throughout the Regime III at the end of which finite chain extensibility prevents further changes

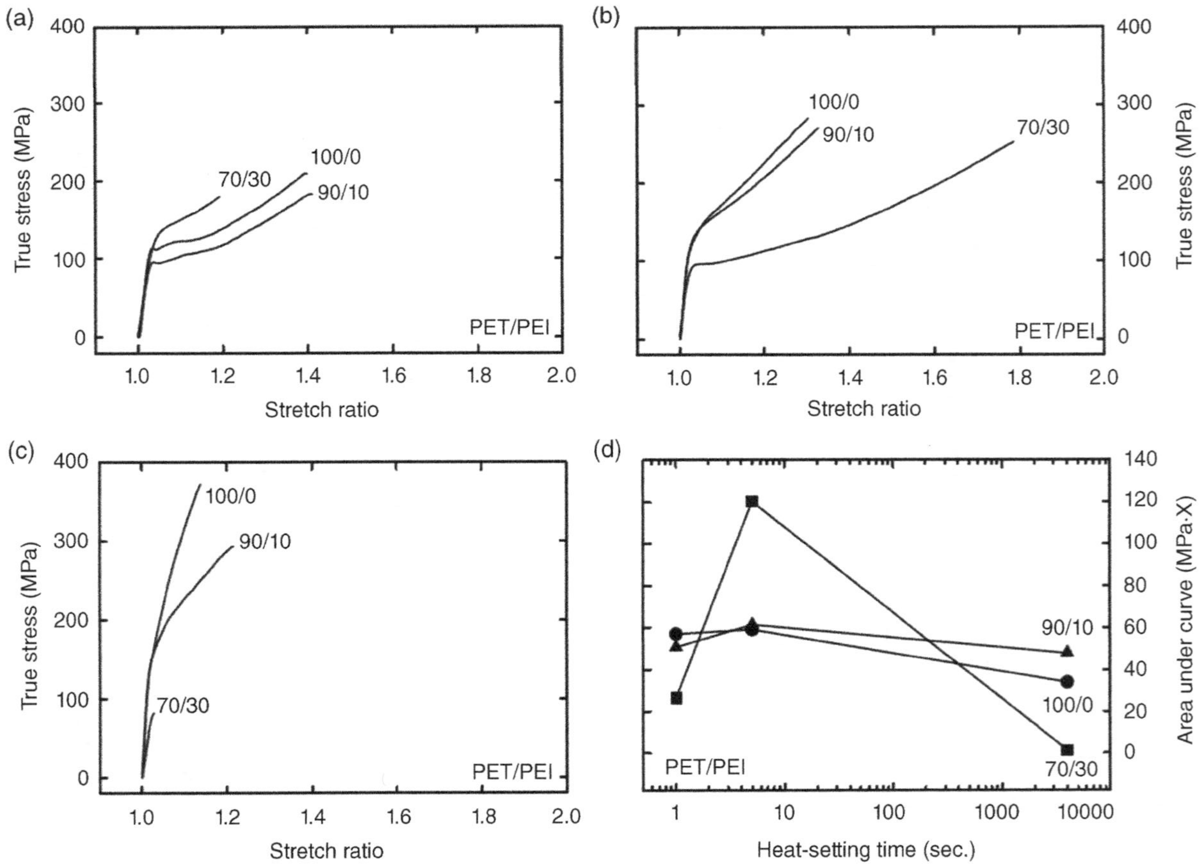


**Figure 5.33** True stress vs. stretch ratio of PET/PEI films, stretched by 5× and heat-set for (a) 1 s, (b) 5 s, (c) 4000 s, and (d) area under the curves. Stretch rate: 300%/min. Stretch temp.: Room temp.

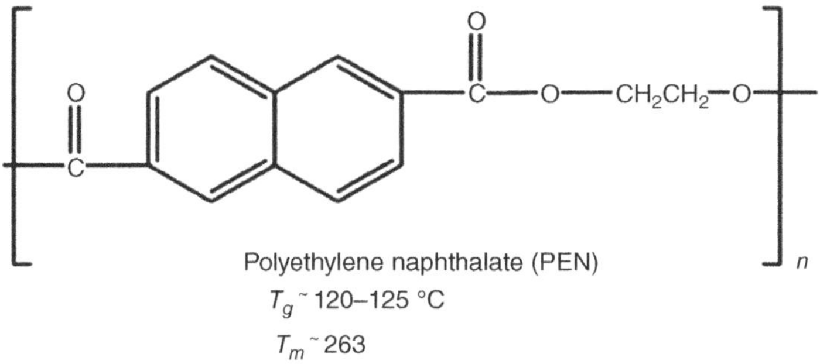


**Figure 5.34** Chemical formula of PEN.

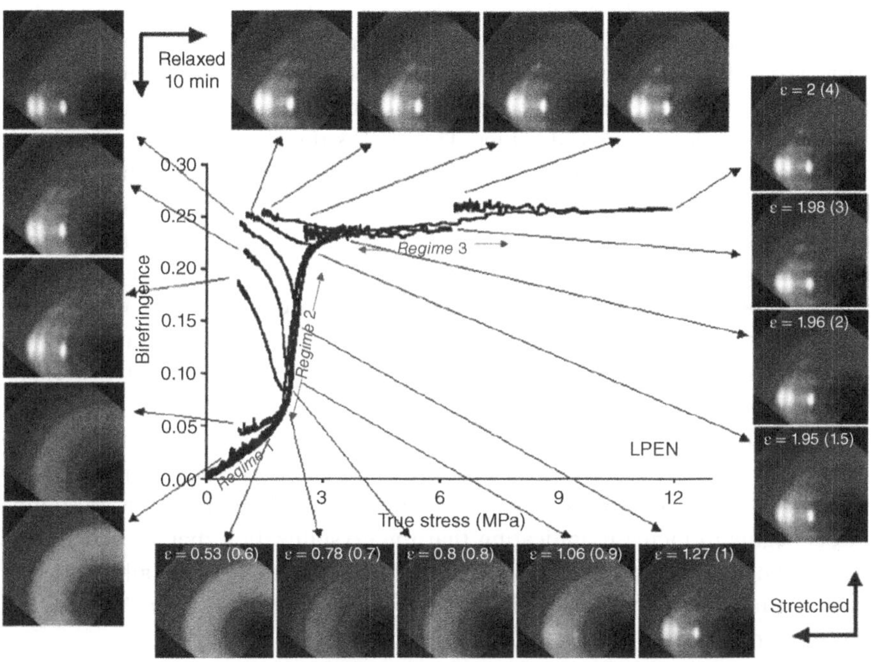


**Figure 5.35** Mechanooptical behavior of LPEN during stretching and relaxation processes with corresponding WAXD pattern. Samples were stretched at 140 °C and 0.01 $s^{-1}$ and relaxed for 10 min. The values indicated in the pictures are referring to the true strain and engineering strain (in parentheses) at relaxation time zero.

in the structure. A model representing the structural mechanisms followed from the amorphous to the crystalline state was suggested by the authors.

The development of crystallinity is governed by the strain applied to the material. This can occur either during the stretching stage or as a spontaneous deformation during the relaxation stage as shown in Figure 5.35. The latter is typically observed after stretching to intermediate deformation levels. Once strain hardening occurs, very small change in the true strain is observed; hence, the saturation behavior observed at the high-strain end of the plot. Relaxation at intermediate states of deformation (relaxation within Regime II) accompanies substantial structural changes. To detail the structural evolution mechanism in this regime during relaxation, different relaxation times were further applied and investigated by Martins et al. The results are shown in Figure 5.36.

Relaxation of uniaxially stretched PEN was found to depend exclusively on the regime from which the relaxation is initiated (Figure 5.36). If the relaxation is initiated from within Regime I, the stress and birefringence decrease traces back the linear stress-optical behavior while the material remains amorphous(not shown). On the other hand, relaxation from the early stages of Regime II deviates from a linear stress-optical behavior and the birefringence eventually begins to increase while the stress declines. This marks the formation of highly localized, highly oriented, ordered regions with low translational ordering along the chain axis in the material. This unexpected behavior during relaxation was attributed to the formation of a long-range physical network that helps arrest the relaxation of orientations gained during stretching. The decrease of birefringence in the first four minutes is accompanied by the appearance of extremely sharp equatorial peak of R(010). Following the appearance of this peak, the birefringence levels off for a while (about six minutes) while the stress relaxation continues.

A crystalline structure with relatively low orientation and no three-dimensional ordering is formed, as seen by the absence of off-equatorial peaks in the WAXD pattern. At two hours relaxation, the azimuthal distribution of the crystalline peaks begins to increase, indicating that the crystallization that is occurring as part of the long-term relaxation process involves reduction in crystalline orientation. The most likely reason for this is the crystallization of polymer chains which has relaxed to a certain extent during this process.

At the start of relaxation within Regime II there already are highly oriented-albeit with considerable translational disorder – crystalline domains, as shown in bottom left and bottom right of Figure 5.37. These two cases correspond to deformations between 90 and 100%, respectively. The initial plateau observed in the birefringence at the early stage of relaxation eventually disappears at 100% prestrained samples where birefringence begins to increase as soon as the stretching stage is terminated. At this point, the material approaches a crystalline state

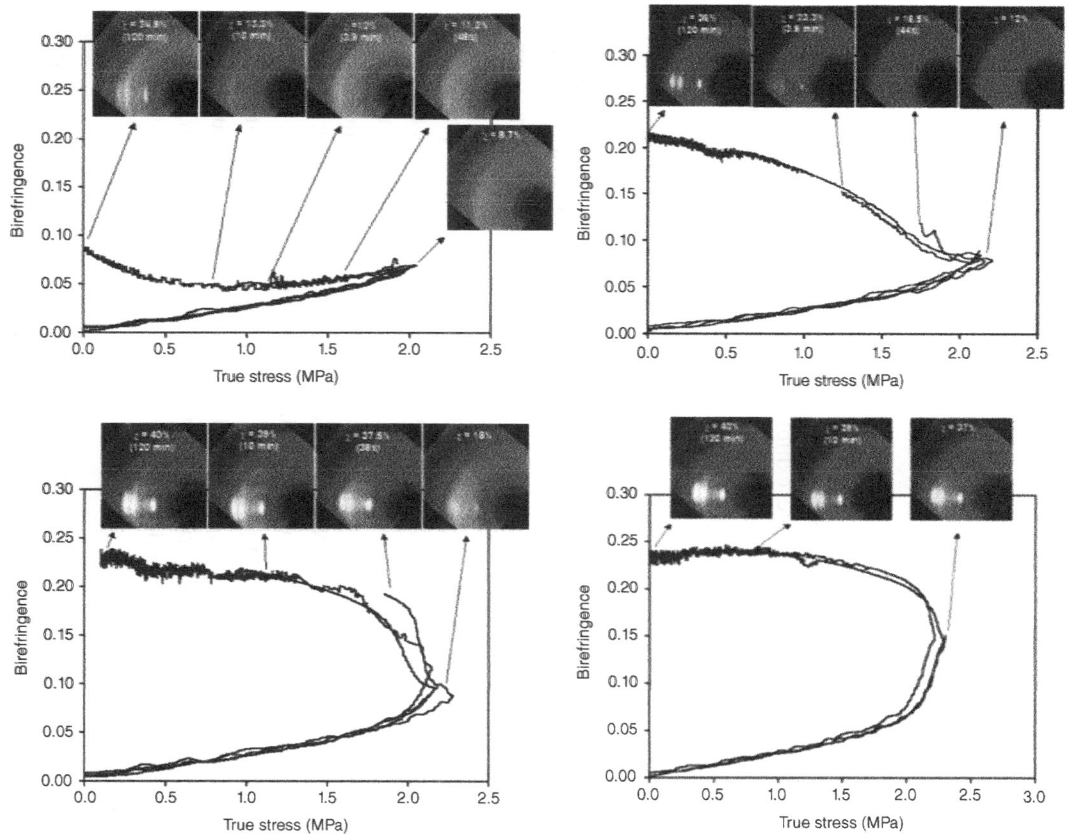


**Figure 5.36** Detailed mechano-optical behavior and WAXD patterns of LPEN during relaxation from 70% strain (early stages of regime 2) (LEFT-TOP), from 80% strain (RIGHT-TOP), from 90% strain (LEFT-BOTTOM) and from 100% strain (RIGHT-BOTTOM).

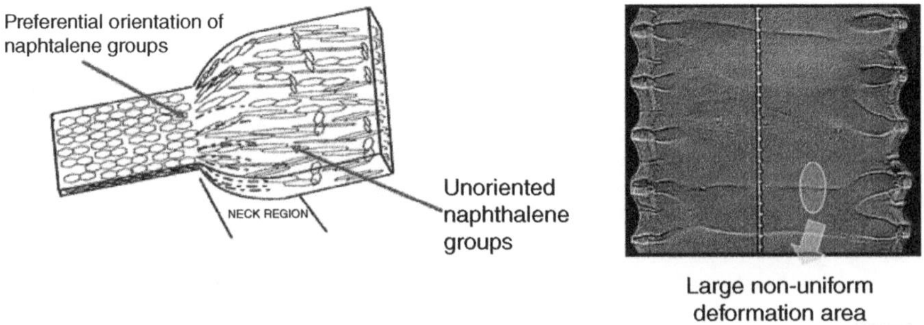


**Figure 5.37** Localized cooperative orientation of the napthalene planes parallel to the surface of the film and the resulting uneven thickness distribution.

where three-dimensional ordering starts to form, as diffused off-equatorial peaks appear while the overall crystalline orientation declines.

This clearly follows earlier observations that the crystalline regions that develop during the relaxation stage exhibit low orientation not because the highly oriented domains that formed during stretching disoriented but because the crystals that formed during the later stages of relaxation simply exhibit lower orientation.

The example above demonstrates the detailed structural phenomena one can deduct from measurement of real-time rheo-optical properties.

Rheo-optical properties of polyesters have been studied by Cakmak to develop dynamic phase diagrams that are useful as a quick guide to understand what structure the polyesters would attain in the final state given the processing conditions (rate, temperature, type of deformation) and molecular weight. PEN/PEI and PEN/Nanoclay blends were two polyester systems for which Cakmak and co-workers developed detailed dynamic phase diagrams that display various stress optical profiles and structures as a function of stretching rate and temperature, with the ultimate goal of avoiding necking without having to resort to high draw ratios. Necking naturally occurs in PEN upon stretching from the amorphous state above the glass transition temperature [91]. It was shown to be caused by a highly localized cooperative orientation of the napthalene planes parallel to the surface of the film as shown in Figure 5.37. In film processing, this neck formation causes a reduction in the processing window by limiting the production to high draw ratios (uniaxial or biaxial), where films become more uniform in thickness once the necks in the films are eliminated.

In the case of PEN/PEI blends, PEI helps avoid necking by increasing the friction forces between the PEN chains by introducing the stiff and bulky PEI chains.

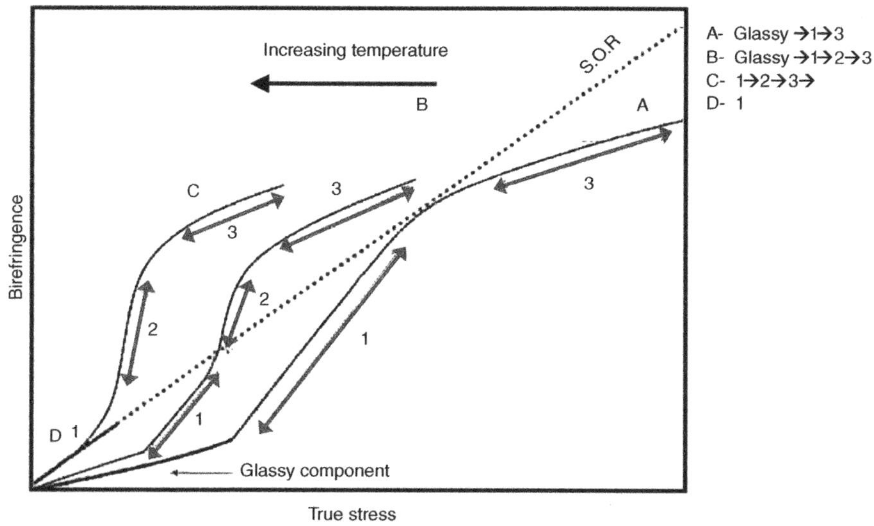


**Figure 5.38** Different stress-optical behaviors (a–d) of PEN/PEI blends.

Molecular level dispersion of PEI is possible as PEN and PEI exhibit miscibility in the amorphous state at all compositions. Kanuga and Cakmak studied rheo-optical behavior of PEN/PEI blends in detail and developed dynamic phase diagrams that show the type of stress-optical behavior over a wide range of temperatures and rates for PEI blends at 5, 10 and 20% [92]. Four types of stress optical behaviors (A, B, C, D) were identified in these phase diagrams based on the stretching temperatures and rates in Figure 5.38.

As the temperature is increased the stress-optical behavior changes from A to B to C and finally to D at very high temperatures. These stress-optical behaviors seen in slow crystallizing polyesters including PEN/PEI blends are discussed below.

Stress-optical behavior A starts with a glassy region characterized by a rapid increase of stress with minimal to no increase in birefringence. With further stretching, this initial glassy regime is followed by a rubber-like deformation Regime 1 wherein birefringence increases linearly with stress, and the material is almost entirely in the amorphous phase. As the stretching continues, Regime 1 is followed by Regime 3 which is characterized by a birefringence/stress curve that negatively deviates from Regime 1 and forms a nematic structure where the chains are highly oriented, but the structure lacks a 3D crystalline order. Nematic structures are explained in detail in the next section for polylactic acid (PLA), which is more prone to nematic structure since PLA has longer relaxation time than PET and PEN due to its higher molecular weight and hence higher entanglement density. In the current PEN/PEI discussion, nematic structures are attained with increasing PEI content because stiff PEI chains in PEN suppress the orientation relaxation, which prevents 3D registering of chains to form oriented crystalline structures.

Stress optical behavior B is characterized by a glassy region followed by linear amorphous Regime 1 similar to that seen in stress optical behavior A. However, linear Regime 1 is followed by a positively deviating Regime 2, where oriented crystals form with further stretching. Regime 3 follows Regime 2 where the oriented crystals reach their finite extensibility limits exhibiting a negative deviation. Final structure is oriented crystalline at the end of stretching.

Stress optical behavior C is the same as B, except the initial glassy region disappears because the chains are above their liquid–liquid transition temperature ($T_{ll}$) where the material transforms from a "structured liquid" to a "true liquid" devoid of the segmental correlations. The liquid–liquid temperature were described by Boyer [93] and later on Frenkel [94]. The liquid–liquid transistion temperature ($T_{ll}$) occurs ~1.08 $T_g$ (°K). In between $T_g$ and $T_{ll}$, the material, althought above the $T_g$, is not truly in a rubbery liquid state because there are segmental crystals that act as additional junction points in a physical network made up of entanglements and prevent the material from behaving as a true liquid. According to Boyer $T_{ll}$ represented a temperature above which the entire polymer chain is free to move as a unit and below which the material has some structure and is represented as a "hindered liquid" or a "fixed liquid state." Frenkel finally associated the concept of "segmental melting" at $T_{ll}$, above which the entropy of the polymer chains gives the driving force to free itself from the segmental restraints. They have shown with experimental evidences the $T_{ll}$-$T_g$ relationship as well as the frequency dependence of $T_{ll}$ exhibiting that $T_{ll}$ is not a thermodynamic transition but has a kinetic nature. Although this field has rather been controversial in the last few decades, the evidences presented in Cakmak and coworkers' work support the presence of liquid–liquid transition.

As the temperature is further increased, stress-optical Behavior D can be observed. This is a linear region where the material stays entirely in amorphous state due to high levels of orientation relaxation.

Dynamic Phase diagrams of PEN/PEI blends as a function of temperature and engineering strain rate are shown in Figure 5.39 along with critical transition regions and stress optical behavior that one can expect in various processing windows. For brevity, only the 5 and 20% PEI/PEN blends are shown.

Three critical structure transitions are observed. (i) Nematic-crystalline transition, (ii) Glassy to liquid ($T_{ll}$) transition and (iii) crystalline to amorphous transition. In the films containing 5% PEI, there is no nematic order at low temperatures. Nematic phase appears at 10% (not shown) and becomes quite clear at 20% PEI concentration. As mentioned earlier, the presence of this nematic phase with

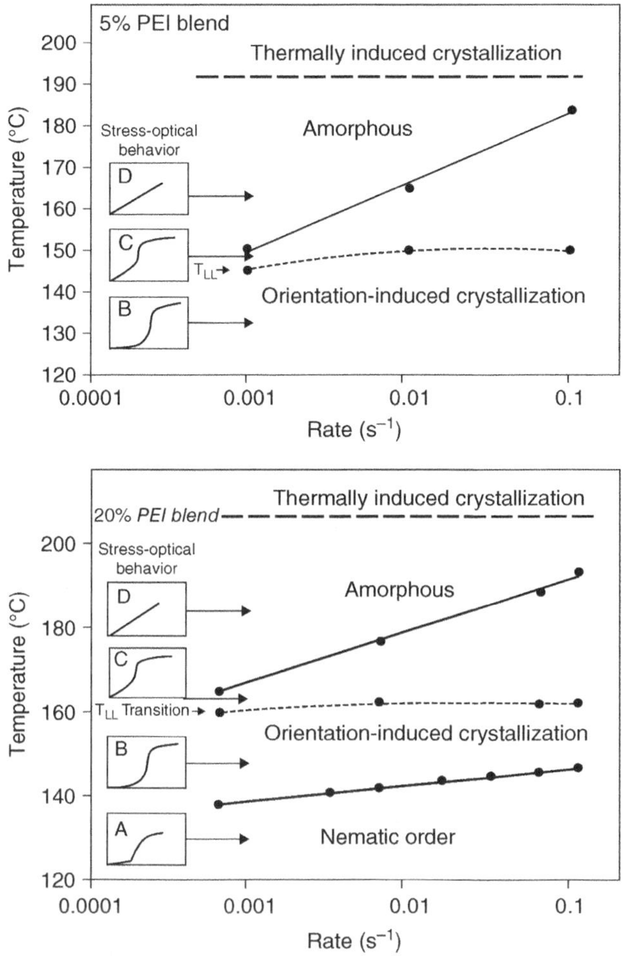


**Figure 5.39** Dynamic Phase diagrams of PEN/PEI blends as a function of temperature and engineering strain rate.

increasing PEI content is attributed to the suppression of orientation relaxation in the presence of stiff PEI chains in PEN. At low stretching temperatures, the films remain in a nematic-like state that converts into orientation-induced crystallization at intermediate temperatures. Another interesting point is that at high enough temperatures, there is a temperature window that yields a completely amorphous structure due to significant orientation relaxations coupled with the absence of thermal activated crystallization. The phase boundary between the orientation-induced crystallization and amorphous phase has stronger rate dependence than $T_{ll}$ transition.

Another solution that was proposed by Kanuga et al. to eliminate necking of PEN was through the incorporation of 0.5–2% montmorillonite (MMT) nanoclay particles in PEN [32]. MMT nanoparticles increase the inter-chain friction in between the naphthalene groups and pre-orient them parallel to the surface of the film during melt casting process prior to rubbery state deformation. Since naphthalene groups are already oriented, necking does not occur during rubbery state stretching.

Dynamic phase diagram of PEN/2% MMT is shown in Figure 5.40. Different from the dynamic phase diagrams of PEN/PEI blends only two critical transitions are observed; (i) Nematic-crystalline transition and (ii) Glassy to liquid ($T_{ll}$) transition. Crystalline to amorphous transition that was observed in PEN/PEI blends does not occur. The structure in PEN/nanoclay blends goes from orientation-induced crystallization to thermal induced crystallization as the temperature of stretching is raised. Authors attributed this to local orientation amplification effect induced by the nanoparticles, which prevents the material from going into a completely amorphous state at high stretching temperatures even though relaxation is dominant. In addition, glassy to liquid ($T_{ll}$) transition was dependent on rate and nanoclay loading which meant that it was a kinetic property rather than a thermodynamic property. As seen in Figure 5.40, the $T_{ll}$ transition occurs at 135 °C at very low stretching rate of 0.001 s$^{-1}$. As the stretching rate is increased, the transition shifts consistently to higher temperature, reaching 147 °C, for the highest stretching rate of 0.1 s$^{-1}$.

Finally increasing the temperature above 178 °C for about 20 minutes of preheating time results in thermally induced crystallization as the nucleation and chain diffusion rates begin to become significant. This marks the point above

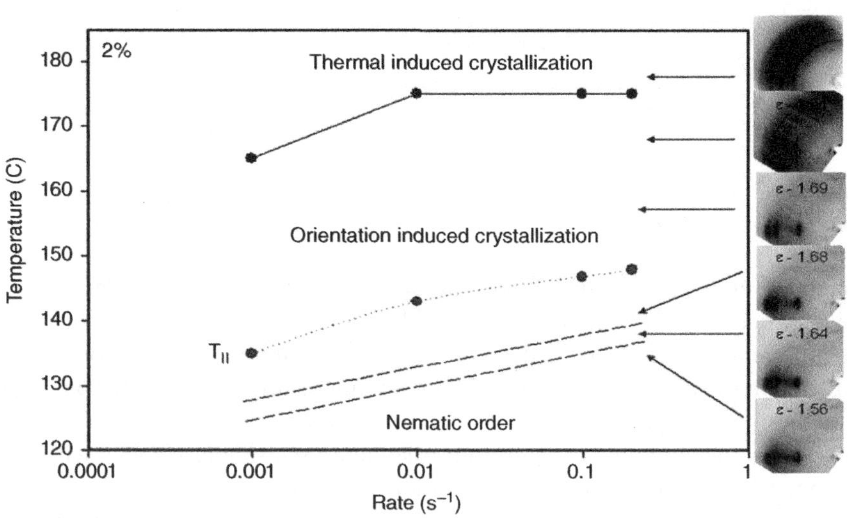


**Figure 5.40** Dynamic phase diagram for PEN with 2 wt% nanoparticle concentration.

which the material thermally crystallizes. This transition point is shown as a function of stretching rate in the dynamic phase diagram occurring at lower temperatures for lower rates. As the sample is exposed to these temperatures longer during slow rate deformation, thermally activated crystallization becomes appreciable due to this prolonged thermal exposure. The authors also pointed out that the increase of clay concentration (not shown) broadened the gap between the nematic transition and $T_{ll}$ at all stretching rates as these samples exhibit wider temperature range where the glassy behavior remains effective.

Nanoclay was shown to reduce the amount of crystallinity as well amorphous orientation but increase crystalline orientation. This drop in crystallinity and the subsequent slow rise in birefringence was explained by the decrease in orientation relaxation in the presence of nanoparticles. For crystallization to occur effectively, some degree of relaxation is needed for the oriented polymer chains to become aligned parallel to each other. The decrease in mobility of the polymer chains in the presence of hard clay surfaces decreases their relaxation substantially to become parallel to each other leading to substantial decrease in their strain crystallizabilities. The decrease in crystallization growth also results in the formation of a weaker long range load bearing network as there is lesser production of crystallites that form the nodes of the network. As a result of malformed crystalline network, orientation levels observed in amorphous regions become lower. This combined effect of diminished crystallinity as well as lower amorphous orientation explains the decrease in birefringence as the nanoparticle concentration is increased in PEN. On the other hand, the crystalline orientation factor increases with the addition of nanoparticles. This is contrary to the effect nanoparticles have on the bulk amorphous orientation. This shows that the nanoparticles promote the alignment of polymer crystallites even though the amorphous orientation decreases with the addition of nanoparticles. Nanoparticles have been shown to act as nucleation agent as the increase in orientation in the presence of these particles decreases the cold crystallization temperature $T_{cc}$. Authors suspected that the crystallites develop in the vicinity of the nanoplatelets, and the relaxation of polymer chains is suppressed in the proximity of the nanoparticles leading to sufficient levels of orientation which leads to oriented crystallization. The effect of nanoclay loading on the crystalline and amorphous orientation factors determined from Wilchinsky's rule as functions of true strain is shown in Figure 5.41.

## 5.7 Poly (Lactic Acid)

Mulligan et al. studied dynamic phase behavior of amorphous cast films of Polylactic acid (PLA) during uniaxial stretching in the rubbery state using online mechano-optical tester shown in Figure 5.12 [95]. PLA is a slow-crystallizing

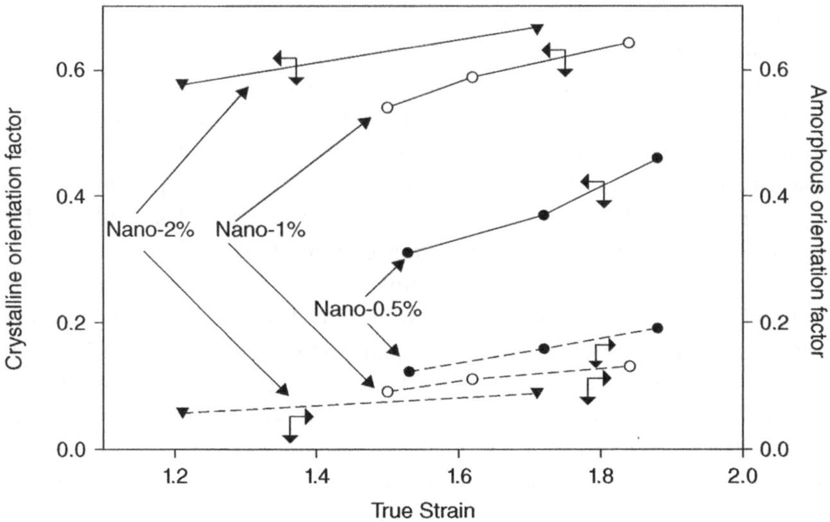


**Figure 5.41** Crystalline and amorphous orientation factors as functions of true strain.

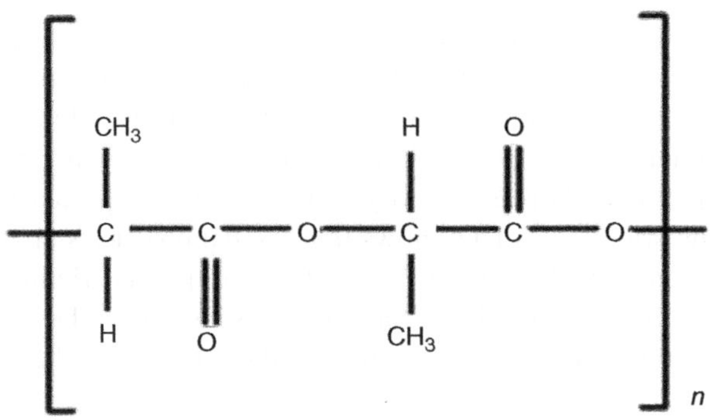


Poly lactic acid (PLA)

$T_g$~60 to 70 °C depending on crystallinity and isomer content

$T_m$~ 150 TO 170 °C depending on isomer content

**Figure 5.42** Chemical Formula of PLA.

polyester that possesses rubbery state deformability and strain-induced crystallization, similar to PET. Its chemical formula and basic thermal properties are shown in Figure 5.42.

PLA has a higher MW and entanglement density than PET, and hence, allows for enhanced ability to suppress orientation relaxation making it easier for the

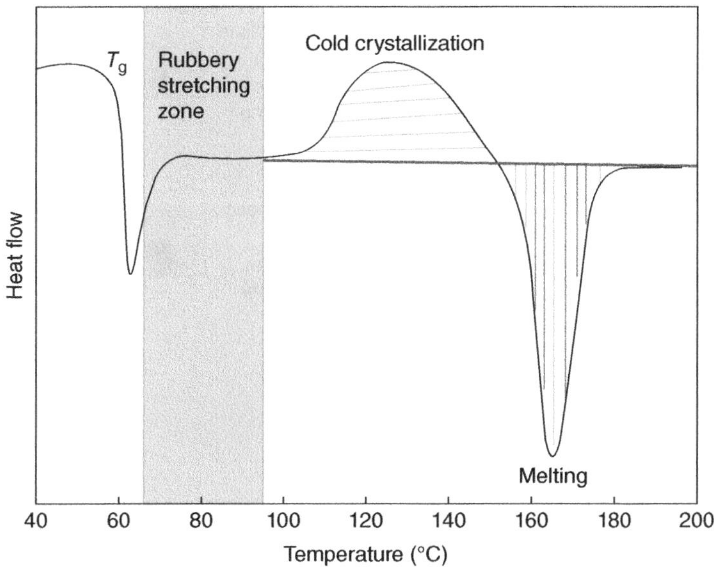


**Figure 5.43** DSC of As-cast PLA films.

chains to be frozen with almost no alteration to the overall structure. DSC of as cast films of PLA (D- isomer content 2%) studied by Mulligan et al. is shown in Figure 5.43. The cast film exhibits the glass transition temperature at 60 °C, and it is in the rubbery state in a wide temperature range between 60 and 95 °C. The areas under the cold crystallization and melting peaks are the same, indicating that the original sample prior to scanning was essentially amorphous.

As-cast amorphous PLA films were stretched at various temperatures between 65 and 90 °C with rates ranging from 60, 240, and 960%/min. The authors found out for all stretching conditions regardless of rate or temperature, the initial Regime I in the Birefringence-True Stress follows the linear stress optical rule with a constant of 3.1 $GPA^{-1}$ (Figure 5.44). As the deformation proceeds beyond Regime I, authors detected two types of structures, nematic and crystalline, that show distinct regime behavior as a function of strain rate and temperature of stretching. Higher stretching rates and lower temperatures favored nematic structure with Regime I-III behavior (negative deviation from linearity), and lower rates and higher temperature favored crystalline structure development with Regime I-II-III behavior (positive deviation from linearity) as depicted in Figure 5.44. Nematic structures can easily be differentiated from crystalline structures in WAXD images. WAXD patterns of nematic structures lack the 3D crystalline order evidenced by the absence of crystalline diffraction spots. Instead, WAXD intensity in nematic structures is fuzzy and concentrated at the equator

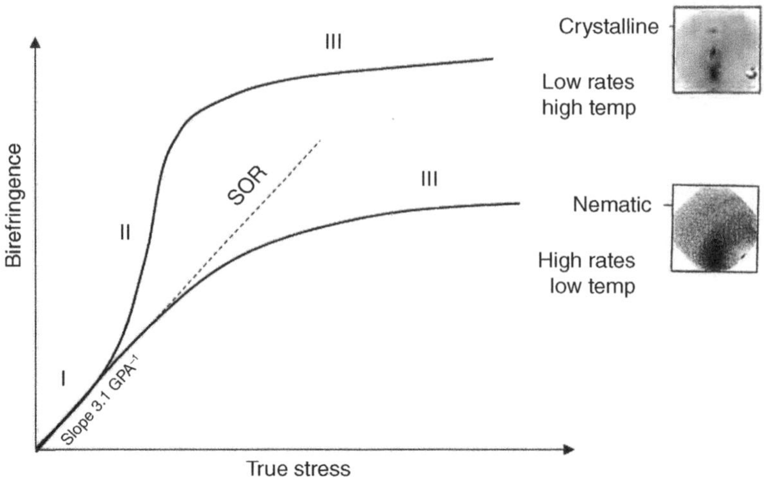


**Figure 5.44** Birefringence- true stress regime behavior for crystalline and nematic phases with representative WAXD images.

indicating that the polymer chains are well-oriented in the stretching direction but without 3D order. This fuzzy intensity concentration in nematic PLA WAXD images occurs at $2\theta \sim 18°$ that represents the interchain correlation. Stress-induced crystallinity in PLA, on the other hand, result in distinct equatorial and off-equatorial diffraction patterns.

Presence of nematic order has previously been observed first by Bonart [96] and then by others through synchrotron studies in PET and other polyesters during stretching from amorphous precursors. Nematic structures are difficult to preserve at the end of stretching for PET chains but much easier for PLA since entanglement densities are higher in PLA and relaxation is suppressed. Mulligan and Cakmak demonstrated PLA chain sluggishness and stability of their nematic PLA structures by relaxation experiments. The time needed for relaxation to turn a nematic structure into a crystalline one was over five minutes which was quite long time needed for the oriented chains to rearrange into a lattice structure with the added relaxation. This implied that quenching the films at the end of stretching was quite sufficient to lock in the nematic structures.

The offline WAXD patterns of samples stretched at a high rate of 960%/min and at a low temperature of 72 °C (left column-a) and a high temperature of 78 °C (left column-b) and rapidly frozen at various stages of deformation are shown on the Birefringence-True stress plots along with corresponding DSC curves in Figure 5.45. After early stages of deformation is complete (Regime I), stretching at a high rate coupled with low temperature of 72 °C induces nematic structures that are progressively oriented at further stages of deformation displaying negative

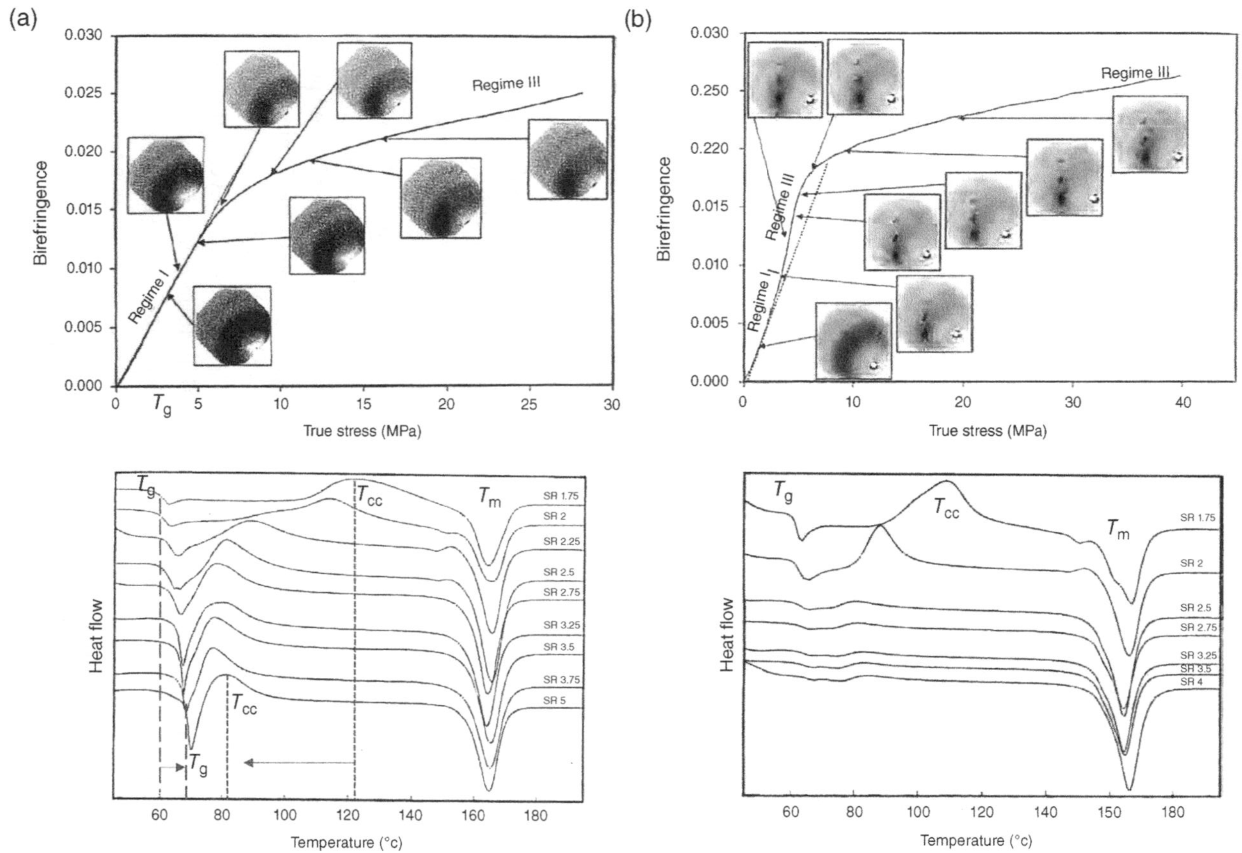


**Figure 5.45** Birefringence-true stress plots of PLA films with WAXD images and DSC graphs taken at various stages of stretching at a rate of 960% min at (a) 72 °C (left) forming nematic structure and (b) 78 °C (right) forming strain crystallized structure.

deviation from linearity of Regime I (Regime I–III behavior). Although not shown here, higher rates and lower temperatures causes an earlier negative deviation from Regime I into Regime III. Here, crystallization does not cause the deviation from Regime I stress optical rule behavior, but the deviation is due to the non-Gaussian segmental distribution characteristic of polymer chains as they approach their finite extensibility. When the temperature of stretching is increased to 78 °C, the same initial stress optical rule behavior (same stress optical constant) seen in curve stretched at 72°C exhibit positive deviation from Regime I into Regime II. This is likely due to the higher propensity for crystallization to occur at higher temperatures due to partial local relaxation of the chains, thereby allowing them to register with each other, establishing the crystalline network. In turn, this leads to a faster increase in birefringence over a small range of stress. This is very similar to rapid formation of physical crosslinks, manifesting here in the form of crystalline domains containing polymer chains that exhibit near-perfect orientation as evidenced by the WAXD measurements. With further deformation, a negative deviation from Regime II to III is observed due to the resistance caused by the increased density of this crystalline physical network, preventing further orientation of the polymer chains. The authors also provided evidence for the absence of crystallization in nematic structures through the DSC scans. At 78 °C, crystallinity increases with deformation and reaches a plateau at a point corresponding to the deviation into Regime III deformation behavior (Figure 5.46). No stress-induced crystallization was observed when the samples were stretched at 72 °C as evidenced by WAXD images (nematic structure).

DSC scans also show that the glass transition temperature increases as the films are stretched since the chains become wellpacked and less mobile as they become aligned in the stretching direction. The cold crystallization peak becomes sharper and shifts to lower temperature, reflecting the decrease in entropy due to an increased orientation in the amorphous regions as the deformation proceeds.

In further analysis of the birefringence true stress curves, the authors fit a power law function of Regime II SOC (slope of Regime II) to the temperature of stretching $T$ through SOC $= KT^n$, where SOC is the stress-optical coefficient, $K$ is the power law coefficient, and $n$ is the power law index. After determining the power law fit, the point at which the Regime II SOC becomes that of the Regime I SOC (i.e. when Regime II first disappears) was taken as the demarcation between Regime I + II + Ill behavior and Regime I + Ill. This was used to develop the quantitative phase diagram for the structure formation during the uniaxial deformation of PLA films in the rubbery state as shown in Figure 5.47. The phase diagram shows that low temperatures and high rates favor the formation of a nematic-like phase during deformation, while high temperatures and low rates favor crystallization while deformation proceeds. As the temperature increases, the rate at the transition increases exponentially.

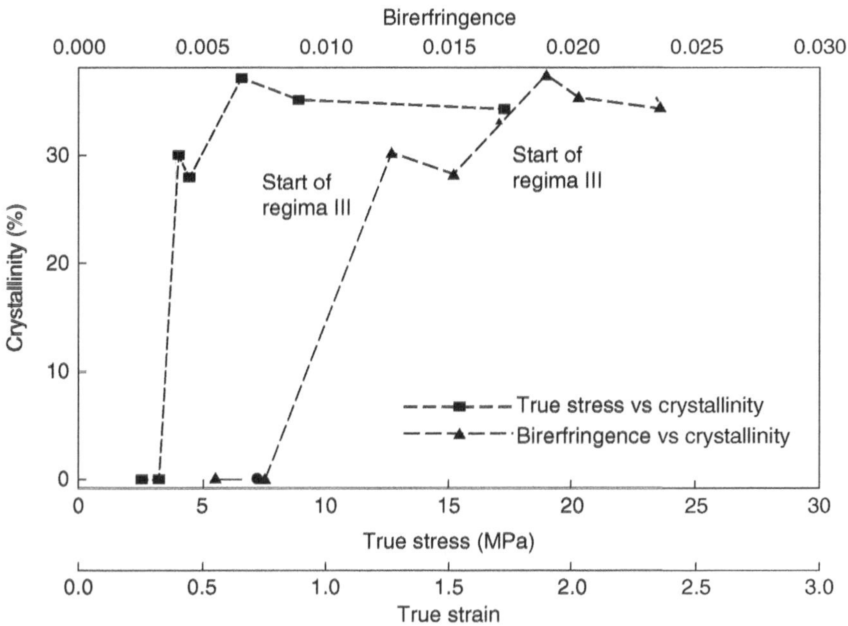


**Figure 5.46** Crystallinity as a function of birefringence and true stress.

## 5.8 Summary

In this chapter, rheo-optical behavior of polyesters during rubbery state processing such as in tenter-frame stretching were reviewed. The focus was on real-time monitoring of birefringence and stress/strain fields during stretching, relaxation, as well as heat-setting modes to provide insight into the instant structural changes and molecular ordering in various types of polyesters.

Rheo-optical properties were presented by analyzing the birefringence-true stress/strain behavior in distinct regimes during deformation, relaxation, or heat-setting stage of the process. All three processes are affected by the stretching rate, ratio, temperature, and the type of polyester. However, some generalizations can be made.

### 5.8.1 Stretching Stage

There is a critical degree of orientation required to achieve strain-induced crystallization in the stretched films. This critical degree of orientation is affected by the stretching rate and ratio. Too low of a rate does not allow for having this degree of orientation, while using high rates do not enable the chains to have a high degree

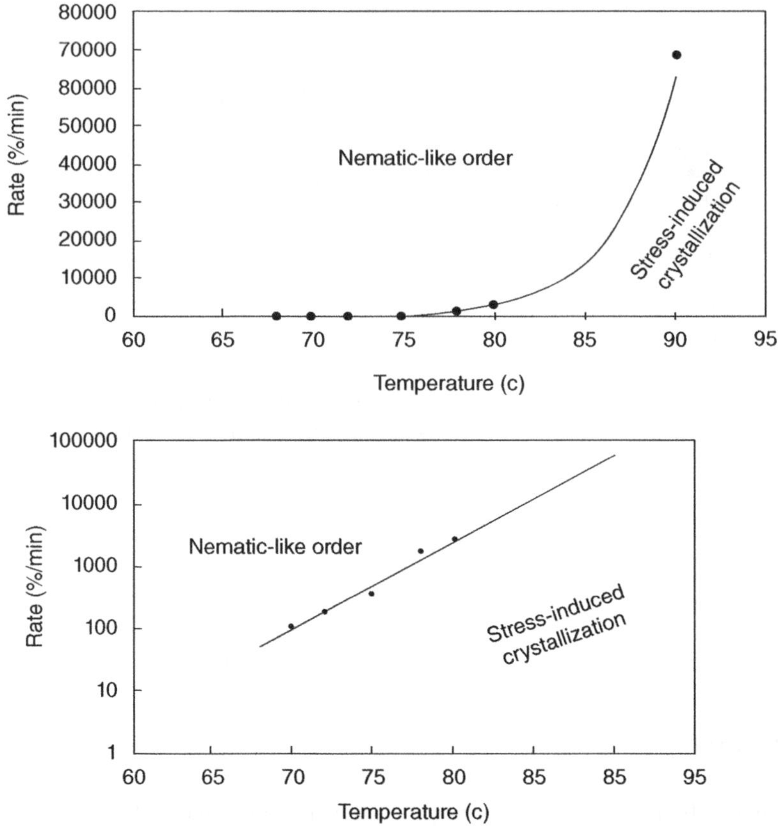


**Figure 5.47** Dynamic phase diagram for structure formation during the uniaxial deformation of PLA, regular (top) and logarithmic scale (bottom) graph.

of relaxation and hence register in crystalline domains. Regime I is a low-strain region where the material is still amorphous. In this regime, the stress is in a linear relationship with the birefringence obeying the stress-optical rule (SOR) (linear rise of stress with birefringence). All characterization techniques point to amorphous if not very low crystalline content in Regime I. Regime II starts with the strain-induced crystallization and the deviation from the SOR where both observations coincide. When the molecular chains are stretched near their finite extensibility limits, Regime III appears and manifests itself by a high increase in the stress compared to the birefringence increase. This causes the decreased slope in the stress-birefringence slope, higher crystallinity, and better alignment in the crystallographic planes.

### 5.8.2 Relaxation Stage

During relaxation in constrained state upon completion of stretching, polyesters can exhibit different mechano-optical behavior. At low stretch ratios, the material remains amorphous in Regime 1 not only during stretching and but also during relaxation stage. At intermediate deformation levels (Regime II), the polymer chains may remain amorphous initially upon completion of stretching, but they begin to exhibit strain-induced crystallization during relaxation stage as the adjacent oriented amorphous chains become locally aligned relative to each other to form crystalline regions. This phenomenon appears as a demarcation point on the evolution of birefringence where the previously decreasing birefringence start to increase while stress continue to decrease. When the relaxation starts after a high stretch ratio where the material has developed a well-connected network during deformation (Regime III), the birefringence does not experience any relaxation but starts to increase from the point of relaxation. A minimum degree of orientation is required for the chains to start crystallizing during relaxation.

### 5.8.3 Heat-Setting Stage

The heat-setting (annealing) step, when used, works on increasing the crystallinity of the stretched films. The kinetics of the process depends on the thickness, orientation, and percentage crystallinity of the pre-stretched films. Similar to stretching and relaxation, the behavior of the films during the process can be categorized into three regimes according to the birefringence trend; either decreasing, decreasing then increasing, or increasing continuously until the end. The intermediate stretch ratio provides the most favorable condition for crystallinity increase upon heating for a long time.

Finally, this chapter reviews studies that demonstrates the use of dynamic phase diagrams as a quick tool to determine what structure the polyesters would attain in the final state given the processing conditions (rate, temperature, type of deformation) and molecular weight.

Such quantitative relationships between the process dynamics such as stress and strain and the structural parameter, birefringence, particularly at large deformation levels, are needed to build constitutive relationships that are essential for predictive modeling of material behavior in shaping/forming processes.

## References

1 Fakirov, S. Handbook of Thermoplastic Polyesters: Homopolymers, Copolymers, Blends, and Composites. 2002 |DOI:https://doi.org/10.1002/3527601961 Wiley-VCH Verlag GmbH & Co. KGaA.

2 John Scheirs, Timothy E. Long Modern Polyesters: Chemistry and Technology of Polyesters and Copolyesters. 2004 |DOI:https://doi.org/10.1002/0470090685 Wiley.

3 Dodiuk, H. (2013). *Handbook of Thermoset Plastics*, Plastics Design Library Series, 3e. Elsevier.

4 Meyer, R.W. (1987). *Handbook of Polyester Molding Compounds and Molding Technology*. https://doi.org/10.1007/978-1-4613-1961-0. New York, NY. Chapman and Hall: Springer.

5 Wilkes, G.L. and Stein, R.S. Physicochemical approaches to the measurement of molecular anisotropy. In: *Structure and Properties of Oriented Polymers*, 2e (ed. I.M. Ward). Dordrecht: Springer. ISBN 978-0-412-60880-3.

6 Kunihiro Osaki,Tadashi Inoue, Eui-Jeong Hwang, Hirotaka Okamoto, Osamu Takiguchi. Dynamic birefringence of amorphous polymers. *J. Non-Cryst. Solids*. Volumes 172–174, Part 2, 1 September 1994, Pages 838–849.

7 Inoue Tadashi Strain-induced birefringence of amorphous polymers and molecular design of optical polymers, *ACS Appl. Polym. Mater.* 2021, 3, 5, 2264–2273 DOI:https://doi.org/10.1021/ACSAPM.1C00149.

8 Iwata, S., Tsukahara, H., Nihei, E.N.E., and Koike, Y. (1996). Compensation for Birefringence of Oriented Polymers by Random Copolymerization Method. *Jpn. J. Appl. Phys.* 35: 3896.

9 Shuheil, W., Satoh, Z., Shafiee, H. et al. (2012). Design and synthesis of zero–zero-birefringence polymers in a quaternary copolymerization system. *Polymer* 53: 3287–3296.

10 Akihiro, T., Ohkita, H., Harada, T. et al. (2006). Zero-birefringence optical polymers. *Macromolecules* 39 (8): 3019–3023.

11 Akihiro, T. and Koike, Y. (2012). Compensation and control of the birefringence of polymers for photonics. *Polym. J.* 44: 306–314.

12 Onogi, S., Keedy, S.A., and Stein, R.S. (1961). The measurement of dynamic birefringence. *J. Polym. Sci.* 153: 515–516.

13 Stein, R.S., Onogi, S., and Keedy, D.A. (1962). Dynamic birefringence of high polymers. *J. Polym. Sci.* 57: 801–821.

14 Offenbach, I., Gupta, S., Chung, T.C.M. et al. (2015). Real-time infrared-mechano-optical behavior and structural evolution of polypropylene and hydroxyl-functionalized polypropylene during uniaxial deformation. *Macromolecules* 48 (17): 6294–6305.

15 Asada, T., Onogi, S., and Yanase, H. (1984). A rheo-optical study on the reformation of structure in racemic poly(γ-benzyl glutamate) liquid crystals. *Polym. Eng. Sci.* 24 (5): 355–360.

16 Galay, J. and Cakmak, M. (2001). Online monitoring of birefringence development in heat-setting polymer films with a fast dual-wavelength optical technique. I. Uniaxially oriented poly(ethylene naphthalate). *J. Polym. Sci. B Polym. Phys.* 39: 1107–1121.

17 Galay, J. and Cakmak, M. (2001). Online monitorin of birefringence development in heat-setting polymer films with a fast dual-wavelength optical technique. II. Uniaxially oriented PEN\PEI films. *J. Polym. Sci. B Polym. Phys.* 39: 1147–1159.
18 Yang, H.H., Chouinard, M.P., and Lingg, W.J. (1982). *J. Polym. Sci. Polym. Phys.* 20: 981.
19 Hongladarom, K., Burghardt, W.R., Baek, S.G. et al. (1993). Molecular alignment of polymer liquid crystals in shear flows. 1. Spectrographic birefringence technique, steady-state orientation, and normal stress behavior in poly(benzyl glutamate) solutions. *Macromolecules* 26: 772.
20 Hongladarom, K. and Burghardt, W.R. (1993). Molecular alignment of polymer liquid crystals in shear flows. 2. Transient flow behavior in poly(benzyl glutamate) solutions. *Macromolecules* 26: 785.
21 Hongladarom, K., Secakusuma, V., and Burghardt, W.R. (1994). Relation between molecular orientation and rheology in lyotropic hydroxypropylcellulose solutions. *J. Rheol.* 38: 1505. https://doi.org/10.1122/1.550556.
22 Hongladarom, K. and Burghardt, W.R. Measurement of the full refractive index tensor in sheared liquid crystalline polymers. *Macromolecules* 27: 483.
23 Serhatkulu, T. (2000). Crystallization and birefringence studies on fast structural changes followed by non-contact spectral birefringence and raman spectroscopy techniques. Ph.D. dissertation. The University of Akron.
24 Serhatkulu, T., Kokturk, G., and Cakmak, M. (1999). Real time monitoring of fast birefringence chances during crystallization of preoriented PETAnd PLA films using spectral. *SPE ANTEC Tech. Pap.* 54: 1645.
25 Serhatkulu, T.F. and Cakmak, M. (1999). The application of new spectral birefringence technique to characterize the structural changes that take place in heat setting of preoriented PET and PETG. *SPE ANTEC Tech. Pap.* 1645.
26 Kokturk, G., Serhatkulu, T.F., Kozluca, A. et al. (1999). Evolution of structural hierarchy in uniaxially deformed poly lactic acid films as followed by spectral birefringence technique and others. *SPE ANTEC Tech. Pap.* 45: 2190.
27 Kokturk, G., Serhatkulu, T.F., and Cakmak, M. (2000). Evolution of structural hierarchy in uniaxially deformed branched poly lactic acid films as followed by spectral birefringence technique and others. *SPE ANTEC Tech. Pap.* 58 (2): 1737–1741.
28 DeMeuse, M.T. (2011). *Biaxial Stretching of Film Principles and Applications*. Woodhead Publishing. ISBN 978-1-84569-675-7.
29 Cakmak, M., Hassan, M., Unsal, E., and Martins, C. (2012). A fast real time measurement system to track in and out of plane optical retardation/ birefringence, true stress, and true strain during biaxial stretching of polymer films. *Rev. Sci. Instrum.* 83 (12): 123901/1–123901/10.

**30** Jiang, F. and Cakmak, M. (2015). Real time optical and mechano-optical studies during drying and uniaxial stretching of polyetherimide films from solution. *Polymer* 68: 168–175.

**31** Wang, E., Batra, S., and Cakmak, M. (2015). A real time study on drying and the mechano-optical behavior of polyvinyl alcohol films in solid and swollen state. *Polymer* 67: 200–207.

**32** Kanuga, K. and Cakmak, M. (2013). Role of nanoparticles and relaxation on strain-induced crystallization behavior in uniaxially stretched polyethylene naphthalate films: a mechano-optical study. *Macromolecules* 46 (15): 6300–6308.

**33** Hassan, M.K. and Cakmak, M. (2015). Strain induced crystallization during relaxation following biaxial stretching of pet films: a real time mechano optical study. *Macromolecules* 48 (13): 4657–4668.

**34** Hassan, M.K. and Cakmak, M. (2014). Mechanisms of structural organizational processes as revealed by real time mechano optical behavior of PET film during sequential biaxial stretching. *Polymer* 55 (20): 5245–5254.

**35** Hassan, M.K. and Cakmak, M. (2013). Mechano optical behavior of polyethylene terephthalate films during simultaneous biaxial stretching: real time measurements with an instrumented system. *Polymer* 54 (23): 6463–6470.

**36** Choi, J., Serhatkulu, T., and Cakmak, M. (2016). Temporal evolution of structure in uniaxially stretched PET/PEI blends during constrained annealing: a real time birefringence study. *Polymer* 97: 569–579.

**37** Hassan, M.K.; Cakmak, M. Rapid structural changes that take place during constrained annealing of biaxially stretched pet films. *Polym. Eng. Sci.* 2016, 6, DOI: https://doi.org/10.1002/pen.24451.

**38** Unsal, E., Nugay, I.I., Offenbach, I. et al. (2013). Rapid integrated rheo-optical and polarized Fourier-transform infrared spectrometry measurement system for polymer films undergoing chemo-mechanical changes. *Rev. Sci. Instrum.* 84 (7): 073901/1–073901/10.

**39** Gerrard, A. and Burch, J.M. (1975). *Introduction to Matrix Methods in Optics*. New York, NY: Wiley.

**40** Kliger, D.S., Lewis, J.W., and Randall, C.E. (1990). *Polarized Light in Optics and Spectroscopy*. Boston: Academic Press.

**41** Collett, E. (1993). *Polarized Light: Fundamentals and Applications*. New York, NY: Marcel Dekker.

**42** Gupta V.B., Z. Bashir, PET fibers, films, and bottles: sections 1–4.13. Stoyko Fakirov Handbook of Thermoplastic Polyesters: Homopolymers, Copolymers, Blends, and Composites, 2002, pp. 317–361. Wiley-VCH Verlag GmbH & Co. KGaA. https://doi.org/10.1002/3527601961.

**43** Jog, J.P. (1995). Crystallization of polyethyleneterephthalate. *J. Macromol. Sci. Polym. Rev.* 35 (3): 531–553.

**44** Cakmak, M., White, J.L., and Spruiell, J.E. (1986). Spruiell structural characterization of crystallinity and crystalline orientation in simultaneously biaxially stretched and annealed polyethylene terephthalate films. *J. Polym. Eng.* 6 (1–4): 291.

**45** Kang, H.J., White, J.L., and Cakmak, M. (1990). Single and double bubble tubular film extrusion of polyethylene terephthalate. *Int. Polym. Process.* 5 (1): 62–73.

**46** Hegemann, B. A. Kech, U. Göschel, K. Belina & P. Eyerer Biaxial deformation behavior of pet dependent on temperature and strain rate, *Journal of Macromolecular Science, Part B*, 41:4–6, 647–656 (2002), DOI: https://doi.org/10.1081/MB-120013056.

**47** Billon, N., Picard, M., and Gorlier, E. (2014). Stretch blow moulding of PET; structure development and constitutive model. *Int. J. Mater. Form.* 7: 369–378. https://doi.org/10.1007/s12289-013-1131-1.

**48** Iwakura, K., Wang, Y.D., and Cakmak, M. (1992). Effect of biaxial stretching on thickness uniformity and surface roughness of PET and PPS films. *Int. Polym. Process.* 7 (4): 327–333.

**49** Marco, Y., Chevalier, L., Régnier, G., and Poitou, A. (2002). Induced crystallization and orientation of poly(ethylene terephthalate) during uniaxial and biaxial elongation. *Macromol. Symp.* 185 (1): 15–34.

**50** Treloar, L.R. (1958). *The Physics of Rubber Elasticity*. Clarendon: Oxford.

**51** Doi, M. and Edwards, S.F. (1986). *The Theory of Polymer Dynamics*. Clarendon: Oxford.

**52** Gupta, M. and Yeh, G.J. (1978). DRDF studies of glassy and partially crystalline poly(ethylene terephthalate). *Macromol Sci.* B15 (1): 119.

**53** Zaroulis, J.S. and Boyce, M.C. (1997). Temperature, strain rate, and strain state dependence of the evolution in mechanical behaviour and structure of poly(ethylene terephthalate) with finite strain deformation. *Polymer* 38 (6): 1303–1315.

**54** Campbell, R. (1999). Polypropylene film comparisons from sequential and simultaneous tenter bopp processes, *1999 Polymers, Laminations & Coatings Conference* Proceedings, Atlanta, GA, Vol. 2, 753e67.

**55** Breil, J. (1999). S-BOPP-film enhancement by linear motor simultaneous stretching technology, *1999 Polymers, Laminations & Coatings Conference Proceedings*, Atlanta, GA, Vol. 2.

**56** Lu, X. and Hay, J. (2001). Crystallization orientation and relaxation in uniaxially drawn poly(ethylene terephthalate). *Polymer* 42: 8055–8067.

**57** Fan, Q. and Qian, R. (1997). The state of high global chain orientation but nearly random segmental orientation of amorphous polymers. *Macromol. Symp.* 124: 59.

**58** Adams, A.M., Buckley, C.P., and Jones, D.P. (2000). Biaxial hot drawing of poly(ethylene terephthalate): measurements and modelling of strain-stiffening. *Polymer* 41: 771.

**59** Ryu, D., Inoue, T., and Osaki, K. (1998). A birefringence study of polymer crystallization in the process of elongation of films. *Polymer* 39 (12): 2515e20.

**60** Venkatesvaran, H. and Cakmak, M. (2001). Kinetics of structural evolution during crystallization of preoriented PET as followed by dual wavelength photometric birefringence technique. *Poly. Eng. Sci.* 41 (2): 341–357.

**61** Werner, E., Janocha, S., Hopper, M.J., and Mackenzie, K. (1987). Polyester films. In: *Encyclopedia of Polymer Science and Engineering*, vol. 12 (ed. H. Mark, N. Bikales, C. Overberger, and G. Menges), 195. New York: Wiley.

**62** Petermann, J. and Rieck, U. (1987). Morphologies and mechanical properties of pet films crystallized under high strain rates. *J. Polym. Sci. Polym. Phys.* 25: 279.

**63** Mitsuisni, Y. and Tonami, H. (1964). Changes in Structure and Properties of PET fibers by heat treatment. *Sen-i Gakkaishi (J. Soc. Text. Cellulose Ind. Jpn.)* 20 (3): 140.

**64** Oswald, H.J., Turi, E.A., Harget, P.J., and Khanna, Y.P. (1977). Development of a middle endotherm in DSC thermograms of thermally treated drawn PET yarns and its structural and mechanistic interpretation. *J. Macromol. Sci.-Phys. B* 13 (2): 231.

**65** Peszkin, P.N., Schultz, J.M., and Lin, J.S. (1986). Kinetics of fiber heat treatment. II. Poly(ethylene terephthalate) fibers. *J. Polym. Sci. Polym. Phys.* 24: 2591.

**66** Gupta, K.M., Motz, H., and Schultz, J.M. (1983). Microstructure rearrangement during the heat-treatment of melt-drawn poly(ethylene terephthalate) fibers. *J. Polym. Sci. Polym. Phys.* 21: 1927.

**67** Fischerand, E.W. and Fakirov, S. (1976). Structure and properties of polyethyleneterephthalate crystallized by annealing in the highly oriented state. *J. Mater. Sci.* 11: 1041.

**68** Fakirov, S. and Evstatiev, M. (1990). New routes to polyethylene terephthalate with improved mechanical properties. *Polymer* 31: 431.

**69** Buckley, C.P. and Salem, D.R. (1987). High-temperature viscoelasticity and heat-setting of poly(ethylene terephthalate). *Polymer* 28: 69.

**70** Buckley, C.P. and Salem, D.R. (1990). Heat setting of oriented poly(ethylene terephthalate): effects of deformation mode, strain level, and heating medium. *J. Appl. Polym. Sci.* 41: 1707.

**71** Gupta, V.B. and Kumar, S. (1981). The effect of heat setting on the structure and mechanical properties of poly(ethylene terephthalate) fiber. IV. Tensile properties other than modulus and their dependence on structure. *J. Appl. Polym. Sci.* 26: 1897.

**72** Itoyama, K. (1987). Structure and properties of poly(ethylene terephthalate) crystallized by prolonged annealing in the highly oriented state. *J. Polym. Sci. Part C Polym. Lett.* 25: 331.

**73** Huisman, R. and Heuvel, H.M. (1989). The effect of spinning speed and drawing temperature on structure and properties of poly(ethylene terephthalate) yarns. *J. Appl. Polym. Sci.* 37: 595.

**74** Gupta, V.B., Ramesh, C., and Gupta, A.K. (1984). Structure–property relationship in heat-set poly(ethylene terephthalate) fibers. I. Structure and morphology. *J. Appl. Polym. Sci.* 29: 3115.

**75** Mitsuishi, Y. and Ikeda, M. (1966). Thermal retraction mechanism of highly oriented, amorphous polyethylene terephthalate filament. *Kobunshi Kagaku* 23 (253): 319–326.

**76** Chultz, J.M.S. and Fakirov, S. (ed.) (1990). *Solid State Behavior of Linear Polyesters and Polyamides*. Englewood, NJ: Prentice-Hall.

**77** Fakirov, S. (1985). *Structure and Properties of Polymers*. Sofia: Sofia Press.

**78** Maruhashi, Y. and Asada, T. (1992). Structure and physical properties of biaxially stretched polyethylene terephthalate sheet after heat-setting. *Polym. Eng. Sci.* 32 (7): 481.

**79** Cakmak, M., Spruiell, J.E., White, J.L., and Lin, J.S. (1987). Small angle and wide angle x-ray pole figure studies on simultaneous biaxially stretched poly(ethylene terephthalate) (PET) films. *Polym. Eng. Sci.* 27 (12): 893.

**80** Shimizu, Y., Toba, E., Sekiguchi, S., and Konda, A. (1988). The structure of highly drawn PET film with transverse lines and its polarized light scattering. *Sen'i Gakkaishi* 44 (3): 111.

**81** Vallat, M.F., Plazek, D.J., and Bhusan, B. (1988). Effects of thermal treatment on biaxially oriented poly (ethylene terephthalate) III. Creep behavior following various thermal histories. *J. Polym. Sci. Polym. Phys.* 26: 555–567.

**82** Klement, J.J. and Geil, P.H. (1971). Deformation and annealing behavior. I. Polyethylene terephthalate films. *J. Macromol. Sci.-Phys. B* 5 (3): 505.

**83** Perkin, W. (1988). Effect of molecular weight and annealing temperature on the oxygen barrier properties of oriented PET film. *Polym. Bull.* 19: 397.

**84** Koenig, J.L. and Cornell, S.W. (1967). Characterization of biaxially drawn polyethylene terephthalate by polarized infrared spectroscopy. *J. Macromol. Sci.- Phys. B* 1 (2): 279.

**85** Matsumoto, K., Ieki, H., and Imamura, R. (1971). Relationship between the structure and the surface state of the simultaneously biaxially stretched poly (ethylene terephthalate) Films. *Sen'i Gakkaishi* 27 (12): 516.

**86** Matsumot, K.O., Izumi, Y., and Imamura, R. (1972). Behavior of crystalline of the biaxially stretched PET films prepared by dry process. *Sen'i Gakkaishi* 28 (6): 189.

**87** Matsumoto, K., Izumi, U., and Imamura, R. (1972). Structure and mechanical properties of the biaxially stretched PET films by dry process. *J. Fiber Sci.* 28 (6): 179.

**88** Martins, C.I. and Cakmak, M. (2007). Control of the strain-induced crystallization of polyethylene terephthalate by temporally varying deformation rates: a mechano-optical study. *Polymer* 48 (7): 2109–2123.

**89** Choi, J. and Cakmak, M. (2016). Morphological evolution during thermal and strain induced crystallization in poly(ethylene terephthalate)/poly(ether imide) blend films. *Polymer* 84: 10–20.

**90** Martins, C.I. and Cakmak, M. (2005). Large deformation mechano-optical and dynamical phase behavior in uniaxially stretched poly (ethylene naphthalate). *Macromolecules* 38 (10): 4260–4273.

**91** Cakmak, M. and Kim, J.C. (1997). Necking mechanism and its elimination in uniaxially drawn films of poly(ethylene naphthalate) (PEN)/polyetherimide (PEI) blends. *J. Appl. Polym. Sci.* 65 (11): 2059–2074.

**92** Kanuga, K. and Cakmak, M. (2005). Nonlinear mechanooptical behavior of poly (ethylene naphthalate)/poly (ether imide) blends. *Dyn. Phase Behav. Macromol.* 38: 9698–9710.

**93** Boyer, R.B. (1987). Polymer yearbook; 1998. In: *Order in the Amorphous "State" of Polymers* (ed. E. Keinath Steven, L. Miller Robert, and K. Rieke James), 234e337; And also see. New York: Plenum Press.

**94** Lobanov, A.M. and Frenkel, S.Y. (1980). The nature of the so-called "liquid-liquid" transition in polymer melts. A review. *Polym. Sci. U.S.S.R.* 22 (5): 1150–1163.

**95** Mulligan, J. and Cakmak, M. (2005). Nonlinear mechanooptical behavior of uniaxially stretched poly (lactic acid): dynamic phase behavior. *Macromolecules* 38: 2333–2344.

**96** Bonart, R. (1966). Parakristalline Strukturen in Polyäthylenterephthalat (PET). *Kolloid Z.Z. Polym.* 213: 1–11.

# 6

# Nano-Layered Films and Foams

*Troy Su, Abdullah Al Faysal, Jianxiang Zhao, and Patrick C. Lee*

*Multifunctional Composites Manufacturing Laboratory (MCML), Department of Mechanical and Industrial Engineering, University of Toronto, Toronto, Ontario, Canada*

## 6.1 Introduction

Polyesters are a class of polymers containing an ester functional group in each repeat unit of the polymer chain. This class of material comprises some of the most common and recognizable polymers found in many widely used consumer products. Some examples of polyesters include polylactic acid (PLA), polyethylene terephthalate (PET), polycaprolactone (PCL), and poly(methyl methacrylate) (PMMA). These materials are utilized in industries ranging from textiles, packaging, healthcare, automotive, aerospace, power distribution, and consumer electronics. Understanding how to reinforce polyesters can lead to cheaper, stronger, and more environmentally friendly products. The properties of a given component are not only a function of its constituent materials but also depend on the particular microstructure generated during processing. In the field of material science and engineering, the correlation between performance and intrinsic properties of a given processed component can be expressed by material-processing-structure-property relationships. Understanding these relationships can prompt the need to focus on processing techniques in conjunction with novel material or material combinations to solve modern technology problems.

Micro-/nano-layered (MNL) coextrusion is a processing technique used to form multilayered composites using passive elements called layer multipliers (LM) to iteratively divide, stack, and recombine extruded streams of molten polymers [1]. This setup is combined with two or more extruders connected in parallel to a

*Polyester Films: Materials, Processes and Applications*, First Edition.
Edited by Miko Cakmak and Jehuda Greener.

feedblock, each containing a different polymer. Doing so allows the formation of two (or more) material-layered composites. In addition, one (or more) extruders in an MNL coextrusion process can be configured to inject gas/supercritical fluid (SCF) into the melt, creating selectively foamed layers. By placing several LMs in series, the extruded stream can easily contain thousands of layers, with layer thickness decreasing in inverse proportion, down to the nanoscale [2,3]. This type of technology was pioneered by Dow Chemical Company in the 1960s, whose original feed block and LM designs persist to this day in MNL coextrusion processes [1].

Utilizing MNL coextrusion, layered composites can be formed quickly and effectively. Layer multiplication is of interest to many industries because it can offer improved mechanical, barrier, or dielectric properties of polyester composites, while reducing cost, percolation threshold of additives, and processing complexity. In some cases, completely novel properties can be attained in MNL films that cannot be achieved by other composite processing technologies. For example, films with tunable shape-memory properties can be made using a polyester, such as PCL, that previously did not exhibit any shape-memory properties by combining it with one that does, such as poly(butylene succinate), due to the inherently high cocontinuous morphology of MNL films [4,5]. This type of structure ensures that the composite maintains the same strain across distinct phases in the direction of deformation. Furthermore, in the last two decades, with the growing awareness of the environmental impact associated with petroleum-based plastics, MNL has been proposed as a reinforcing mechanism for bio-based polyesters without compromising their green nature [6–8]. This chapter is sub-divided into four major subcategories: the processing, microstructure, and properties of MNL coextrusion processed multilayered films, and multilayered film/foam structures. These sections will provide an overview of the technology covering the corresponding microstructures and the resultant properties. The pertinent applications of each type of structure will be outlined, and several brief examples will be provided for each section. MNL films have a wide variety of applications in packaging, biomedical components, and dielectric elements, whereas MNL foam/film structures have usage mostly in thermal and sound insulation. The wider adoption of this technology presents an exciting and promising step toward creating multifunctional, stronger, lightweight, and low-cost polyester products.

## 6.2 Processing of MNL Films

### 6.2.1 Manufacturing Technologies of Multilayered Films

Due to the many advantages of multilayered films, many processing methods have been proposed for preparing such films. These methods can be divided into two categories: self-assembly and forced assembly. Self-assembly methods, such as for block copolymer, can fabricate multi-layer films with layer thicknesses down to several nanometers [9–11]. However, the scaling-up of self-assembly technology is still a challenge. In contrast, forced assembly methods have been used in industrial manufacturing for a long time. The most common forced assembly methods for polymeric materials include melt processing methods such as multi-layer coextrusion and solution processing methods such as dip coating, spin coating, and spray coating [3, 12]. Compared with solution processing methods, which use solvents that are often not recyclable and might pose environmental challenges, melt processing has the advantage of being more cost-effective and solvent-free.

Typical melt processing forced assembly technologies are blown film MNL coextrusion and cast film MNL coextrusion [13]. Blown film coextrusion is utilized extensively in packaging applications. For example, two- or three-layer films for trash bags or five- to nine-layer structures for flexible and semi-rigid packages. This represents the largest commercial application for coextruded films and sheets to date. The typical type of blown film coextrusion system uses spiral mandrel dies, where layers are formed by separate spiral manifolds present at different radial distances. The melt from each manifold joins each other near the exit of the die to create a layered structure. With an increasing number of layers, this version's die geometry tends to be very large and complex, which increases fabrication cost and material residence time within the die.

Cast film coextrusion is another typical melt-processed forced assembly technology. It combines two or more extruders through a multi-channel layered feedblock where separate streams of polymer materials are combined into parallel layers in the feedblock before exiting the die. Advanced feedblock designs, such as the ones designed by Cloeren Inc., can directly produce more than 1000 layers in a single unit by redistributing the incoming melt streams from extruders into hundreds or thousands of layers. However, such designs are usually complex and expensive. In contrast, the layer multiplication system utilizing a feedblock and a series of LMs first developed by the Dow Chemical Company [1] is relatively simple and cost-efficient. It can generate 2–4096 layers with layer thicknesses as low as 10 nm using only 10 LMs in series with carefully controlled flow rates through each extruder-gear pump set [2, 3]. A typical layer multiplication process with three initial layers is shown in Figure 6.1. First, a conventional feedblock is used to extrude a two to five layered polymer melt. When the polymer melt goes through the LM, the melt flow is divided in half as it passes through a vertically oriented slicer. It is then directed by parallel flow channels and reoriented such that they are stacked on top of one another horizontally. In doing so, the number of layers is doubled. By using a series of LMs, the number of layers can be doubled many times in sequence. For example, consider a processing line with two initial layers containing $n$ LMs in series. The final number of layers, $N$, in an extruded

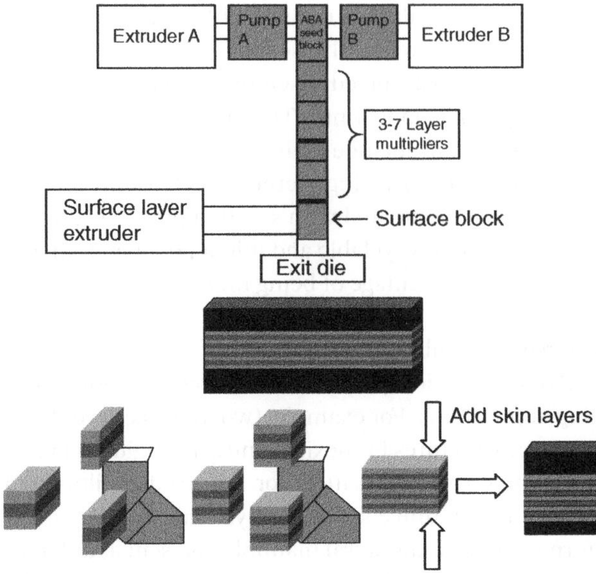


**Figure 6.1** Schematic of a typical layer multiplication process, with three initial layers (A/B/A) and skin layers added at the surface block by surface layer extruder. *Source:* Zhang et al. [14]/ with permission from Elsevier.

stream that has passed through the MNL line is given by Eq. (6.1). Accordingly, a system with three initial layers (A/B/A), a series of $n$ LMs, and a surface layer block and extruder, such as the one shown in Figure 6.1, will have its final number of layers given by Eq. (6.2).

$$N = 2 \cdot 2^n \tag{6.1}$$

$$N = 3 + 2 \cdot 2^n \tag{6.2}$$

### 6.2.2 MNL Structures Produced by Layer Multiplication Technology

Compared with conventional coextrusion technologies, layer multiplication technology can be more flexible and can easily be modified to create interesting layer structures. For example, by changing the LM configuration, MNL films with unique structures can be created. As shown in Figure 6.2, LMs can unevenly divide the flow, allowing for the creation of gradient layer structures that have an unequal weighting of adjacent layer thicknesses [15]. By changing the number of LMs and altering the arrangement of "even" and "uneven" LMs, the final layer number and gradient distribution can be carefully controlled. Likewise, a careful arrangement

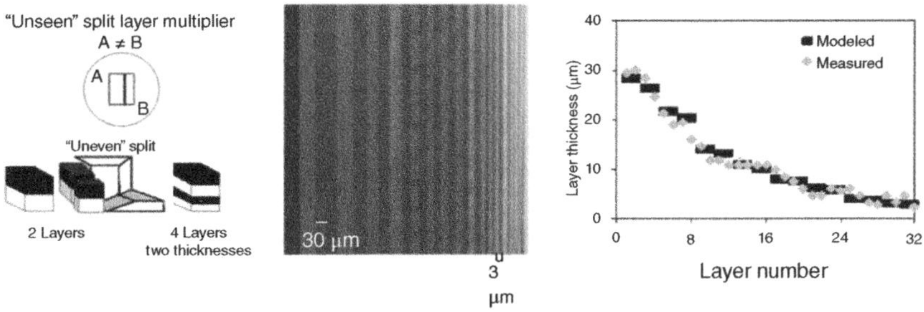


**Figure 6.2** (Left) Schematic illustrating "uneven" split die multiplication producing two-layer thicknesses. (Center) 32-layer PS/PMMA gradient film by coextrusion through four "uneven" split dies to produce a 10× gradient in layer thickness. (Right) Comparison of the measured and the calculated layer thicknesses for a 32-layer PS/PMMA gradient film. *Source:* Ponting et al. [2]/ with permission from John Wiley & Sons.

of LMs can be used to fabricate micro-/nanofibers instead of micro-/nanolayers, as shown in Figure 6.3. Consider a layered polymer stream rotated by 90° in the plane normal to the machine direction. As a result, the layers will be oriented from left to right, rather than from up to down. A skin layer, used to separate individual fibers, can be added around the new melt before it passes through several more LMs. At the end of the line, the extrudate will consist of isolated fibers oriented along the machine direction embedded in an isolating matrix. The fibers can be extracted with additional post-processing to etch away the matrix polymer.

In addition, the thickness ratio of alternating layers can be easily modulated by adjusting the flow rate of the two (or more) streams entering the feedblock. For example, in a two-component multilayered coextrusion system containing two initial layers of polymer A and B, the layer thickness ratio of the A/B composite can be modified to a certain degree by adjusting the flow rate of their respective gear pumps and extruders. Further, the material choice of each layer is flexible, and a multitude of tools can be employed to increase layer compatibility. For example, if the bonding between layers A and B needs enhancement, a tie layer, T, can be added by way of an A/T/B/T/A initial structure. This can be achieved by adding one more extruder-gear pump set with an appropriate feedblock.

### 6.2.3 Flow Instabilities in Layer Multiplication Technology

Though layer multiplication technology has clear benefits over other forms of composite forming techniques, there are several limitations that deserve mention. The sequential nature of layer multiplication technology results in some contact times between the polymer and dies, and between layers. Further, high level of

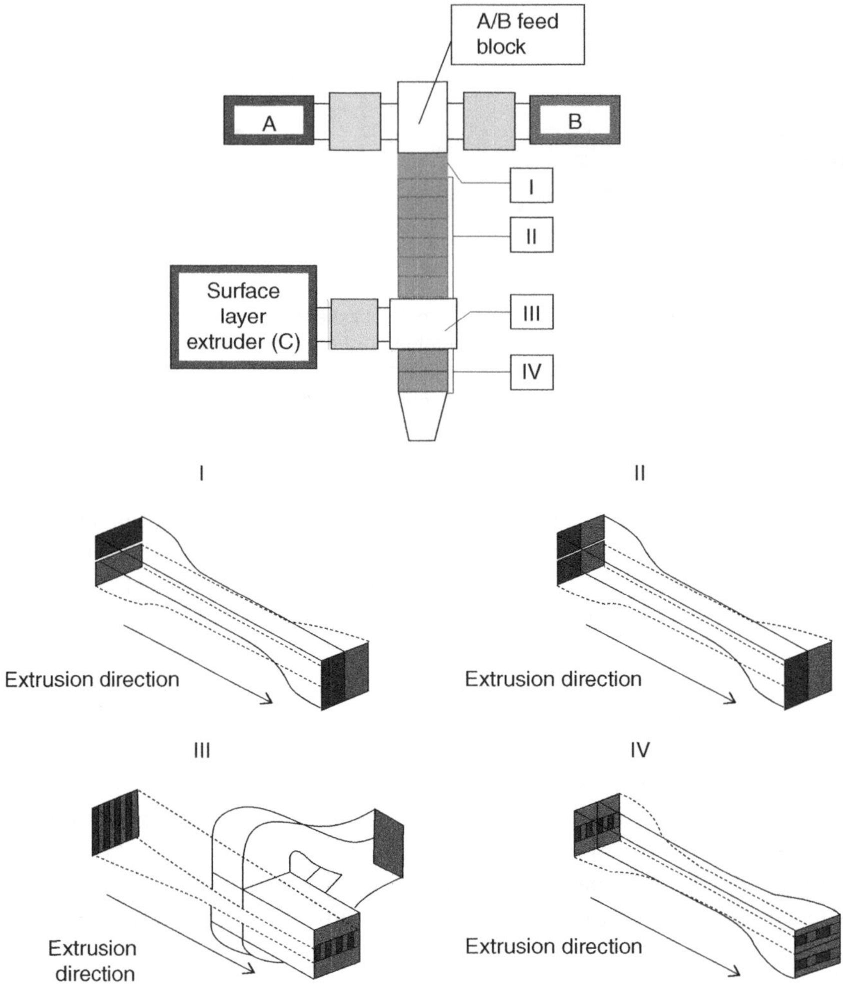


**Figure 6.3** Diagrammatic example of an LM setup to produce fibers embedded in a matrix. *Source:* Wang et al. [16]/ with permission from Elsevier.

applied shear force in the LMs can make the flow susceptible to instabilities and nonuniformities. This typically results in layer thickness nonuniformity, layer thickness variation, irregular polymer intermixing at interfaces, and film thickness variation [13]. This issue is one that researchers and industry experts have investigated for decades and currently poses one of MNL coextrusion technology's greatest challenges.

There are typically four kinds of flow instability that cause layer nonuniformities during MNL coextrusion. First, layer nonuniformity induced by interfacial instability. Interfacial instabilities are a transient phenomenon in which the interface location between two adjacent layers varies locally. This interface distortive behavior causes layer thickness nonuniformities in multilayered structures. Two types of interfacial instabilities are common in coextrusion: zig-zag instabilities and wave instabilities [17–19]. Zig-zag instabilities display extremely chaotic and intermixed interfaces where the onset of instability is often correlated with a critical interfacial shear stress. In contrast, wave instabilities have larger amplitudes and more regular patterns than zig-zag instabilities, as illustrated in Figure 6.4. Wave instabilities are assumed to be related to the extensional rheological properties of the polymers in coextruded structures. Although practical remedies from the accumulated experience exist to reduce or eliminate these instabilities, no complete understanding nor predictive theory is available for these complex flow phenomena.

Second, layer nonuniformity can arise due to viscoelastic secondary flow. The effect of polymer viscoelasticity on layer thickness uniformity has been studied by

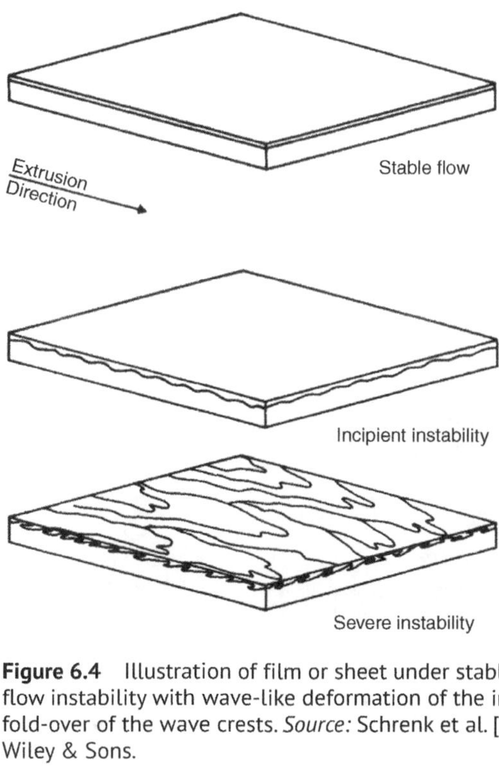


**Figure 6.4** Illustration of film or sheet under stable flow conditions, incipient interfacial flow instability with wave-like deformation of the interface, and severe instability with fold-over of the wave crests. *Source:* Schrenk et al. [20]/ with permission from John Wiley & Sons.

various researchers [21–23]. It has been demonstrated both experimentally and numerically that polymers that are comparatively high in elasticity produce secondary flows normal to the primary flow direction, which can distort the layer interface. The imbalance of normal stresses is known to cause elastic layer rearrangement behavior as shown in Figure 6.5. The effect is typically not observed in radially symmetric channels such as tubular dies.

Third, layer nonuniformity can also arise due to encapsulation. Encapsulation was first observed and studied in capillary flow of two polymers in side-by-side bicomponent fiber processing [25–27]. The viscosity ratio between the two components is a dominant factor controlling the interface conformation in coextrusion; the lower viscosity polymer encapsulates the higher viscosity polymer, while elasticity ratio has relatively little effect on the interface distortion [28]. An attempt to explain this phenomenon is based on the principle of minimum viscous dissipation where a less viscous material moves to the area of highest stress (near the wall) to minimize the energy to flow [29]. It has been observed that the higher the viscosity difference between two polymers, the faster viscous encapsulation occurs. Furthermore, the viscous encapsulation velocity is not constant along the flow channel; it is highest near the entry of the channel and slows down along the channel.

Last, layer nonuniformity can be caused by interfacial slip. Significant slip can occur in the flow of a blend of two immiscible polymers due to reduced entanglements at their interface. The slip is of practical importance because of its effect on morphology and adhesion in, for example, disordered two-phase blends or multilayered films. Interfacial slip can be quantified using two polymer pairs each with closely matched viscosity and elasticity but different

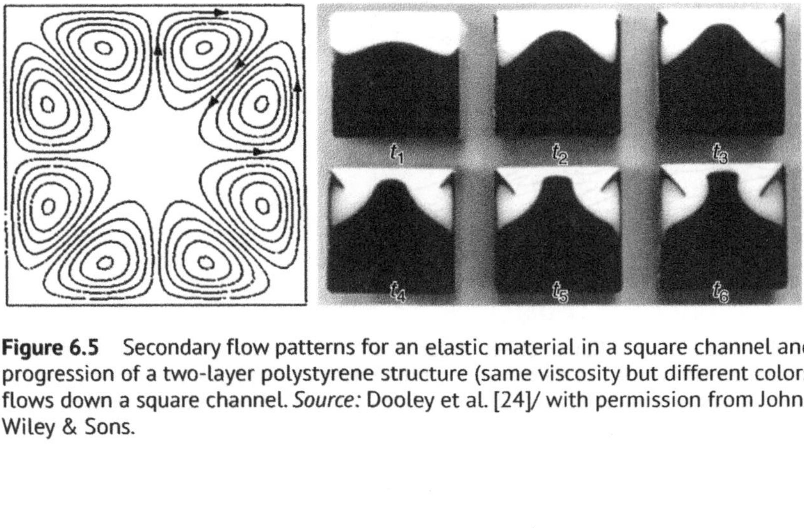


**Figure 6.5** Secondary flow patterns for an elastic material in a square channel and the progression of a two-layer polystyrene structure (same viscosity but different colors) as it flows down a square channel. *Source:* Dooley et al. [24]/ with permission from John Wiley & Sons.

miscibility [30, 31]. The nominal viscosity, as well as normal shear stresses of the multi-layer samples decrease with the number of layers (i.e. more interfacial areas). From the viscosity reduction, an apparent interfacial slip velocity can be calculated. The detectable slip, about 0.001–100 mm $s^{-1}$, occurs at the shear stress ranges of the performed experiments. However, the slip velocity did not show a sudden increase at a critical shear stress, as predicted by Brochard-Wyart and de Gennes, but rather increases the velocity of slip with different power law indices depending on the polymer pairs. It was shown that encapsulation can be affected by the degree of interdiffusion and/or chemical reactions that take place at the interface [32].

To improve the stability of layer multiplication technology and expand the range of materials that can be coextruded, researchers have been developing newer generations of LMs [33–35]. Dubbed the "Second Generation Layer Multiplier Die," a proposed newer LM structure utilizes a constant cross-sectional area and balanced symmetrical flow paths to reduce flow instabilities [36]. It can better handle mismatched viscosities and high-elasticity materials.

However, due to the non-Newtonian melt flow property differences between adjacent polymer layers that cause flow instability, the thickness of layers fabricated through layer multiplication technology can not decrease indefinitely despite improved system design. For example, PMMA/polystyrene (PS) and PCL/PS systems layered using currently available technology, layer breakdown occurs when the nominal thickness decreases to around 30 to 40 nm. There is a critical layer thickness around 10 nm, below which the layers break up despite optimizing processing parameters due to material characteristics such as the small interfacial perturbations that are amplified by Van der Waals disjoining forces [37, 38].

## 6.3 Microstructure of MNL Films

### 6.3.1 Interphase Formation

Typically, a composite is formed between two or more materials out of a desire to leverage properties or characteristics of each constituent. The Rule of Mixtures postulates that the degree to which the desired properties manifest is a function of the fractional contribution its parent makes to the composite. The use of MNL coextrusion technology presents a new way to tune composite properties without altering the blend composition, and only by decreasing layer thickness between the alternating materials. In contrast, traditional blend processing techniques rely on altering the blend composition to achieve different resulting properties.

Liu et al. were able to measure the glass transition temperature ($T_g$) of an MNL coextruded structure of PMMA and polycarbonate (PC) at equal volume ratios [39]. At relatively large layer thicknesses, two distinct $T_g$ were observed.

These $T_g$s correspond to those of PC and PMMA, 144 °C and 112 °C, respectively. As the layers got thinner, the two $T_g$s gradually shifted to converge with one another, ultimately creating only one observable $T_g$ for the entire multilayered composite at 122 °C [39]. It is also interesting to note that the convergent $T_g$ was asymmetrically located between the two constituents $T_g$s, approximately 22 °C below that of PC but only 10 °C above that of PMMA. This complements traditional understandings of composite property tuning because the concentration of PC and PMMA remained constant. Leveraging the formation of properties that are tunable by altering the microstructure of the composite alone opens new doors to creating novel engineering materials for targeted solutions. To understand this mechanism, the micro-morphologies within the layers and/or at the interfaces must be scrutinized as they have significant effect on the overall composite properties.

When processed through MNL coextrusion, the boundary between two materials may not always be as clear and crisp as diagrams may imply. At the melt interface, the polymer chains of both materials have an opportunity to intermingle and blend, creating what is known as an "interphase" material. This is shown diagrammatically in Figure 6.6 [40]. The thickness of this blend ranges from 5 to 10 nm but can contribute significantly to the bulk properties of the composite film when their concentration becomes significant. Interphase material concentration is directly proportional to the number of layer interfaces as each layer interface presents a new opportunity for polymer chains to intermingle. As such, interphase concentration increases as the number of layers increase. Attempts to study the interphase have proven to be difficult due to the thinness of this "layer," the low concentration of interphase, and the shielding afforded by the original two materials. To circumvent these obstacles, thin films can be created with thousands of layers, making each layer less than 20 nm thick and entering the range of interphase layer thicknesses. It was shown that as layer thicknesses entered this regime, the interphase fraction increased dramatically due to the creation of more boundaries, but also the reduction of bulk material, which was absorbed to form the interphase [41]. Once composites were created with sufficiently high interphase fractions, they were studied using traditional characterization methods, such as mechanical, thermal, and microscopy analysis tests, to understand the properties of the interphase material. It is worth noting at this stage that the interphase material was not expected to possess the same properties as either of its constituent materials.

Helfand et al. set forth to theoretically determine the thickness of the interphase at a two-material boundary. They postulated that the interphase thickness is a function of the interaction parameter ($\chi$), and the statistical segment step length ($b$), which is a theoretical measure of the number of degrees of freedom along a polymer chain when considering rotation-inhibiting structures such as functional

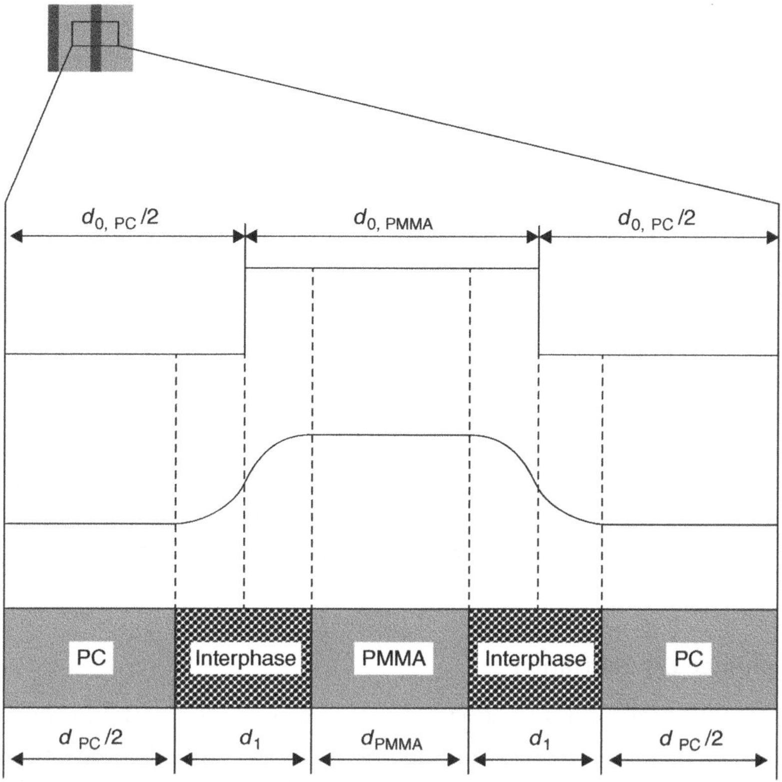


**Figure 6.6** Diagrammatic example of the formation of an interphase layer in a multilayered PC/PMMA MNL coextruded composite. *Source:* Liu et al. [40]/ with permission of American Chemical Society.

groups. This equation can be seen in Eq. (6.3), with $\chi$ defined in Eq. (6.4) as a function of the materials' solubility parameters (subscripts 1 and 2 denote the two materials used in the layered composite). Though the equation is simple, it is remarkably accurate and able to estimate interphase thicknesses to within 10% of the experimentally reported values for a PC/PMMA sample [42–44]. Note that $\chi$ is a property specific to each material combination and provides insight into the interaction between the constituent molecules at an interface [45].

$$d_1 = \frac{2b}{\left(6\chi_{1-2}\right)^{0.5}} \tag{6.3}$$

$$\chi_{1-2} = \frac{v}{\mathrm{RT}}\left(\delta_1 - \delta_2\right) \tag{6.4}$$

Returning to the idea of property blending in composites, a more rigorous understanding can be established, which explains the existence of a converging composite property as layer thicknesses decrease. The resultant singular blended property is that of the interphase material, which can be treated as a distinct material from the two-component materials of the original composite. As layer thicknesses fall below 10 nm (the typical thickness of the interphase layer), the relative fractional composition of interphase in the structure increases dramatically, causing its properties to become dominant over that of the two original materials. In a way, a third material has been introduced to the composite with properties that fall between those of the materials originally used in the composite [39]. From this example, it is clear to see that the glass transition temperature of the interphase is not necessarily the average of those of the original materials, even when they are used in equal parts. Instead, the resultant value depends upon on polymer chain miscibility, molecular weight, factors, by proxy through the prevalence of the interphase material [46]. Understanding the formation of interphase material and how MNL coextrusion can be used to create a composite whose properties are distinct to either of its constituent parts introduces countless opportunities for synergistic combinations of materials for targeted solutions.

### 6.3.2 Confined Crystallization

When allowed to grow with no restriction, some polymers will exhibit isotropic spherulitic growth of lamellae. These randomly oriented lamellae structures can be observed in bulk melts, where the lamellae are considered crystalline volume due to the tight packing and alignment of polymer chains. Within MNL coextruded composites, obstruction of crystal growth by layer interfaces results in the one-dimensional confinement of lamellae growth if layers thicknesses are on the order of the thickness of a single lamella. The obstruction of crystal growth can be caused by a variety of factors due to the molecular and bulk properties of the adjacent layers. For example, polymer chain mismatch may prevent neat crystals from forming near the interfaces, or the two materials may have significantly different crystallization/glass transition temperatures such that they are not simultaneously crystallizing/vitrifying at the same time during cooling, among other causes. This can manifest as anisotropic growth behavior if the layer thicknesses are low enough to effectively constrain crystal growth [47]. When confined to planar growth, lamellae orientation falls between two mutually perpendicular extremes: edge-on and in-plane. Which of the two orientations a polymer may exhibit depends on a variety of factors including its crystallinity, melting temperature, the respective interaction parameters, and bulk crystal morphology [46,47]. Polymers that exhibit edge-on lamellar growth when subject to confined crystallization include polyethylene (PE) [48–51], isotactic polypropylene (PP) [52], polyamide-6 [53], poly(ethylene naphthalate) [54, 55], and PET [56]. In contrast, poly(ethylene oxide) (PEO) [57, 58], isotactic PS [59], syndiotactic PP [60], PCL [61], and PLA [62] all have been reported to exhibit in-plane lamellar growth. It is clear to see that polyesters tend to exhibit in-plane growth of lamellae under confined crystallization. The use of MNL coextrusion as a facilitator for lamellae growth manipulation can dramatically alter the final composite's bulk properties. This will be pertinent in the manufacturing of gas/moisture barrier films.

As layer thicknesses decrease, the in-plane and edge-on orientations are favored over isotropic growth, creating plate-like structures that extend between layer interfaces. In the case of in-plane orientation, lamellae growth is parallel to the layer interfaces, while edge-on orientation refers to perpendicular growth of lamellae with respect to the layer interfaces. This threshold layer thickness which triggers this transition is typically around the same order of magnitude as the thickness of lamellae, which range from 10 to 20 nm [63]. Even with layer thicknesses of just 100 nm, strong orientation of lamellae can be observed. As layer thicknesses approach small integer multiples of the lamellae thickness, the degree of lamellae orientation increases [4]. This phenomenon is known as confined crystallization and can be seen in Figure 6.7. How the lamellar growth of a crystalline polymer in confined systems proceeds is determined by a variety of factors including, but not limited to, whether the confining polymer is crystalline or amorphous, the thermal properties of each polymer, the interaction properties between the confining and confined polymers (as measured by the interaction parameter), the presence of additives and stabilizers, and recrystallization or quenching conditions [46, 47].

The tunability of lamellae orientation by modulating the number of layers in an extruded stream of constant thickness is a powerful tool that can be used to directly influence the mechanical, barrier, and many other properties of the

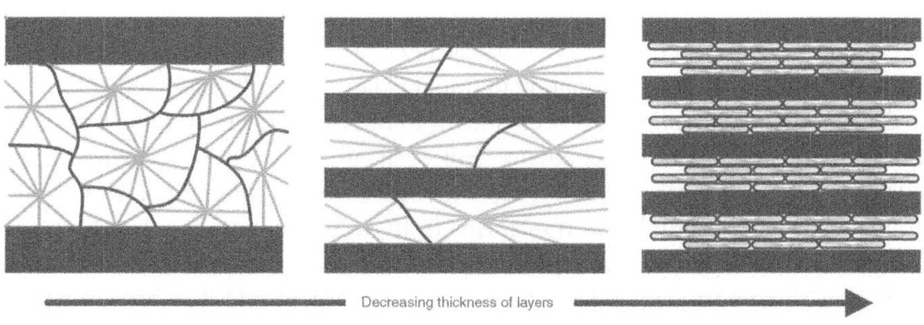


**Figure 6.7** Diagrammatic representation of confined crystallization where in-plane lamellae orientation is preferred as layer thickness decreases. Note that some polymers will not exhibit in-plane lamellae orientation, instead exhibiting random or edge-on orientation as layer thicknesses decrease.

resultant film. Crystalline films are most commonly used for their enhanced gas barrier properties [4,5,47,63,64]; however, these films can also be exploited for their enhanced stiffness when compared to their amorphous counterparts.

Wang et al. demonstrated that PEO could be one-dimensionally confined using ethylene acrylic acid (EAA). They created films using an MNL coextrusion setup capable of producing films with 33, 257, and 1025 layers alternating between PEO and EAA. Furthermore, they varied the EAA/PEO volume ratio between 50/50, 70/30, 80/20, and 90/10 to observe the effect of confined crystallization in PEO with thinner layers than what would be possible with their experimental set up had the volume ratios been held constant. Using this procedure, the team was able to create PEO layers as thin as 8 nm. As PEO layer thicknesses decreased below 110 nm, they observed that crystallization occurred as stacks of 3–5 lamellae oriented in-plane. Further decreasing the PEO layer thickness to 20nm caused in-plane lamellar growth of PEO crystals across the entire layer thickness (i.e., stacked 1 lamella high) [47,63,64]. The group found these results to correlate with improved barrier properties, which will be discussed in Section 6.4.2.

As mentioned earlier, one of the determining factors for lamellar morphology of a crystalline polymer when confined is the interaction parameter between the two polymers. To study the effect of $\chi$ on the resultant lamellar morphology, Ponting et al. observed the crystal orientation of PCL as a function of layer thickness for three different systems: PCL/PS, PCL/PMMA, and PCL/PC [65] [66]. The pertinent interaction parameters are as follows: $\chi = 0.168$ (immiscible), $\chi_{PCL/PMMA} = 0.017$ (intermediate), and $\chi_{PCL/PC} = 0$ (miscible) [47,65]. In the cases of PCL/PS and PCL/PMMA, the PCL crystals exhibited spherulitic morphology and tended toward in-plane orientation with decreasing layer thickness. Though these trends are identical, it was noted that PMMA/PCL exhibited less orientation than PS/PCL at the same layer thickness, with their degrees of lamellar orientation converging at low layer thicknesses (below 100nm). Interestingly, PCL/PC exhibited no lamellar orientation as a function of layer thickness. The researchers concluded from these results that the degree of chemical interaction, intrinsically related to the interaction parameter, can slow or even block the formation of oriented lamellae in PCL [47,65].

Confined crystallization has been shown to not only affect crystal morphology but also the nucleation mechanism. Sun et al. utilized a poly(4-vinylpyridine) (P4VP)/PCL multilayered structure to show that when confined below a critical thickness, PCL crystallization transforms from occurring homogeneously to heterogeneously at layer interfaces [67]. The critical thickness for this transformation fell below the reported lamellae thickness for PCL of 15 nm [47], at just 6 nm [67]. The authors explained that typically, PCL crystals favor thick crystalline stems with minimal chain folding and hypothesized that the one-dimensional confinement below the threshold thickness made this type of structure less energetically favorable such that interfacial heterogeneous nucleation became the dominant nucleation mode. Understanding the mechanics of heterogeneous interfacial nucleation in multilayered composites may help better control resultant crystallization behaviors of multilayered films.

### 6.3.3 In-Situ Fibrillation of a Secondary Phase

The LMs used in MNL co-extrusion processing impart high shear and elongational forces due to their converging nature and long contact times with the extrudate. These forces will tend to stretch the composite along the machine direction. If a spherical secondary phase is present in the blend at the multiplication stage, these LMs can be used to plastically deform this phase along with the matrix, forming in-situ fibers. However, it should be noted that the material compatibility and properties such as viscosity and elasticity ratios are important factors in ensuring the formation of a fibrillar secondary phase. As the number of LMs in series increases, this phenomenon becomes more and more prominent. Wang et al. demonstrated that a PC/PE microfibrillar composite could be made using just four LMs. The formation of these fibers using four LMs was also shown to improve tensile strength by 20% in the machine direction compared to the blends processed without any LMs at the same composition [68]. Though this technique presents a unique and novel way to create fibrillar composites, one of its major drawbacks is that only microfibers have been reported to have been made using this technique. Fiber diameters of 1.11μm were reported by Shen et al., who used a polyamide 6 (PA6)/isotactic PP blend [69].

### 6.3.4 Orientation of Fillers and Fibers

When subjected to the shear and elongational forces in an MNL coextrusion system, a melt containing non-deformable anisotropic fillers will tend to orient them in the direction of flow. Though this phenomenon affects all nonspherical fillers, the effect becomes more obvious as aspect ratio increases due to the increase in magnitude of the applied torque. Due to the amplification effect of aspect ratio, this is a particularly useful phenomenon to exploit for fibers or other rigid oblong fillers, which have inherently large aspect ratios [70].

Several studies have examined the ability of MNL co-extrusion systems in tuning the orientation of fibers in a melt, demonstrating that the mechanical benefits imparted by the presence of fibers can be tuned in accordance with their average angle of orientation compared to the machine direction. For example, Embabi et al. demonstrated that a PP/fibrillar PET system subjected to MNL extrusion saw a decrease in the average angle of orientation, representing closer alignment between the fibers to the machine direction. By coupling the use of multipliers to impart shear and elongation, and varying the flow rate through the multipliers by increasing the extruder and gear pump RPMs, the group was able to see a reduction of average orientation angle from 41° to 23.9° with respect to the flow direction when going from zero to six multipliers [71]. In this study, 45° represents the random orientation of fibers. Embabi et al. were further able to show that the tunability of the fiber orientation ultimately causes the tensile properties of the composite to be tunable as well. They concluded that 45 and 46% improvements were observable in the elastic modulus and yield strength, respectively, under the most optimal processing conditions [71], as seen in Figure 6.8.

### 6.3.5 Selective Spatial Confinement of Phases and Additives

Additives can be used to mechanically reinforce polymers or enhance their functional properties, such as electrical conductivity, thermal conductivity, and ultraviolet shielding, among others. Generally, these additives are more expensive than the polymer matrix, such that it is desirable to minimize the amount of additives as much as possible. Further, with less additives in the melt, the overall viscosity decreases, changing the processability of the composite. Owing to its ability to selectively place or orient additives within the layered structure by confining it to a single phase, MNL coextrusion technology has been explored as a method of decreasing the level of additives in a matrix needed to achieve desired bulk properties. Furthermore, MNL coextrusion can be used to separate and selectively confine whole phases relative to others. This is particularly useful if the desired product is a single-phase fiber, for example.

Certain properties, such as electrical conductivity, require a "network" through which electrons can travel, to be formed from one end of the material to the other. This pathway need not be physical; however, filler material must be dispersed such that electrons may "hop" from particle to particle. As such, to statistically ensure that a complete conductive pathway is formed by a randomly scattered filler, a high concentration of filler must be added. MNL extrusion can be used to limit the amount of filler necessary to create a conductive pathway by spatially selectively locating the material containing the conductive filler. To do this effectively, LMs can be used to spatially confine the phase containing the conductive filler, and those phases that do not. LMs come in a variety of orientations and types and can be combined in series to form complex structures. One such configuration that is useful in creating conductive pathways utilizes a series of vertical and horizontal multipliers to create a checkerboard-like alternating pattern of the two phases. One of these phases can contain a conductive filler while the other does not, allowing for the significant reduction of additive content.

The degree to which an additive's concentration is reduced while still exhibiting a particular property can be characterized in terms of the percolation threshold, which represents the minimum concentration of an additive needed to manifest a property. This concept originates in mathematics to describe long-range

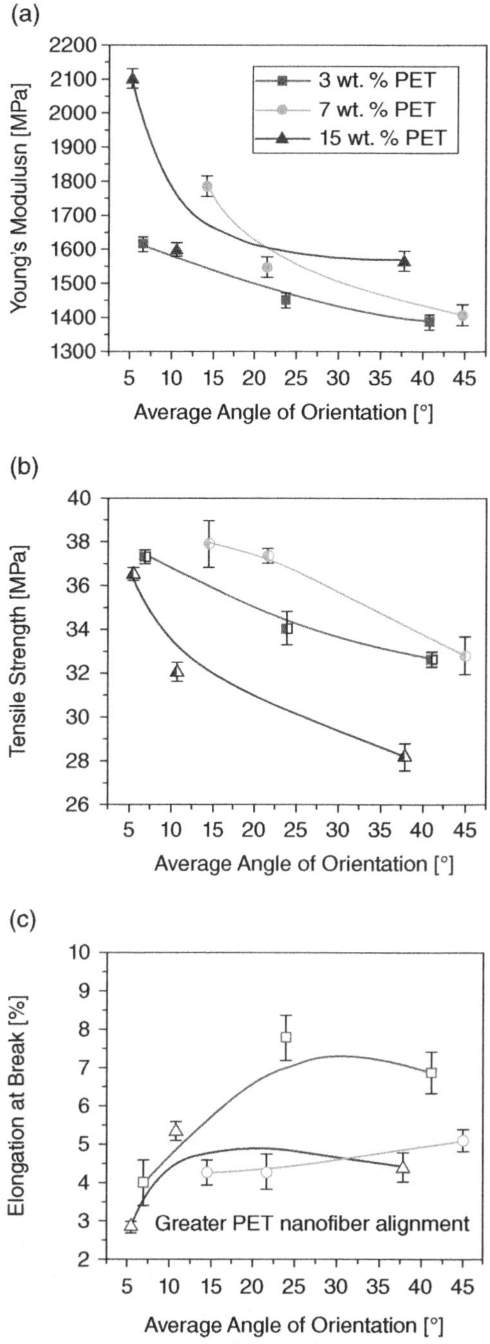


**Figure 6.8** Data showing tunable tensile properties by varying average angle of orientation of fibrillar PET in a PP matrix processed through MNL extrusion. These graphs show variations of (a) elastic modulus, (b) tensile strength, and (c) elongation at break with respect to average angle of orientation. *Source:* Embabi et al. [71]/MDPI/CC BY 4.0.

connectivity in statistically random systems. When processed through MNL, the selective spatial confinement of additives reduces the statistical randomness of the system, allowing for long-range connectivity to form more frequently. This has the effect of reducing percolation threshold and minimizing the amount of additive needed to achieve conductivity.

The idea of selective spatial confinement can be extended from just two dimensions to three. Injection molding processes have been utilized in conjunction with splitting serpentine channel molds to create small components that possess complex, 3D multilayered structures. These types of molds were originally developed to promote high efficiency (low Reynolds number) mixing by repeatedly stretching, cutting, folding, and stacking the stream while modulating the channel width [72]. Neerincx et al. have exploited and adapted their preliminary splitting serpentine channel design to create extremely fine layered structures [73]. Further, the folding and stacking nature of these molds can allow for the formation of hierarchical fractal structures within the melt. As the name suggests, fractal structuring refers to the formation of structures that repeat across a range of length scales. These complex patterns give rise to interfacial areas between phases that are exponentially higher than that of a two-dimensional multilayered structure made using LMs [73]. Further, this work represents some of the first single-step processes capable of producing 3D structures in the melt. Though this technology is still in its early stage, the potential to create advanced fractal structured polymeric and hybrid composites with high interfacial surface area could be hugely beneficial for applications that require high energy density abilities, shape memory abilities, and mechanical enhancements.

## 6.4 Mechanical and Functional Properties of MNL Films

### 6.4.1 Mechanical Properties

As concerns about environmental issues caused by petroleum-derived plastics grow, polyester biopolymers such as PLA and poly(propylene carbonate) (PPC) that are produced from renewable resources, and other biodegradable polyesters such as PCL, are drawing increasingly more attention. However, polyester biopolymers have not replaced petroleum-derived plastics yet partially because their mechanical properties are often not as good as those of traditional petroleum-derived plastics such as PP, PE, and PS. For example, neat PLA and PPC materials suffer from brittleness, and their applications are thus limited [74].

When discussing PLA, it is important to note that there exist two varieties: poly(l-lactide) (PLLA) and poly(d-lactide) (PDLA). Ikada et al. found that an equimolar blend of PLLA and PDLA would create a stereocomplex crystal, which has dense and strong intermolecular hydrogen bonding and greatly benefits the overall mechanical properties of PLA [75]. Unfortunately, a relatively large amount of PDLA is needed to achieve such improvement if using traditional melt blending, and the production of PDLA is more costly than PLLA. To address this problem and improve PLA's scratch resistance, Yi et al. [76] studied a method to enhance the mechanical performance of PLA material by inducing the coexistence of transcrystallinity and stereocomplex crystals via the multilayered assembly of PLLA and PDLA. By using a layer multiplying coextrusion process, Yi et al. fabricated a multi-layer architecture consisting of alternating PLLA and PLLA+PDLA blend layers. They observed that with the multiplication of layers, the spherulites tended to disappear and the whole specimen exhibited a transition from isotropic to highly oriented crystalline morphology. When the number of layers reached 1024, the heterogeneous nucleation of stereocomplex crystal caused the oriented lamellae to grow from the layer interfaces. They tended to occupy the PLLA layers, creating a specific transcrystallinity (PLLA layers)/stereocomplex crystal (PLLA+PDLA blend layers) alternating multilayered structure. As a result of this unique microstructure, the toughening effect of stereocomplex crystals was observed in the PLLA matrix with only 2.5 wt % of PDLA. This would previously only manifest when over 10 wt % PDLA was incorporated using traditional melt blend processing methods. This significant reduction in percolation threshold can be attributed to the multiple interfaces formed between PLLA+PDLA blend layers and PLLA layers, and the benefits to crack deflection and termination leading to high ductility. In addition, the deflection of transcrystalline crystals can continuously dissipate energy during the uniaxial stretching process, leading to highly improved toughness. In their work, tensile strength and elongation at break of the 1024 layered sample were increased by 17 and 1528%, respectively in comparison to those of the conventional blend specimen containing the same content of PDLA (~2.5 wt %), providing an effective and green strategy for fabricating high-performance PLA materials.

Comparable to the above, Guo et al. [77] adopted long-chain branched PLA to achieve high-ductility PLA sheets without compromising its strength and sustainable nature. By compounding linear PLA with branched PLA and processing the blend through a series of LMs, the growth of a compact and uniform network-like precursor from the scaffold-network nuclei was promoted through the pre-orientation and homogeneity of the PLA melt. From this, a compact and uniform stress conduction network was constructed inside the PLA based on the scaffold-network nuclei from an entanglement network of long-chain branched PLA. This compact and uniform stress conduction network dramatically increased the elongation at break from 8 to 229% and simultaneously enhanced the yield strength by 5% compared with neat PLA, showing a combination of strength and ductility. Evidenced by the deformation of the network-like precursor, the toughening mechanism was proposed to be the stress transfer of the network-like precursor as a stress-conducting network. The micro-fibrillation and micro-sized voids formed during deformation also contributed to energy dissipation during tensile tests. This work also utilizes the MNL coextrusion-based method to enhance the mechanical properties of PLA materials for potential industrial manufacturing.

Boufarguine et al. [78] blended PLA with 10% Poly (3-hydroxybutyrate-co-3-hydro-xyvalerate) (PHBV) by way of multi-layer coextrusion, creating alternating layers of PLA and PHBV, which exhibited improved ductility compared with neat PLA and traditional melt blending methods. Furthermore, Boufarguine et al. observed that for a critical number of theoretical layers, the PHBV layers would break into thin and long lamellae. As such, a well-developed lamellar morphology with thin and long lamellae of highly crystalline PHBV in the PLA matrix can be obtained. Films with improved mechanical properties can be obtained by balancing the aspect ratio and crystallinity of these lamellae and by controlling their dispersion within the PLA matrix. The sample's elongation at break increased threefold without compromising the Young's modulus compared to neat PLA. Adding PHBV, a brittle polymer, into the PLA matrix made the multilayered coextruded samples more ductile than neat PLA, which is rather counterintuitive. The authors attribute this result to the good dispersion of PHBV lamellae within the PLA matrix formed by multi-layer coextrusion. When PHBV issues dispersed in long thin lamellae, the increased interfacial surface area led to a reduction of the glass transition temperature of PLA. This reduction is more pronounced at the interface between the PLA matrix and the PHBV lamellae, leading to the formation of softened PLA regions that can deform and absorb energy.

Though many researchers are focusing their attention on biopolymers, traditional polymers will not be easily replaced. Therefore, lots of studies on enhancing the mechanical properties via LM technology are still based on traditional polyesters such as PET and polyvinyl acetate (PVA). For example, Embabi et al. [71] used MNL coextrusion to reprocess PET nanofibers prepared by a spun-bond system in a PP matrix. Due to shear and elongational flow fields in the LM channels, the degree of orientation of the PET nanofibers along the machine direction increased with the increasing number of LM and flow rate, contributing to a 45 and 46% enhancement in elastic modulus and yield strength, respectively. This study provides a new technique to enhance the tensile properties of polymer composites by changing the degree of orientation. Jordan et al. [79] further studied the structural evolution during mechanical deformation in poly(vinylidene fluoride-co-tetrafluoroethylene) (PVDF-TFE)/PET multi-layer films. PVDF-TFE is confined between alternating layers of PET through layer multiplication technology. Postprocessing techniques, including biaxial orientation and melt recrystallization, were used to tune the crystal orientation of the PVDF-TFE layers and achieve crystallinity in the PET layers through strain-induced crystallization and thermal annealing during the melt recrystallization step. They observed an enhancement in the modulus from 730 MPa in the as-extruded state (isotropic crystal) to 840 MPa in the biaxially oriented state (edge-on crystal) to 2230 MPa in the melt-recrystallized state (in-plane crystal). The authors further correlated the mechanical response (e.g. strain hardening) of the films to evolved crystal structures and highlighted the ability to tailor the mechanical response.

The use of MNL coextrusion to create a layered composite presents an alternative failure mode in layer delamination, where some methods of composite processing may not exhibit this type of behavior. For example, blend processing techniques may not produce distinct phases which can delaminate from one-another. This is significant because the composite's overall toughness can be improved by leveraging the relationship between the dominant mode of failure and the layer thickness [80–82]. Furthermore, in MNL composites, as layer thicknesses decrease below a critical value, the failure mode shifts from two-phase behavior to single-phase behavior [80]. Two-phase behavior is characterized by the localization of shear bands in an individual layer or several individual layers, noting in particular that shear bands do not pass between layer interfaces. In contrast, single-phase behavior is characterized by the passage of shear bands originating in one layer into an adjacent layer. In essence, the layered composite is thought to behave as a single bulk material as its layer thicknesses decrease past the critical threshold layer thickness. This can significantly improve the mechanical properties of certain polymer blends, which may contain materials whose toughness are drastically different from each other [83]. For example, a PC/ styrene-acrylonitrile (SAN) blend exhibits a transition between single and two-phase behavior as PC and SAN layer thicknesses decreased beyond the critical thickness for this system of 1.7μm [80]. In such a multilayered polymer blend, whose layer thickness is larger than the critical layer thickness, thus exhibiting single-phase failure behavior, the composite toughness will simply be that of the less tough phase. By transitioning towards single-phase failure behavior through the reduction of layer thickness below the critical layer thickness, the composite toughness increases. The reason for the increase in composite toughness during the transition is discussed below. Layer multiplication technology offers an alternative way to enhance the mechanical strength of a polymer without the need to compromise on cost, integrity, or biodegradability [84–88].

Ebeling et al. were able to demonstrate that the failure mode of a PC and SAN layered composite was dependent on the relative thicknesses of the PC and SAN layers using a series of T-peel tests. The researchers expected that decreasing layer thicknesses have an effect on the peel strength due to the restriction of the plastic zone that forms at the crack tip. Furthermore, the significance of polymer miscibility should not be understated as the interaction between the polymer chains of neighboring layers is what facilitates or prevents the transfer of stress between layers [80, 89]. Four different modes of failure were observed: single-layer SAN crazing, single-layer delamination, multi-layer crazing, and multi-layer delamination as shown in Figure 6.9 [80].

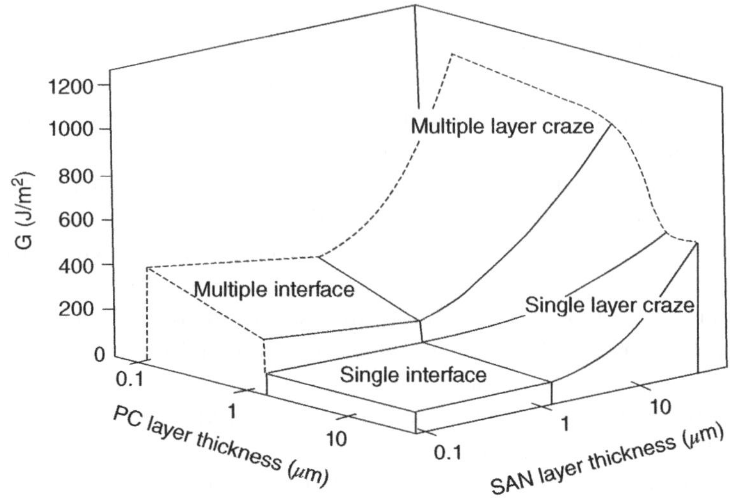


**Figure 6.9** Data showing the peel toughness of a PC/SAN MNL coextruded multilayered composite with respect to the PC and SAN layer thicknesses with corresponding failure mode. *Source:* Ebeling et al. [80]/ with permission from John Wiley & Sons.

Whether the composite exhibited two-phase or single-phase failure (irrespective of failure by delamination or crazing) was determined by the PC thickness. As PC layers decreased in thickness, the stresses were increasingly transferred to adjacent layer interfaces, causing the formation of secondary delamination cracks. The threshold PC layer thickness was found to be 1.7 μm, below which the stresses would induce multiple layer failure [80]. Conversely, whether the composite failed by delamination at PC/SAN layer interfaces or by crazing (irrespective of single or multiple layer failure) was found to be determined by the SAN layer thickness as the confinement by neighboring PC layers may have restricted the formation of the plastic zone at the crack tip. As a result, SAN layers thinner than 1.5 μm caused the dominant failure mode to transition from crazing to layer delamination [80].

In addition to modulating the failure mode, the use of MNL structures can reinforce the composite toughness through the introduction of micro-shear bands (MSBs) confined to a single layer, which serve to improve the overall toughness. Shear bands are a form of plastic energy dissipation that occurs when the crazing strength exceeds that of the yield strength in a material. This is particularly common in polymeric materials with high levels of molecular entanglement. In this scenario, as stress is gradually applied to the material, instead of undergoing crazing, the polymer chains begin to slip past one another and align parallel to the direction of maximum shear. The formation of these shear bands is a natural response to the applied stress whose purpose is to dissipate energy. In the case of an MNL coextruded composite, the crazing that occurs in one layer can subsequently induce the formation of MSBs in another layer as the composite naturally tries to dissipate energy. This idea can be further extended into the aforementioned regime where crazing occurs in multiple layers to conclude that as more crazing occurs, more localized MSBs will form, creating an overall tougher composite as the number of layers increases even if its material composition remains unchanged [90]. In instances of high applied stress, prior to component failure, coalescence of MSBs into shear bands can be observed.

Adhikari et al. observed these MSBs in PC layers originating from crazed regions of PET in a PC/PET multilayered composite. It was demonstrated that as crazes formed in the PET and approached the PC/PET interface, MSBs would extend into the PC, dissipating energy. As the thickness of individual layers decreased below 1 μm, the deformation mechanism changed from two-phase behavior to single-phase behavior. Upon further decreasing the layer thickness below 100 nm, the composites exhibited homogeneous plastic deformation [88].

Im et al. showed that a similar phenomenon was present in a PC/SAN multilayered structure. Moreover, at fixed PC concentrations, impact strength also increased significantly simply by increasing the number of layers in the structure [91]. This further supports the earlier idea that more layers allow for increased number of crazes when the layer thickness causes the failure mode to be in the multi-layer crazing regime. Each of those crazes can spawn new MSBs when they reach layer boundaries. These MSBs dissipate energy from the crack tip and increase the toughness of the composite [91].

This phenomenon, though extensively studied for nonpolyester materials, has also been investigated for a PC/PMMA multilayered structure. Kearns et al. demonstrated that due to the interfacial adhesion between PC and PMMA layers, delamination was not the dominant failure mode [90]. Instead, as stress increased, crazing occurred in the PMMA layers, propagating outwards toward the layer interfaces. As the cracks propagated into PC layers, the MSBs coalesced into shear bands, ultimately resulting in the failure of the sample. These experiments were carried out using PMMA/PC layered structures of 32, 256, 1024, 2048, and 4096 layers.

Based on a similar mechanism, the fatigue performance of the composite also changes with the number of layers while keeping composition constant in an MNL film. As number of layers increase and layer thicknesses decrease, there is a noted transition in failure modes, as discussed above. This phenomenon was demonstrated by Ma et al. when it was observed that a PC/SAN multilayered composite exhibited improved fatigue crack propagation resistance when layer thicknesses were decreased [92]. Though the two samples had similar fatigue lifetimes ($9087 \pm 472$ for the 49-layer sample and $8853 \pm 388$ cycles for the 776-layer sample), the 49-layer sample experienced a longer crack initiation phase ($8302 \pm 211$ cycles) and shorter propagation phase ($707 \pm 136$ cycles) compared to the 776 layer sample which had more than a 250% longer propagation phase (1867 ± 201 cycles) [92]. The result of a longer propagation phase was a longer stable crack length, measuring 13.3 ± 1.2 mm for the 776 layer sample compared to 9.7 ± 1.8 mm for the 49 layer sample [92]. This type of improvement can be useful in engineering applications because the propagation phase under fatigue loading is when cracks can be identified, and components can be removed from service prior to failure.

### 6.4.2 Barrier Films

Barrier films can be found in a variety of applications ranging from food packaging to pharmaceutical products, and filters. Their key function is to regulate the passage of gas or moisture across the film in order to preserve specific conditions that either slow or prevent contamination. Due to the rapid growth of the packaging industry, the search for a commercially competitive green packaging film that exhibits significant barrier properties while not compromising on mechanical strength or thermal stability has been ongoing. Researchers have begun to apply MNL coextrusion in search of a suitable process capable of producing films that meet the requirements of such a product. By leveraging the effect of confined crystallization in MNL structures, vastly superior barrier properties can be attained compared to a nonlayered structure [47]. Additionally, the use of MNL coextrusion technology in creating multilayered barrier films has advantages over other multi-layering assembly methods owing to its continuous melt processing nature, lack of solvents, flexibility in polymer choice, and nearly defect free confinement [47]. The use of MNL coextrusion as a way of producing layers with confined crystallization presents opportunities for carefully tuning the permeability properties of multilayered composites [4, 5, 47, 63, 64]. In particular, the in-plane growth of lamellae can result in the formation of a tortuous path through a layer, forcing gas or water molecules to travel around closely packed crystalline domains. This is shown in Figure 6.10. Several systems have been examined for their potential barrier property enhancements when subjected to confined crystallization including: PEO/EAA [63], PS/PEO [4], polyvinylidene fluoride (PVDF)/PEO [64], PCL/PS [2], [70], PMMA/PC [2], [42], PMMA/PCL [65, 66], Polyamide 12 (PA12)/PCL [65], and polypropylene-graft-maleic anhydride (PP-g-MA)/PCL [65] systems.

As noted above, Wang et al. showed that with a PEO/EAA MNL coextruded layered structure, oxygen gas transmission rates decreased by two orders of magnitude when layer thicknesses decreased to around 20nm. Further decrease in layer thickness was shown to compromise layer integrity and adversely affect the transmission rate [4, 63]. Building upon this research, Ponting et al. demonstrated that crystalline PCL confined by either PS or PMMA exhibited a 2.5 order of

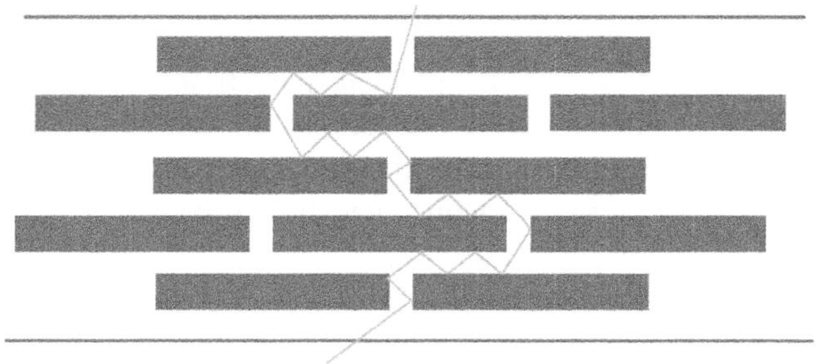

**Figure 6.10** Illustration of the tortuous path of a gas or water molecules diffusing around the in-plane oriented crystal lamella of a single layer.

magnitude improvement in oxygen gas barrier properties at minimum PCL layer thickness [65]. As mentioned earlier, the PCL/PS film showed stronger in-plane orientation of PCL lamellae compared to the PCL/PMMA film, thus requiring less confinement at only 5 μm PCL layers, compared to 1 μm PCL layers when confined by PMMA to achieve the same barrier property improvement. The researchers also investigated a PCL/PC blend but found that the strong miscibility of the two polymers prevented strong in-plane orientation of lamellae and thus did not improve barrier properties [66].

Similar results were observed by Mackey et al. who performed a similar experiment using PC/PVDF and PC/PVDF-TFE layered films. Additionally, an isothermal recrystallization procedure was utilized to further promote the growth of in-plane oriented crystals. These films were left to isothermally recrystallize for varying times (0.5, 5, 96 hours), and at varying temperatures (PC/PVDF – 135 and 145 °C; PC/PVDF-TFE – 110, 115, and 120 °C) to observe the effect of crystallization time and temperature on barrier properties across a range of layer thicknesses (530, 225, 112, 80, 47, and 28 nm) before cooling at a rate of 10 °C/min to room temperature. Their results suggest that increasing recrystallization temperature and decreasing layer thickness lead to overall improvement in oxygen barrier properties of the film. Comparing PVDF and PVDF-TFE performance, shows that the barrier performance of the thinnest layers increased 30-fold for PVDF whereas the PVDF-TFE films showed only a 7-fold improvement between the thinnest and thickest layers [93]. The authors postulated that this decrease in performance of PVDF-TFE compared to pure PVDF is due to the fact that copolymers crystallize more poorly compared to homopolymers [94]. Though this study does not use polyesters, it would be interesting to study how the use of an isothermal recrystallization procedure in polyesters can contribute to the improvement of film barrier properties. The use of this technique for a polyester multilayered film composite may be an interesting path to pursue for achieving better barrier performance.

### 6.4.3 Dielectric Films

Polymeric film capacitors are used extensively in various industrial applications including consumer electronics, microelectronic devices, power systems, hybrid electric vehicles, and grid converters [95]. One of the most widely used polymer capacitors are biaxially oriented PET films, having a low dielectric loss and high energy density compared to other available polymer capacitors, such as biaxially oriented polypropylene (BOPP) [96]. However, the intrinsically low energy densities of these polymers limit their use in high-performance applications having more stringent power requirements [96]. As a result, superior materials with higher dielectric constants, higher breakdown strengths, and enhanced energy storage capabilities are necessary to produce compact capacitors. Joel et al. reported that a multilayered film with improved dielectric and breakdown performance can be manufactured by combining two polymers with contrasting dielectric properties [97]. In addition, it was found that, increasing the dielectric constant and/or breakdown strength of a material increases its energy storage capability [98]. Based on these ideas, several studies were carried out to improve the overall dielectric performance of different polymeric materials. Kim et al. prepared a composite containing phosphonic acid (PA), a polymer with high breakdown strength and high permittivity, and barium titanate. Though the dielectric constant of the composite was improved, the breakdown strength was impacted negatively [99].

Recently, biaxially-stretched PET films have been widely studied for use in dielectric applications in various industries [96]. The idea of combining dissimilar dielectric polymers was adopted for the fabrication of PET dielectric films. In this method, a PVDF copolymer with a high dielectric constant and high breakdown strength is coextruded with PET to produce a multilayered film structure. The layered film is then biaxially stretched to form thin PET layers and copolymer nanolayers. In the study carried out by Deepak et al., PVDF–TFE was coextruded with PET to produce a multilayered film structure with 32 layers of alternating PVDF-TEF and PET polymers [96]. In the coextruded films, the PVDF-TFE crystals were randomly oriented, while the PET layers were in an amorphous state. Biaxial stretching of this extruded film at 105 °C reduced its overall thickness from 200 to 10 μm, while oriented fibrils were formed in the PET layers. These PET/PVDF-TFE coextruded films exhibited enhanced breakdown strength compared to the PET and PVDF-TFE control films. The biaxial stretching approach further enhanced the breakdown strength by 25 to 40% [97]. The structures obtained by the multilayer coextrusion and biaxial stretching method are schematically represented in Figure 6.11 [97]. Jordan et al. used the same material combinations to enhance the mechanical properties through biaxial stretching and melt recrystallization, as this had a direct impact on the use of these materials in dielectric applications [79]. Yin et al. prepared a 3-component multilayered dielectric film containing PET,

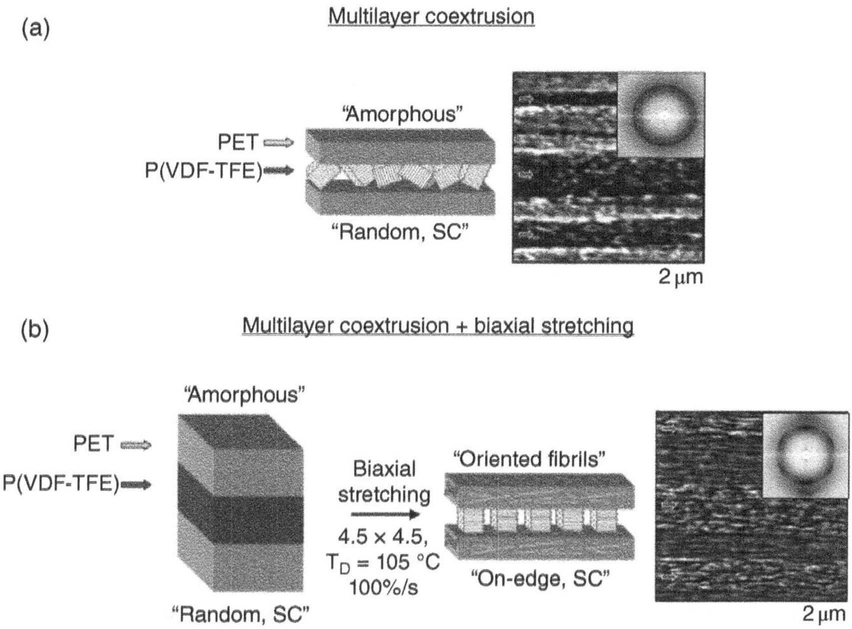


**Figure 6.11** Structures obtained in the PET/PVDF-TFE multi-layer system through (a) multi-layer coextrusion and (b) biaxial stretching of thick PET/PVDF-TFE multi-layer films at high draw ratios. *Source:* Carr [97]/ with permission from John Wiley & Sons.

PMMA, and poly-(vinylidene fluoride-co-hexafluoropropylene) (PVDF-HFP) and reported to achieve an enhanced energy density of 17.4 $Jcm^{-3}$ at breakdown when biaxially stretched. The improvement here is about 150% compared to the as-extruded PET/PVDF-HFP films [95]. Thus, these multilayered films are promising candidates for various high-energy density film capacitor applications.

### 6.4.4 Shape-Memory Films

Shape-memory polymers have the unique ability to recover from a temporary deformation shape to its original shape in response to the application of an external stimulus such as heat, light, voltage, solvent, or water vapor. These materials could be used for flexible electronics, smart packaging, biomedical instruments, aerospace structures, self-deployable devices, sensors, actuators, and drug delivery [101]. Polyester materials such as PMMA and PCL are widely used to fabricate thermally responsive shape-memory polymers due to their thermal and phase transition properties. Furthermore, a cocontinuous phase morphology is preferred for achieving the highest shape-memory performance because the parallel layers are capable of maintaining the same strain along the deforming direction. As such, they possess the maximum ability to recover from temporary deformation and trigger their recovery to original form through interfacial shearing and high phase continuity along the deformation direction [102]. Therefore, MNL films are a suitable for enhancing shape-memory performance due to the inherent cocontinuous morphology they exhibit [103, 104].

Recently, a significant amount of research has been focused on this area, and promising results have been obtained. For example, Zheng et al. fabricated shape-memory polymeric materials based on multilayered and cocontinuous PBS/PCL [105, 106], thermoplastic polyurethane (TPU)/PCL [107], and multilayered multi-walled carbon nanotube (MWCNTs) filled TPU/PCL [102] composites. Based on the PBS/PCL composite, they found that the multilayered structure has higher degree of shape recovery ratio (measure of the capability of shape memory materials to recover their permanent shape) than if the polymers were to simply be blended at the same concentration. The shape recovery ratios were 85% in dual- and triple-shape-memory processes when the number of layers reached 128, which was difficult to achieve through conventional melt-processing methods. Their study indicates that the assembled parallel layers were capable of maintaining the same strain along the deforming direction and showed the greatest capacity to recover their original form after deformation. These results were attributed to the high phase continuity along the deformation direction and the strong interfacial steering effect promoted by the multilayer structure.

Electro- and thermo-responsive shape-memory capabilities were introduced into a TPU/PCL multilayered system with the addition of MWCNTs [102]. By adjusting processing conditions and number of layers, the layer multiplication process can modify the phase interfaces and electrical conductive pathways in the multilayered structure, which gives the materials tunable thermo- and electro-responsive shape-memory characteristics. Zheng et al. concluded that by increasing the number of layers, the shape-memory effect could be further enhanced. This study also demonstrated that due to the confinement effect of the layered structure, the connection between the MWCNTs is more compact compared with blended composite. However, if the material goes through too many multiplying steps, the conductive network originally formed in TPU tends to break up during layer multiplication which might affect the thermal and electrical properties of the final material. Therefore, the number of layers should be optimized to balance conductivity and shape-memory performance.

Zeng et al. fabricated a hierarchical PPC/TPU alternating multilayered film with body heat responsive shape-memory properties, high transparency, and balanced mechanical properties [108]. The multilayered films exhibited an outstanding body heat responsive shape-memory effect due to the glass transition temperature of PPC, which is roughly 37 °C. Compared with conventional cocontinuous blending, much higher shape fixation and recovery ratios (96.1 and 93.6%)

are achievable. The force generated by shape recovery allowed an automatic expansion of the spiral and lifting of a load 880 times its own weight in 37 °C water.

Polyurethane (PU) is another elastomer that is often used with polyesters to achieve shape-memory effects. For example, Li et al. studied micro-scale surface shape – memory effects on multi-layer films with alternating layers of PVA and PU [109]. They embossed a sharp micro-prismatic pattern and a smooth micro-lens pattern on the surface of PVA/PU multilayered films and programmed the micro-structured surface using compression and shape recovery to switch the corresponding optical properties of the film. Their study showed that the shape-memory effect of the multilayered film depends on the PVA composition relative to that of the PU. The shape-memory effect of the layered PVA/PU films was enhanced compared to the PVA control film. They also conclude that higher recovery temperature and lower compression pressure enhanced the micro-scale recovery rate and recovery ratio. And a smooth microstructure had a higher recovery rate and recovery ratio than the micro-prismatic pattern with its sharp microstructure.

Through MNL coextrusion of PVDF/PMMA multilayered structure, Ji et al. studied the increased interfacial area leading to a broader and more continuous thermal transition temperature from PVDF to PMMA layers [110]. The low and high thermal transition temperatures correspond to the glass transition temperature of PMMA and the melting of PVDF crystals, respectively. Their study shows that compositional diffusion broadened the relaxation distribution of amorphous chains and strengthened the interaction between amorphous and crystalline domains. The results indicate that the 1024-layer specimen has a 3-4x improvement in shape-memory capacity compared to a conventional capacity than a conventional blend. They also concluded that the higher phase continuity plays a significant role in strengthening the shape-fixing and -recovering ability during each shape-memory cycle.

## 6.5 MNL Film-Foam Structures

### 6.5.1 Polymeric Foams and Foaming Process

The use of foamed plastics has increased significantly in the past few decades and is expected to continue increasing [111]. These foamed or cellular plastics are desirable for superior mechanical, thermal, and acoustic properties when compared to their solid counterparts [112]. The improved properties of cellular plastics are attributed to high cell density, small cell size, and uniform cell distribution. However, the low melt strength of some polymer materials poses critical challenges during foaming as the cell walls are not strong enough to support cell integrity, causing cell coalescence, cell wall rupture, and cell ripening/coarsening. The novel approach of multi-layer coextrusion processing of alternating film/foam structures can effectively address these challenges through mechanical property improvement of the film/foam bulk. Further improvement can be achieved by cell confinement and cell uniformity [113].

Different types of multi-layer foam/film structures found in the literature mostly include PP, PE, ethylene octene copolymer, cross-linked polyolefin elastomers, polyolefin blends, and polyester [114–118]. In the coextrusion process, the first polymer contains a blowing agent and/or a nucleating agent necessary for foaming, while the second polymer forms the film layers separating the foam layers. The presence of a blowing agent and nucleating agent facilitates the nucleation, growth, and stabilization of gaseous cells in the melt. A brief description of the general polymeric foaming process, properties of foamed plastics, and the underlying factors affecting the foaming mechanism is discussed in the following section.

### 6.5.2 Fabrication of Microcellular Foam Structures

Cellular plastics or plastic foams generally have a minimum of two phases: a solid polymer matrix and a gaseous phase derived from a blowing agent. In polymer foams, cells are typically generated by the decomposition of chemical blowing agents (CBAs) that release gases, or by the phase separation of dissolved physical blowing agents (PBAs) directly from the polymer melt. During the foaming process, formation of a one-phase polymer-gas solution is the preliminary condition required to produce high quality foams. The maximum amount and saturation time of a PBA that can be dissolved in the polymer melt is governed by the gas solubility and diffusivity, respectively.

Figure 6.12 shows a schematic of a typical foam process that makes plastic foams using a PBA. Several crucial steps are involved: (1) the formation of a uniform polymer-gas solution, (2) the cell nucleation via thermodynamic instability,

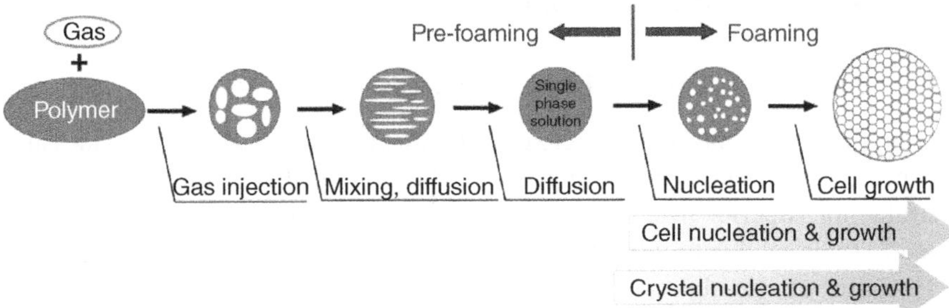


**Figure 6.12** The plastic foaming process by a PBA.

which is induced by either a reduction in pressure [119] or an increase in temperature [120], (3) the cell growth to achieve a desired expansion ratio, and (4) the stabilization of cells to retain the desired cell structure. Efforts have been made to reduce the cell size and enhance the cell size uniformity because plastic foams with a finer cell size and a more uniform distribution exhibit better mechanical and thermal properties [121–126]. These advances have led to the development of so-called "microcellular plastics," which have cell sizes smaller than 10 mm and cell-population densities larger than $10^9$ cells/$cm^{-3}$. These novel foams exhibit high impact strength (up to a five-fold increase over unfoamed plastic [121,124]), high toughness (up to a five-fold increase over unfoamed plastic [127]), high fatigue life (up to a fourteen-fold increase over unfoamed plastic [125]), high heat-insulation properties [128], high sound-insulation properties [129–131], high thermal stability [126], and high light reflectability [132]. Applications of foamed plastics range from household commodities (e.g., packaging, and insulation materials) to advanced engineering products (e.g., airplane, and automotive parts with high strength-to-weight ratios). Furthermore, foamed plastic products can reduce material usage, which typically accounts for up to 70% of the total production cost. Due to these unique properties, microcellular plastics have attracted the attention of numerous plastics manufacturers. To improve foaming technologies and products, intense research has been undertaken using small-scale industrial foam processing equipment such as foam extrusion or injection molding systems. However, the cell nucleation and growth phenomena cannot be easily observed in these systems, often resulting in a trial-and-error approach to optimize the processing strategies, and conditions.

Recently, some researchers have prepared multilayered foam sheets through multi-layer coextrusion technique [114,115,133–140]. Their work revealed that the foam morphology, sound absorption efficiency, and mechanical properties of the multilayered foam sheets were improved over their single-layer counterparts. These studies have indicated the feasibility of the functional structure design of multilayered polymeric systems for desired applications. However, these studies used CBAs to produce foams with large cells and anisotropic cell shapes. It is critical to use PBAs with the proper process parameters to produce desirable foam morphologies, such as nanocellular and/or gradient foam structures. The three main steps of polymeric foaming process namely, cell nucleation, growth, and stabilization are briefly discussed in the following sections.

#### 6.5.2.1 Cell Nucleation

Cell nucleation in polymer foaming can be explained as a function of different processing parameters. The spontaneous growth or collapse of the nucleated cells depends on a state parameter known as the critical radius ($R_{cr}$) [141].

Theoretically, $R_{cr}$ is a function of the thermodynamic state that can be expressed as Eq. (6.5) [112].

$$R_{cr} = \frac{2\gamma_{lg}}{P_{bub,cr} - P_{sys}} * \tag{6.5}$$

where, $\gamma_{lg}$ is the interfacial energy at the polymer-gas interface, $P_{bub,cr}$ is the pressure of a critically sized cell, $P_{sys}$ is the system pressure, and the term ($P_{bub,cr} - P_{sys}$) represents the supersaturation level of the polymer-gas solution. Additionally, by assuming that the polymer-gas solution is a weak solution, $P_{bub,cr}$ can be estimated using Henry's law and Eq. (6.5) can be rewritten as Eq. (6.6)

$$R_{cr} = \frac{2\gamma_{lg}}{HC - P_{sys}} \tag{6.6}$$

where, $H$ is the Henry's law constant and $C$ is the gas concentration dissolved in the polymer. From this equation, it is clear that by increasing the dissolved gas concentration, the critical radius is reduced [112].

Based on the classical nucleation theory derived from continuum thermodynamics, the degree of super-saturation ($\Delta P$), which is the main force for cell nucleation, can be expressed as Eq. (6.7) [111]:

$$\Delta P = P_{bub,cr} - \left(P_{sys} + \Delta P_{local}\right) \tag{6.7}$$

where, $\Delta P_{local}$ is the difference between $P_{sys}$ and the local pressure at the nucleating site. If the local region is under a compressive stress, $\Delta P_{local}$ is positive. Conversely, if the local region experiences an extensional stress, $\Delta P_{local}$ is negative. If $\Delta P_{local}$ is negative, then the super-saturation level will increase, which will reduce the critical radius for cell nucleation as well as the free energy barrier for homogeneous and heterogeneous cell nucleation. From this, it is apparent that compressive stresses tend to discourage cell nucleation, whereas tensile stresses encourage the formation of cells. Equation (6.5) can then be rewritten as Eq. (6.8).

$$R_{cr} = \frac{2\gamma_{lg}}{P_{bub,cr} - \left(P_{sys} + \Delta P_{local}\right)} \tag{6.8}$$

When the pressure drop rate is increased, the degree of supersaturation increases proportionally, reducing the critical nucleation radius and free energy barrier required for nucleation. As a result, the nucleation rate and the final cell density will also increase [142]. Simultaneously, the gas depletion rate will also increase, reducing the nucleation time. On the other hand, if the concentration of gas is increased, the surface tension will decrease. With the increase in gas concentration and decrease in surface tension, the critical nucleation radius will also decrease. As a result, the onset time for nucleation, nucleation rate, and final cell density will increase [143].

#### 6.5.2.2 Cell Growth

Cell growth depends on several physical parameters. At the initial phase of cell growth, the cell pressure for the cells with smaller cell radius is higher. This higher cell pressure leads to a higher initial growth rate. As the cell gradually grows in size, the cell pressure reduces accordingly. As a result, the growth rate also gradually reduces [144]. On the other hand, cells with a larger initial radius will have a lower growth rate [144]. Another parameter, called the influenced radius, is a measure of the volume around the cell from which gas can be used to fuel cell growth. As cells nucleate around each other, they impinge on each others' influenced radii, reducing the amount of gas available to fuel growth for all cells in the region. When the influenced radius around a cell is bigger, the corresponding cell grows into a bigger-sized cell at the end because of the higher amount of available gas or SCF for that individual cell to grow [144]. However, the effect of influenced radius is not apparent at the initial stage of cell growth, as the amount of gas around the cell remains constant until the later stages of growth. A schematic diagram of the nucleated cell and the surrounding polymer-gas solution is shown in Figure 6.13 [144].

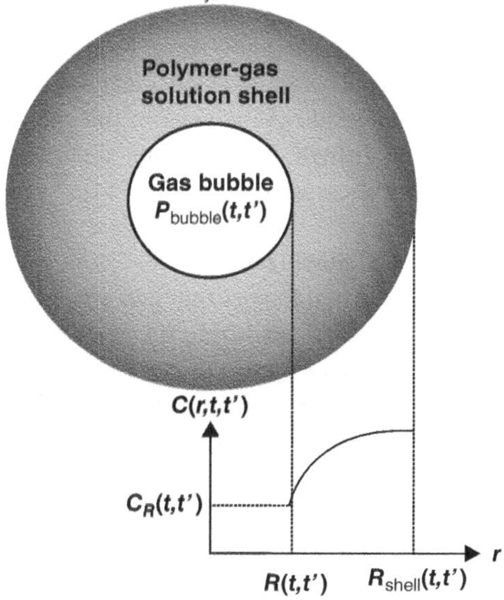


**Figure 6.13** A schematic diagram of the nucleated cell and its corresponding polymer-gas solution shell. *Source:* Leung et al. [144]/ with permission of American Chemical Society.

Gas diffusivity is another important parameter that can significantly influence cell growth. A higher gas diffusivity will enable gas to move through the matrix at a higher rate into nucleated cells during cell growth [112]. As a result, the cell growth rate will increase at higher diffusivity. However, sometimes even with a higher diffusivity, cells may not grow past a certain size due to the limited availability of gas [144]. To account for this, the gas solubility must also be considered. Gas solubility represents the capability of the polymer to dissolve a certain amount of gas before becoming saturated. With more gas in the matrix, more gas will be available for the cells to grow faster and bigger [112]. On the other hand, when the gas solubility is low, the cell growth and cell size will be limited by the amount of available gas content.

Furthermore, surface tension of the cell must be considered as well, since it represents the energy needed to create new surfaces or increase the interfacial area between two phases. A larger surface tension will result in a lower growth rate at the very early stage of the cell growth process because of the greater retarding force imposed by the matrix onto the growing cell [144]. Additionally, a larger surface tension will yield a larger initial cell radius and a lower cell pressure during the early cell growth stage, which ultimately translates to a lower cell growth rate. However, with the increase of cell size, the effect of surface tension becomes less significant and can eventually be considered a constant as opposed to a function of cell radius [144].

Another parameter that affects the cell growth phenomenon is the relaxation time, which is relevant because of the viscoelastic nature of the polymer melt. A longer relaxation time means that the stress caused by the deformation of the matrix by the cell will take a longer time to relax and accumulate [144]. Therefore, with a shorter relaxation time, small cells will be formed due to the high rate of stress accumulation and slow cell growth rate [144]. Lastly, shear viscosity of the polymer melt takes into account the elastic nature of the polymer melts and also affects cell growth. With a higher shear viscosity, the growth rate is decreased due to a larger stress accumulation [144].

#### 6.5.2.3 Cell Stabilization

There are several factors that can affect the stability of a nucleated cell. Cell deterioration takes place during the growth stage of foaming [143]. The deterioration is mainly affected by two different mechanisms: cell coalescence and cell coarsening/ripening [145]. During the growth of nucleated cells, adjacent cells tend to undergo coalescence because of the reduction of total free energy and surface area. If coalescence occurs during growth, the cell density is negatively affected [112]. The shear field generated during the shaping stage stretches the nucleated cells and further accelerates the cell coalescence. This occurs because the contact surface area of two elongated cells is higher compared to their spherical counterparts. Cell coalescence reduces the cell density and cell uniformity of the foamed polymer, negatively impacting the foamed polymer as a whole. The coarsening of cells is a process driven by gas diffusion between adjacent cells of different sizes that result in the growth of some cells and the shrinkage and collapse of other cells as the larger, more energetically favored cell absorbs the gas in the smaller cell [146]. This can occur if the gas diffusivity is too high or if the polymer is unable to cool and stabilize quickly enough. In this way, bigger cells continue to grow while smaller cells collapse until they disappear completely. Cell coarsening will ultimately reduce the total cell density and uniformity of cell size.

Cell coarsening/ripening mechanism can be understood as a force imbalance, or an energy imbalance [147]. From a force imbalance perspective, the pressure inside a smaller cell is higher than that of a larger cell. When gas diffuses from the smaller cell with high-pressure region to the bigger cell with low-pressure region, the smaller cell shrinks while the larger one expands. From an energy imbalance perspective, the critical cell nucleation radius increases as the gas concentration decreases and surface tension increases [141]. As the larger cells continue to grow further, gas depletion occurs, and the surface tension increases even further, meaning that larger cells will outgrow the critical radius and grow spontaneously. Conversely, smaller cells continue to collapse because they are not energetically favorable.

### 6.5.3 Processing of MNL Film-Foam Structures

Typically, a coextrusion system is used for producing a multilayered film/foam structure. A polymeric material is combined with a blowing agent and/or a nucleating agent and extruded through one set of extruder and gear pump, while the other polymer is extruded through the second set of extruder and gear pump. The melt flow rates through each of the components determine the relative volumetric flow rates, and ultimately the layer thickness ratios of the film and foam layers [3]. When a CBA is used for foaming, the blowing agent decomposes in the extruder at elevated temperatures and releases gas which immediately dissolves into the melt. In the case of a PBA, the blowing agent is directly pumped into the extruder by way of a gas injection port and is mixed with polymer melt to form a polymer-gas mixture. At the die exit, the foaming process starts to occur and continues as discussed in previous sections. During the cell growth and stabilization phases, the integrity and stability of the film/foam structure are determined primarily by the stress produced by the cell nucleation/growth phases and stiffness of the film layer [3, 148]. As such, polymers with high melt strength are generally suitable for multilayered film/foam structures. It is well known that smaller and more uniform cell sizes can significantly improve the overall composite properties of a film/foam structure. Chen et al. reported that, the average cell size of a 16-layer

PC/PMMA film/foam structure can be reduced from 13.6 to 2.1 μm by increasing the number of layers to 512. More uniform cell morphologies were observed in this study when the number of layers were gradually increased from 16 to 512 layers [148]. The schematic of a typical two-component film/foam coextrusion system is shown in Figure 6.14 [115].

In an MNL film/foam structure, the confining film layers provide active heterogeneous nucleation sites at the interfaces between the film and foam layers. The stress variation at the interfaces is also increased due to the stiff film layers. Both mechanisms help in reducing the free energy barrier and critical cell nucleation radius during the nucleation phase. As a result, heterogeneous nucleation of cells is significantly promoted in the film/foam structures in this polymeric foaming process [148].

As cells are heterogeneously nucleated and growing at the film/foam interfaces, instant expansion of the foam layers occurs. However, the film layers do not experience internal expansion like the foam layers, the expansion difference between film layers and foam layers induces huge extensional stress at the film/foam interfaces, producing large local stress variations and pressure variations, $\Delta P_{local}$. Consequently, both the critical radius, $R_{cr}$ and the free energy barrier

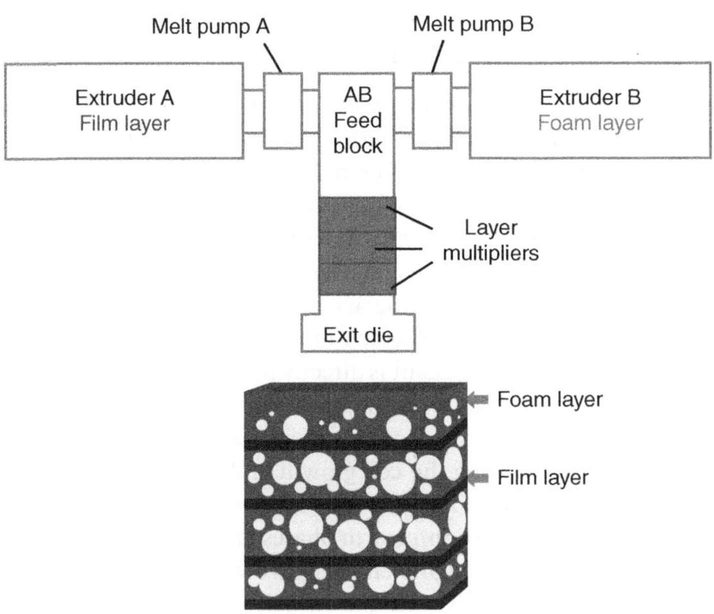


**Figure 6.14** Schematic of a two-component film/foam coextrusion system. *Source:* Wu et al. [115]/ with permission from American Chemical Society.

decreases drastically [149]. Thus, more layers, expressed as the layer density, results in more enhanced heterogeneous cell nucleation and higher cell densities [148, 149].

### 6.5.4 Properties and Applications of Foamed Polyesters

#### 6.5.4.1 Micro- and Nanocellular Foams

The commonly used polymers in foaming include PS, PMMA, polyvinyl chloride (PVC), PE, and PET [148, 150–153]. PET foams have competitive advantages over the other thermoplastic polymers due to its high elastic modulus, thermal stability, high service temperature, as well as high impact resistance [154]. Despite having good mechanical properties, foaming of common polyester materials like, PET, polybutylene terephthalate (PBT), or PLA is highly challenging due to the low melt strength, low shear viscosity, and low elongational viscosities of these materials [153]. Enhancing the melt strength by chemical modification has been found to be effective in achieving uniform cellular morphology using polyester materials [154].

Extrusion foaming is a common technique for polyester foaming, as chemical modification of the polymer chains can be easily done during the extrusion process. Typically, chain extenders are added to the polymer melt. This lengthened polymer chain increases the molecular weight, and in turn improves the melt viscosity and strength, which are favorable parameters for foaming [154]. Thus, by using suitable chain extenders and favorable processing parameters, micro-/nanocellular polyester foams can be achieved. Today, due to the availability of high molecular weight linear PET, studies related to PET foaming have increased significantly [153–155].

Sorrentino et al. investigated the foamability of two high-molecular-weight PETs and reported achieving high cell density and uniform cell morphology when $N_2$ was used as a PBA. This was attributed to the higher diffusivity of $N_2$ compared to $CO_2$. The results suggested the potential use of these foamed PETs in food packaging industries [153]. The flame-retardant properties of a bottle-grade PET foam were investigated by Bethke et al. In this study, a chain extender and different fire retardants (FRs) were added in several samples to assess the foamability of the polymer. It was found that the addition of FRs not only improved the fire retarding ability, but also the mechanical properties of the foams compared to neat PET foams. Additionally, uniform microcellular morphology was achieved in this study [153]. PLA is another commonly used foamed polyester material that has received increased attention in recent years due to its biodegradable nature. The foamability of PLA is dependent on the ratio of L and D contents. In general, increased D content makes the polymer more amorphous and the foam processing gets easier, but the mechanical properties are negatively affected. However, in semi-crystalline PLA foams, mechanical properties are increased, while the foam processability decreases [156]. Nofar et al. studied the effect of nanosized additives in PLA foaming using $CO_2$ as a PBA [157]. This study reported that well-dispersed nanoparticles affected the crystallization behavior, which ultimately enhanced the foam expansion ratio and the cell density of these foams. Julien et al. investigated the physical and morphological characteristics of extruded PLA foams using chain extenders [158]. The incorporation of chain extenders decreased the average cell size and open cell content, while the tensile strength was increased from 17.8 to 27.5 MPa. Due to having reduced weight and enhanced structural properties, the foamed PLA materials have become one of the most promising materials for green packaging applications. The physical properties and the usability of the foamed materials can be further enhanced through MNL film/foam structures discussed in the next section.

#### 6.5.4.2 MNL Film-Foam Structures

Introduction of selective foaming in MNL polymeric structure can enhance the properties of the materials in several ways. Continuously coextruded MNL film/foam composites exhibit improved mechanical and physical properties when compared with traditional foams [116]. One of the recent studies in this area was carried out by Chen et al., who studied the cell nucleation and growth mechanisms in advanced MNL structures with alternating PC or PS film layers and PMMA foamed layers [148]. It was found that the cell nucleation and growth phenomenon in MNL systems is governed by the synergy of two categories of parameters: morphological parameters, such as film and foam layer thicknesses and the number of layer interfaces, and material parameters, such as material stiffness and compatibility with neighboring layers. The presence of adjacent film layers can significantly increase the cell density through two mechanisms: promoting heterogeneous cell nucleation and preventing cell deterioration. The confinement produced by adjacent film layers can also decrease the average cell size. The influence of film layers varied in different layer thickness regions and interface densities. The use of a stiffer and more compatible film layer enhanced the overall foam quality by taking advantage of the above-mentioned mechanisms, which was observed in PC/PMMA samples. On the other hand, in the case of a soft and less compatible film layer, the cell nucleation and growth were less affected compared to PC/PMMA case. As a result, by increasing the stiffness of the PC layers in PC/PMMA samples through $CO_2$-induced crystallization, the cell morphology was further improved. This study reported to achieve MNL structures with an average cell size of 2.1 μm and a normalized cell density of $1.74 \times 10^{10}$ cells $cm^{-3}$. Figure 6.15 shows an SEM image of a coextruded PC film/PMMA foam sample.

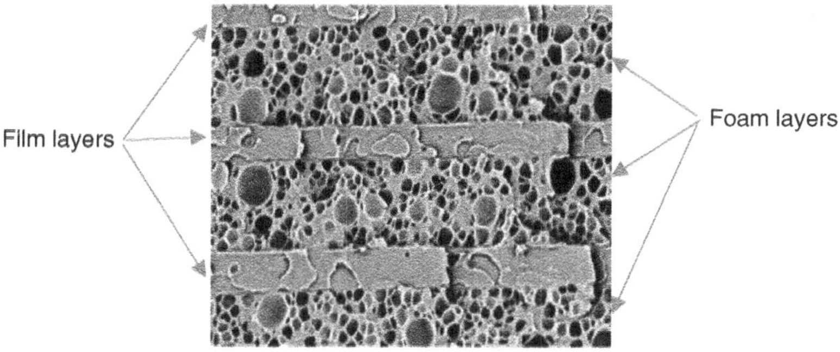


**Figure 6.15** SEM image of a coextruded film (PC)/foam (PMMA) sample. *Source:* Chen et al. [148]/ with permission from Elsevier.

The confining film layers in a film/foam structure can effectively control the growth of cells and reduce cell wall rupture during foaming [136]. Benning et al. reported that strong foam materials with much higher mechanical properties can be achieved if the orientation of the cells can be maintained after processing [159]. As such, highly oriented cellular structures are advantageous for mechanical property improvement in film/foam structures. These composites exhibit improved toughness, tear resistance, and puncture resistance [160].

## 6.6 Summary and Perspective

MNL coextrusion is a simple, yet effective way of creating layered composites that exhibit enhanced, or completely novel properties compared to their constituent polymers via precise spatial control of each phase on the micro-/nanometer scales. As noted above, the industries which benefit from the use of layer multiplication technology is diverse and growing. Ranging from food packaging to smart electronics, MNL coextruded films can contribute in innovative ways through the phenomenon that arise from the coontinuous layers as thin as 10nm. Furthermore, the use of LMs can be combined with foaming technology, producing film-foam structures. These types of structures leverage properties of both foams and films to produce a composite that has improved mechanical and physical properties compared to traditional foams but are less dense than solid film structures. Lastly, because of the ubiquity of polyesters in consumer products, the use of MNL manufacturing technology to produce well-engineered multilayered polyester-based composite films is expected to grow.

A growing number of research groups are focusing on processing techniques that leverage the use of unique microstructures made possible by MNL coextrusion. Compared to traditional melt blending techniques for forming polymeric composites, MNL coextrusion offers fundamentally unique material design approaches, such as the ability to confine crystallization and create cocontinuous phase interfaces, to name just a few. There is a particularly large push for producing sustainable food packaging, an industry where polyesters have reigned supreme for decades. Researchers are also utilizing MNL coextrusion to create polymeric capacitors, which promise to be a viable alternative to existing components used in hybrid vehicles and smart grids. Many other applications of MNL films have been explored with promising results. For example, the use of MNL in the healthcare industry to create bio-compatible smart devices capable of self-actuating through their shape-memory properties have been a focus of several research groups. These are just some of the many industries which are working to assess and exploit the potential benefits that this technology can offer.

Though MNL coextrusion technology has been around for several decades, its adoption has been recently growing as its benefits and limitations slowly become unraveled. Some of these limitations include the inability to decrease layer thicknesses below 10 nm due to the onset of layer breakup, or the need to match material viscosities and elasticities, which limits the available materials for use in these novel composite materials. Recent advancements in this field have produced new LMs with unique and novel channel designs targeted at alleviating some of these issues, though they have not been widely adopted as of yet. With advancements in computing technology, complex fluid dynamic simulations can be used in conjunction with artificial intelligence to design more effective LMs. As MNL coextrusion technology is made accessible to industries and research groups all over the world, more polyester-based consumer products may be replaced with stronger, lighter, and smarter multilayered composites.

## List of Abbreviations

| | |
|---|---|
| BOPP | Biaxially oriented polypropylene |
| CBA | Chemical Blowing Agent |
| EAA | Ethylene acrylic acid |
| EVA | Ethylene vinyl acetate |
| FR | Fire retardant |
| LM | Layer Multiplier |
| MNL | Micro-/Nano-Layered |

| | |
|---|---|
| MSB | Microshearband |
| MWCNT | Multi-Walled Carbon Nanotube |
| P-(VDF-HFP) | poly-(vinylidene fluoride-co-hexafluoropropylene) |
| P4VP | Poly(4-vinylpyridine) |
| PA6 | Polyamide 6 |
| PA12 | Polyamide 12 |
| PBA | Physical Blowing Agent |
| PBS | Poly(butylene succinate) |
| PBT | Polybutylene terephthalate |
| PC | Polycarbonate |
| PCL | Polycaprolactone |
| PDLA | Poly(d-lactide) |
| PE | Polyethylene |
| PEO | Poly(ethylene oxide) |
| PET | Polyethylene terephthalate |
| PLA | Polylactic Acid |
| PLLA | Poly(l-lactide) |
| PMMA | poly(methyl methacrylate) |
| PP | Polypropylene |
| PPC | poly(propylene carbonate) |
| PS | Polystyrene |
| PU | Polyurethane |
| PVA | Polyvinyl Acetate |
| PVC | polyvinyl chloride |
| PVDF-TFE | Poly(vinylidene fluoride-co-tetrafluoroethylene) |
| $R_{cr}$ | Critical Radius |
| SAN | Styrene-acrylonitrile |
| SCF | Supercritical Fluid |
| SEM | Scanning Electron Microscopy |
| $T_g$ | Glass Transition Temperature |
| TPU | Thermoplastic Polyurethane |

## References


1 Harder, R.E. (1968). Interfacial surface generator. US Patent 3404869A.

2 Ponting, M., Hiltner, A., and Baer, E. (2010). Polymer nanostructures by forced assembly: process, structure, and properties. *Macromol. Symp.* 294 (1): 19–32.

3 Li, Z., Olah, A., and Baer, E. (2020). Micro- and nano-layered processing of new polymeric systems. *Prog. Polym. Sci.* 102: 101210.

**4** Wang, H.P.P., Keum, J.K.K., Hiltner, A., and Baer, E. (2009). Confined crystallization of PEO in nanolayered films impacting structure and oxygen permeability. *Macromolecules* 42 (18): 7055–7066.

**5** Thellen, C., Schirmer, S., Ratto, J.A. et al. (2009). Co-extrusion of multilayer poly(m-xylylene adipimide) nanocomposite films for high oxygen barrier packaging applications. *J. Memb. Sci.* 340 (1–2): 45–51.

**6** Messin, T., Follain, N., Guinault, A. et al. (2017). Structure and barrier properties of multinanolayered biodegradable PLA/PBSA films: confinement effect via forced assembly coextrusion. *ACS Appl. Mater. Interfaces* 9 (34): 29101–29112.

**7** Messin, T., Follain, N., Lozay, Q. et al. (2020). Biodegradable PLA/PBSA multinanolayer nanocomposites: effect of nanoclays incorporation in multinanolayered structure on mechanical and water barrier properties. *Nanomaterials* 10 (12): 2561.

**8** Messin, T., Marais, S., Follain, N. et al. (2020). Biodegradable PLA/PBS multinanolayer membrane with enhanced barrier performances. *J. Memb. Sci.* 598: 117777.

**9** John, R., Pal, K., Jayan, J.S. et al. (2021). New emerging review on advances in block copolymer based water purification membranes. *J. Mol. Struct.* 1231: 129926.

**10** Kumorek, M., Minisy, I.M., Krunclová, T. et al. (2020). PH-Responsive and antibacterial properties of self-assembled multilayer films based on chitosan and tannic acid. *Mater. Sci. Eng. C* 109 (January 2019): 110493.

**11** Ergün, A., Tümer, E.H., Cengiz, H.Y., and Deligöz, H. (2020). Monitoring the salt stability of layer-by-layer self-assembled films from polyelectrolyte blends by quartz crystal microbalance-dissipation and their ion separation performances. *Polym. Eng. Sci.* 60 (5): 1006–1018.

**12** Richardson, J.J., Björnmalm, M., and Caruso, F. (2015). Technology-driven layer-by-layer assembly of nanofilms. *Science (80-)* 348 (6233): aaa2491.

**13** Langhe, D. and Ponting, M. (2016). Coextrusion processing of multilayered films. In: *Manufacturing and Novel Applications of Multilayer Polymer Films* (ed. D. Jackson), 16–45. Elsevier.

**14** Zhang, G.J., Lee, P.C., Jenkins, S. et al. (2014). The effect of confined spherulite morphology of high-density polyethylene and polypropylene on their gas barrier properties in multilayered film systems. *Polymer (Guildf).* 55 (17): 4521–4530.

**15** Ponting, M., Burt, T., Korley, L. et al. (2010). Gradient multilayer films by forced assembly coextrusion. *Ind. Eng. Chem. Res.* 49 (23): 12111–12118.

**16** Wang, J., Langhe, D., Ponting, M. et al. (2014). Manufacturing of polymer continuous nanofibers using a novel co-extrusion and multiplication technique. *Polymer* 55: 673–685.

**17** Han, C.D., Kim, Y.J., and Chin, H.B. (1984). Rheological investigation of interfacial instability in two-layer flat-film coextrusion. *J. Polym. Eng.* 4 (3): 177–200.

**18** Khomami, B. and Ranjbaran, M.M. (1997). Experimental studies of interfacial instabilities in multilayer pressure-driven flow of polymeric melts. *Rheol. Acta 36* (4): 345–366.

**19** Martyn, M.T. and Coates, P.D. (2013). Pressure variation during interfacial instability in the coextrusion of low density polyethylene melts. *Int. Polym. Process.* 28 (5): 516–527.

**20** Schrenk, W.J., Bradley, N.L., Alfrey, T., and Maack, H. (1978). Interfacial flow instability in multilayer coextrusion. *Polym. Eng. Sci.* 18 (8): 620–623.

**21** Anderson, P.D., Dooley, J., and Meijer, H.E.H. (2006). Viscoelastic effects in multilayer polymer extrusion. *Appl. Rheol.* 16 (4): 198–205.

**22** Debbaut, B., Avalosse, T., Dooley, J., and Hughes, K. (1997). On the development of secondary motions in straight channels induced by the second normal stress difference: experiments and simulations. *J. Nonnewton. Fluid Mech.* 69 (2): 255–271.

**23** Yue, P., Dooley, J., and Feng, J.J. (2008). A general criterion for viscoelastic secondary flow in pipes of noncircular cross section. *J. Rheol. (N. Y. N. Y)* 52 (1): 315–332.

**24** Dooley, J., Hyun, K.S., and Hughes, K. (1998). An experimental study on the effect of polymer viscoelasticity on layer rearrangement in coextruded structures. *Polym. Eng. Sci.* 38 (7): 1060–1071.

**25** Minagawa, N. and White, J.L. (1975). Co-extrusion of unfilled and $TiO_2$-filled polyethylene: influence of viscosity and die cross-section on interface shape. *Polym. Eng. Sci.* 15 (12): 825–830.

**26** White, J.L. and Lee, B. (1975). Theory of interface distortion in stratified two-phase flow. *Trans. Soc. Rheol.* 19 (3): 457–479.

**27** Townsend, P., Walters, K., and Waterhouse, W.M. (1976). Secondary flows in pipes of square cross-section and the measurement of the second normal stress difference. *J. Nonnewton. Fluid Mech. 1* (2): 107–123.

**28** Han, C.D. (1973). A study of bicomponent coextrusion of molten polymers. *J. Appl. Polym. Sci.* 17 (4): 1289–1303.

**29** Korteweg, D.J. (1883). On a general theorem of the stability of the motion of a viscous fluid. *London, Edinburgh, Dublin Philos. Mag. J. Sci.* 16 (98): 112–118.

**30** Lee, P.C., Park, H.E., Morse, D.C., and Macosko, C.W. (2009). Polymer-polymer interfacial slip in multilayered films. *J. Rheol. (N. Y. N. Y)* 53 (4): 893–915.

**31** Park, H.E., Lee, P.C., and Macosko, C.W. (2010). Polymer-polymer interfacial slip by direct visualization and by stress reduction. *J. Rheol. (N. Y. N. Y)* 54 (6): 1207–1218.

**32** Lamnawar, K., Bousmina, M., and Maazouz, A. (2012). 2D Encapsulation in multiphase polymers: role of viscoelastic, geometrical and interfacial properties. *Macromolecules* 45 (1): 441–454.

**33** Harris, P.J., Carson, S.O., Gadley, J.L., and Maia, J. (2015). Multilayer coextrusion of rubber compounds. *Polym. Eng. Sci.* 55 (7): 1520–1527.

**34** Huang, R., Silva, J., Huntington, B.A. et al. (2015). Co-extrusion layer multiplication of rheologically mismatched polymers: a novel processing route. *Int. Polym. Process.* 30 (3): 317–330.

**35** Harris, P.J., Patz, J., Huntington, B.A. et al. (2014). Improved interfacial surface generator for the co-extrusion of micro- and nanolayered polymers. *Polym. Eng. Sci.* 54 (3): 636–645.

**36** Harris, P.J. (2015). *Layered Polymeric Systems: New Processing Methods and Novel Mechanical Design in Extensional Rheology*. Case Western Reserve University.

**37** Bironeau, A., Salez, T., Miquelard-Garnier, G., and Sollogoub, C. (2017). Existence of a critical layer thickness in PS/PMMA nanolayered films. *Macromol. (Washington, DC, United States)* 50 (10): 4064–4073.

**38** Feng, J., Zhang, Z., Bironeau, A. et al. (2018). Breakup behavior of nanolayers in polymeric multilayer systems – creation of nanosheets and nanodroplets. *Polymer (Guildf).* 143: 19–27.

**39** Liu, R.Y.F., Jin, Y., Hiltner, A., and Baer, E. (2003). Probing nanoscale polymer interactions by forced-assembly. *Macromol. Rapid Commun.* 24 (16): 943–948.

**40** Liu, R.Y.F., Ranade, A.P., Wang, H.P. et al. (2005). Forced assembly of polymer nanolayers thinner than the interphase. *Macromolecules* 38 (26): 10721–10727.

**41** Liu, R.Y.F., Bernal-Lara, T.E., Hiltner, A., and Baer, E. (2004). Interphase materials by forced assembly of glassy polymers. *Macromolecules* 37 (18): 6972–6979.

**42** Helfand, E. and Tagami, Y. (1972). Theory of the interface between immiscible polymers. II. *J. Chem. Phys.* 56 (7): 3592–3601.

**43** Helfand, E. and Tagami, Y. (1971). Theory of the interface between immiscible polymers. *J. Polym. Sci. Part B Polym. Lett. 9* (10): 741–746.

**44** Helfand, E. and Sapse, A.M. (1975). Theory of unsymmetric polymer–polymer interfaces. *J. Chem. Phys.* 62 (4): 1327–1331.

**45** Tadros, T. (2013). Interaction parameter. *Encycl. Colloid Interface Sci.* 634–634.

**46** Langhe, D. and Ponting, M. (2016). Gas transport, mechanical, interphase, and interdiffusion properties in coextruded-multilayered films. *Manuf. Nov. Appl. Multilayer Polym. Film.* 46–116.

**47** Carr, J.M., Langhe, D.S., Ponting, M.T. et al. (2012). Confined crystallization in polymer nanolayered films: a review. *J. Mater. Res.* 27 (10): 1326–1350.

**48** Douzinas, K.C. and Cohen, R.E. (2002). Chain folding in ethylene-butylene-ethylethylene semicrystalline diblock copolymers. *Macromolecules* 25 (19): 5030–5035.

**49** Hamley, I.W., Fairclough, J.P.A., Ryan, A.J. et al. (1996). Crystallization of nanoscale-confined diblock copolymer chains. *Polymer (Guildf).* 37 (19): 4425–4429.

**50** Bartczak, Z., Argon, A.S., Cohen, R.E., and Kowalewski, T. (1999). The morphology and orientation of polyethylene in films of sub-micron thickness crystallized in contact with calcite and rubber substrates. *Polymer (Guildf).* 40 (9): 2367–2380.

**51** Mellbring, O., Øiseth, S.K., Krozer, A. et al. (2001). Spin coating and characterization of thin high-density polyethylene films. *Macromolecules* 34 (21): 7496–7503.

**52** Padden, F.J. and Keith, H.D. (1966). Crystallization in thin films of isotactic polypropylene. *J. Appl. Phys.* 37 (11): 4013.

**53** Muratoğlu, O.K., Argon, A.S., and Cohen, R.E. (1995). Crystalline morphology of polyamide-6 near planar surfaces. *Polymer (Guildf).* 36 (11): 2143–2152.

**54** Tsuji, M., Fernando, A., Novillo, L. et al. (1999). Morphology of melt-crystallized poly(ethylene 2,6-naphthalate) thin films studied by transmission electron microscopy. *J. Mater. Res.* 14 (1): 251–258.

**55** Zhang, Y., Mukoyama, S., Hu, Y. et al. (2007). Thermal behavior and molecular orientation of poly(ethylene 2,6-naphthalate) in thin films. *Macromolecules* 40 (11): 4009–4015.

**56** Durell, M., Macdonald, J.E., Trolley, D. et al. (2002). The role of surface-induced ordering in the crystallisation of PET films. *Europhys. Lett.* 58 (6): 844–850.

**57** Zhu, L., Cheng, S.Z.D., Calhoun, B.H. et al. (2000). Crystallization temperature-dependent crystal orientations within nanoscale confined lamellae of a self-assembled crystalline–amorphous diblock copolymer. *J. Am. Chem. Soc.* 122 (25): 5957–5967.

**58** Schönherr, H. and Frank, C.W. (2003). Ultrathin films of poly(ethylene oxides) on oxidized silicon. 2. In situ study of crystallization and melting by hot stage AFM. *Macromolecules* 36 (4): 1199–1208.

**59** Kamdar, A.R., Ayyer, R.K., Poon, B.C. et al. (2009). Effect of tie-layer thickness on the adhesion of ethylene-octene copolymers to polypropylene. *Polymer (Guildf).* 50 (14): 3319–3328.

**60** Bu, Z., Yoon, Y., Ho, R.-M. et al. (1996). Crystallization, melting, and morphology of syndiotactic polypropylene fractions. 3. Lamellar single crystals and chain folding. *Macromolecules* 29 (20): 6575–6581.

**61** Mareau, V.H. and Prud'homme, R.E. (2004). In-situ hot stage atomic force microscopy study of poly(ε-caprolactone) crystal growth in ultrathin films. *Macromolecules* 38 (2): 398–408.

**62** Ho, R.-M., Lin, F.-H., Tsai, C.-C. et al. (2004). Crystallization-induced undulated morphology in polystyrene- b -poly(l -lactide) block copolymer. *Macromolecules* 37 (16): 5985–5994.

**63** Wang, H., Keum, J.K., Hiltner, A. et al. (2009). Confined crystallization of polyethylene oxide in nanolayer assemblies. *Science (80-.)* 323 (5915): 757–760.

**64** Wang, H., Keum, J.K., Hiltner, A., and Baer, E. (2010). Crystallization kinetics of poly(ethylene oxide) in confined nanolayers. *Macromolecules* 43 (7): 3359–3364.

**65** Ponting, M., Keum, J., Freeman, B. et al. (2011). Substrate effects on the confined crystallization of polycaprolactone in coextruded nanolayered films. *Annu. Tech. Conf. Soc. Plast. Eng.* 3: 2328–2332.

**66** Ponting, M., Lin, Y., Keum, J.K. et al. (2010). Effect of substrate on the isothermal crystallization kinetics of confined poly(ε-caprolactone) nanolayers. *Macromolecules* 43 (20): 8619–8627.

**67** Sun, Y.-S., Chung, T.-M., Li, Y.-J. et al. (2006). Crystalline polymers in nanoscale 1d spatial confinement. *Macromolecules* 39 (17): 5782–5788.

**68** Wang, J., Zhang, X., Zhao, T. et al. (2014). Morphologies and properties of polycarbonate/polyethylene in situ microfibrillar composites prepared through multistage stretching extrusion. *J. Appl. Polym. Sci.* 131 (7): 40108. (1–8).

**69** Shen, J., Wang, M., Li, J., and Guo, S. (2011). In situ fibrillation of polyamide 6 in isotactic polypropylene occurring in the laminating-multiplying die. *Polym. Adv. Technol.* 22 (2): 237–245.

**70** He, G., Li, J., Zhang, F. et al. (2014). Effect of multistage tensile extrusion induced fiber orientation on fracture characteristics of high density polyethylene/short glass fiber composites. *Compos. Sci. Technol.* 100: 1–9.

**71** Embabi, M., Kweon, M.S., Chen, Z., and Lee, P.C. (2020). Tunable tensile properties of polypropylene and polyethylene terephthalate fibrillar blends through micro-/nanolayered extrusion technology. *Polymers (Basel).* 12 (11): 2585.

**72** Neerincx, P.E., Denteneer, R.P.J., Peelen, S., and Meijer, H.E.H. (2011). Compact mixing using multiple splitting, stretching, and recombining flows. *Macromol. Mater. Eng.* 296 (3–4): 349–361.

**73** Neerincx, P.E. and Meijer, H.E.H. (2013). Fractal structuring in polymer processing. *Macromol. Chem. Phys.* 214 (2): 188–202.

**74** Wu, F., Misra, M., and Mohanty, A.K. (2021). Progress in polymer science challenges and new opportunities on barrier performance of biodegradable polymers for sustainable packaging. *Prog. Polym. Sci.* 117: 101395.

**75** Ikada, Y., Jamshidi, K., Tsuji, H., and Hyon, S.H. (1987). Stereocomplex formation between enantiomeric poly(lactides). *Macromolecules* 20 (4): 904–906.

**76** Yi, L., Li, D., Xu, Y. et al. (2019). Coexistence of transcrystallinity and stereocomplex crystals induced by the multilayered assembly of poly(L-lactide) and poly(D-lactide): a strategy for achieving balanced mechanical performances. *Ind. Eng. Chem. Res.* 58 (5): 1914–1923.

**77** Guo, Y., Peng, S., Wang, Q. et al. (2020). Achieving high-ductile polylactide sheets with inherent strength via a compact and uniform stress conduction network. *Ind. Eng. Chem. Res.* 59 (26): 12096–12105.

**78** Boufarguine, M., Guinault, A., Miquelard-Garnier, G., and Sollogoub, C. (2013). PLA/PHBV Films with improved mechanical and gas barrier properties. *Macromol. Mater. Eng.* 298 (10): 1065–1073.

**79** Jordan, A.M., Lenart, W.R., Carr, J.M. et al. (2014). Structural evolution during mechanical deformation in high-barrier PVDF-TFE/PET multilayer films using in situ X-ray techniques. *Acs Appl. Mater. & Interfaces* 6 (6): 3987–3994.

**80** Ebeling, T., Hiltner, A., and Baer, E. (1998). Delamination failure mechanisms in microlayers of polycarbonate and poly(styrene-co-acrylonitrile). *J. Appl. Polym. Sci.* 68 (5): 793–805.

**81** Adhikari, R., Seydewitz, V., Löschner, K. et al. (2010). Structure and properties of multilayered PET/PC composites. *Macromol. Symp.* 290 (1): 156–165.

**82** Adhikari, R., Lebek, W., Godehardt, R. et al. (2005). Investigating morphology and deformation behavior of multilayered PC/PET composites. *Polym. Adv. Technol.* 16 (2–3): 95–101.

**83** Scholtyssek, S., Seydewitz, V., Adhikari, R. et al. (2013). Influence of laminate thickness reduction on the deformation mechanism of coextruded multilayered PC/PMMA films. *J. Appl. Polym. Sci.* 127 (6): 4262–4272.

**84** Chen, B., Gao, W., Shen, J., and Guo, S. (2014). The multilayered distribution of intumescent flame retardants and its influence on the fire and mechanical properties of polypropylene. *Compos. Sci. Technol.* 93: 54–60.

**85** Lai, C.-Y., Ayyer, R.K., Hiltner, A., and Baer, E. (2010). Effect of confinement on the mechanical properties of polyethylene oxide in nanolayered films. In: *Annual Technical Conference - Society of Plastics Engineers*, vol. 2, 1580–1584. Society of Plastics Engineers.

**86** Mao, G.L., Andrews, J., Crescimanno, M. et al. (2011). Co-extruded mechanically tunable multilayer elastomer laser. *Opt. Mater. Express* 1 (1): 108–114.

**87** Lai, C.-Y., Ayyer, R.K., Hiltner, A., and Baer, E. (2010). Mechanical properties of confined polyethylene oxide nanolayers. *PMSE Prepr.*

**88** Adhikari, R., Seydewitz, V., Loschner, K. et al. (2010). Structure and properties of multi layered PET/PC composites. *Macromol. Symp.* 290: 156–165.

**89** Ma, M., Vijayan, K., Hiltner, A. et al. (1990). Thickness effects in microlayer composites of polycarbonate and poly(styrene-acrylonitrile). *J. Mater. Sci.* 25 (4): 2039–2046.

**90** Kerns, J., Hsieh, A., Hiltner, A., and Baer, E. (2000). Comparison of irreversible deformation and yielding in microlayers of polycarbonate with poly(methylmethacrylate) and poly(styrene-co-acrylonitrile). *J. Appl. Polym. Sci.* 77 (7): 1545–1557.

**91** Im, J. and Schrenk, W.J. (1988). Coextruded microlayer film and sheet. *J. Plast. Film Sheeting* 4 (2): 104–115.

**92** Ma, M., Im, J., Hiltner, A., and Baer, E. (1990). Fatigue crack propagation in microlayer composites of polycarbonate and poly(styrene–acrylonitrile). *J. Appl. Polym. Sci.* 40 (5–6): 669–684.

**93** Mackey, M., Flandin, L., Hiltner, A., and Baer, E. (2011). Confined crystallization of PVDF and a PVDF-TFE copolymer in nanolayered films. *J. Polym. Sci. Part B-Polymer Phys.* 49 (24): 1750–1761.

**94** Mandelkern, L. (1964). *Crystallization of Polymers: Volume 1, Equilibrium Concepts*, vol. 1. Cambridge University Press.

**95** Yin, K., Zhou, Z., Schuele, D.E. et al. (2016). Effects of interphase modification and biaxial orientation on dielectric properties of poly(ethylene terephthalate)/ poly(vinylidene fluoride-co-hexafluoropropylene) multilayer films. *ACS Appl. Mater. Interfaces* 8 (21): 13555–13566.

**96** Langhe, D. and Ponting, M. (2016). Dielectric and electrical properties of multilayered films. In: *Manufacturing and Novel Applications of Multilayer Polymer Films* (ed. D. Langhe and M. Ponting), 141–189. Elsevier.

**97** Carr, J.M., Mackey, M., Flandin, L. et al. (2013). Effect of biaxial orientation on dielectric and breakdown properties of poly(ethylene terephthalate)/ poly(vinylidene fluoride-co-tetrafluoroethylene) multilayer films. *J. Polym. Sci. Part B-Polymer Phys.* 51 (11): 882–896.

**98** Luo, J., Du, J., Tang, Q., and Mao, C. (2008). Lead sodium niobate glass-ceramic dielectrics and internal electrode structure for high energy storage density capacitors. *IEEE Trans. Electron Devices* 55 (12): 3549–3554.

**99** Kim, P., Jones, S.C., Hotchkiss, P.J. et al. (2007). Phosphonic acid-modified barium titanate polymer nanocomposites with high permittivity and dielectric strength. *Adv. Mater.* 19 (7): 1001–1005.

**100** Sarjeant, W.J., Clelland, I.W., and Price, R.A. (2001). Capacitive components for power electronics. *Proc. IEEE* 89 (6): 846–855.

**101** Lendlein, A. and Gould, O.E.C. (2019). Reprogrammable recovery and actuation behaviour of shape-memory polymers. *Nat. Rev. Mater. 2018 42* 4 (2): 116–133.

**102** Zheng, Y., Zeng, B., Yang, L. et al. (2020). Fabrication of thermoplastic polyurethane/polycaprolactone multilayered composites with confined distribution of mwcnts for achieving tunable thermo- and electro-responsive shape-memory performances. *Ind. Eng. Chem. Res.* 59 (7): 2977–2987.

**103** Ji, S.Z., Wang, J., Olah, A., and Baer, E. (2017). Triple-shape-memory polymer films created by forced-assembly multilayer coextrusion. *J. Appl. Polym. Sci.* 134 (5): 44405. (1–10).

**104** Armstrong, S.R., Du, J., and Baer, E. (2014). Co-extruded multilayer shape memory materials: nano-scale phenomena. *Polymer (Guildf).* 55 (2): 626–631.

**105** Zheng, Y., Ji, X., Wang, Q. et al. (2018). Structural design of polyurethane/ poly(butylene succinate)/polycaprolactone compounds: via a multilayer-assembled strategy: achieving tunable triple-shape memory performances. *RSC Adv.* 8 (74): 42337–42345.

**106** Zheng, Y., Ji, X., Yin, M. et al. (2017). Strategy for fabricating multiple-shape-memory polymeric materials via the multilayer assembly of co-continuous blends. *ACS Appl. Mater. Interfaces* 9 (37): 32270–32279.

**107** Zheng, Y., Dong, R., Shen, J., and Guo, S. (2016). Tunable shape memory performances via multilayer assembly of thermoplastic polyurethane and polycaprolactone. *ACS Appl. Mater. Interfaces* 8 (2): 1371–1380.

**108** Zeng, B., Yang, L., Qin, J. et al. (2020). Nacre-inspired polymeric materials with body heat-responsive shape-memory effect, high optical transparence, and balanced mechanical properties. *ACS Appl. Mater. Interfaces* 12 (46): 52008–52017.

**109** Li, Z.P., Black, T., Rahman, M.A. et al. (2018). Opto-mechanical programming of micro-scale information on transparent multilayer shape memory film. *Polymer (Guildf).* 137: 156–168.

**110** Ji, X., Chen, D., Zheng, Y. et al. (2019). Multilayered assembly of poly(vinylidene fluoride) and poly(methyl methacrylate) for achieving multi-shape memory effects. *Chem. Eng. J. (Amsterdam, Netherlands)* 362: 190–198.

**111** Wong, A., Guo, Y., and Park, C.B. (2013). Fundamental mechanisms of cell nucleation in polypropylene foaming with supercritical carbon dioxide – effects of extensional stresses and crystals. *J. Supercrit. Fluids* 79: 142–151.

**112** Leung, S.N., Wong, A., Guo, Q. et al. (2009). Change in the critical nucleation radius and its impact on cell stability during polymeric foaming processes. *Chem. Eng. Sci.* 64 (23): 4899–4907.

**113** Langhe, D. and Ponting, M. (2016). *Introduction to Multilayered Films*. Elsevier.

**114** Rahman, M.A., Andrade, R., Maia, J., and Baer, E. (2015). Viscosity contrast effects on the structure – property relationship of multilayer soft film/foams. *Polymer (Guildf).* 69: 110–122.

**115** Wu, H., Zhang, J., Zhang, C. et al. (2016). Structure–properties relationship of a novel multilayer film/foam material produced through co-extrusion and orientation. *Ind. Eng. Chem. Res.* 55 (41): 10947–10954.

**116** Lee, S., Morgan, A.B., Schiraldi, D.A., and Maia, J. (2020). Improving the flame retardancy of polypropylene foam with piperazine pyrophosphate via multilayering coextrusion of film/foam composites. *J. Appl. Polym. Sci.* 137 (15): 48552.

**117** Feng, J.X., Li, Z.P., Olah, A., and Baer, E. (2018). High oxygen barrier multilayer EVOH/LDPE film/foam. *J. Appl. Polym. Sci.* 135 (26): 9.

**118** Ranade, A.P. (2007). *Structure Property Relationships in Various Layered Polymeric Systems*. Case Western Reserve University.

**119** Lee, S.-T. (1993). Shear effects on thermoplastic foam nucleation. *Polym. Eng. Sci.* 33 (7): 418–422.

**120** Park, C.B. and Suh, N.P. (1996). Rapid polymer/gas solution formation for continuous production of microcellular plastics. *J. Manuf. Sci. Eng.* 118 (4): 639–645.

**121** Collais, D.I. and Baird, D.G. (1995). Tesile toughness of microcellular foams of polystyrene, styrene-acrylonitrile copolymer, and polycarbonate, and the effect of dissolved gas on the tensile toughness of the same polymer matrices and microcellular foams. *Polym. Eng. Sci.* 35 (14): 1167–1177.

**122** Collias, D.I., Baird, D.G., and Borggreve, R.J. (1994). Impact toughening of polycarbonate by microcellular foaming. *Polymer (Guildf).* 35 (18): 3978–3983.

**123** Doroudiani, S., Park, C.B., and Kortschot, M.T. (1998). Processing and characterization of microcellular foamed high-density polythylene/isotactic polypropylene blends. *Polym. Eng. Sci.* 38 (7): 1205–1215.
**124** Matuana, L.M., Park, C.B., and Balatinecz, J.J. (1997). Processing and cell morphology relationships for microcellular foamed pvc/wood-fiber composites. *Polym. Eng. Sci.* 37 (7): 1137–1147.
**125** Seeler, K.A. and Kumar, V. (1993). Tension-tension fatigue of microcellular polycarbonate: initial results. *J. Reinf. Plast. Compos.* 12 (3): 359–376.
**126** Shimbo, M., Baldwin, D.F., and Suh, N.P. (1995). The viscoelastic behavior of microcellular plastics with varying cell size. *Polym. Eng. Sci.* 35 (17): 1387–1393.
**127** Baldwin, D.F. and Suh, N. (1992). Microcellular poly (ethylene terephthalate) and crystallizable poly (ethylene terephthalate): characterization of process variables. In: *ANTEC 92--Shaping the Future*, vol. 38, 1503–1507.
**128** Glicksman, L.R. (1994). Heat transfer in foams. In: *Low Density Cellular Plastics* (ed. N.C. Hilyard and A. Cunningham), 104–152. Dordrecht: Springer.
**129** Ghaffari Mosanenzadeh, S., Naguib, H.E., Park, C.B., and Atalla, N. (2015). Design and development of novel bio-based functionally graded foams for enhanced acoustic capabilities. *J. Mater. Sci.* 50 (3): 1248–1256.
**130** Mosanenzadeh, S.G., Doutres, O., Naguib, H.E. et al. (2015). A semi-empirical model relating micro structure to acoustic properties of bimodal porous material. *J. Appl. Phys.* 117 (3): 034305.
**131** Ghaffari Mosanenzadeh, S., Doutres, O., Naguib, H.E. et al. (2015). A numerical scheme for investigating the effect of bimodal structure on acoustic behavior of polylactide foams. *Appl. Acoust.* 88: 75–83.
**132** Chen, H., Guan, R., Zhao, J. et al. (2013). Microcellular foaming of plasticized thin PC sheet I. Effects of processing conditions. *J. Wuhan Univ. Technol. Sci. Ed.* 28 (2): 235–242.
**133** Han, T., Wang, X., Xiong, Y. et al. (2015). Light-weight poly(vinyl chloride)-based soundproofing composites with foam/film alternating multilayered structure. *Compos. Part A Appl. Sci. Manuf.* 78: 27–34.
**134** Lee, S., Zhu, L., and Maia, J. (2015). The effect of strain-hardening on the morphology and mechanical and dielectric properties of multi-layered PP foam/PP film. *Polymer (Guildf).* 70: 173–182.
**135** Rahman, M.A., Wang, J., Zhang, C. et al. (2016). Novel micro-/nano- porous cellular membranes by forced assembly co-extrusion technology. *Eur. Polym. J.* 83: 99–113.
**136** Ranade, A.P., Hiltner, A., Baer, E., and Bland, D.G. (2004). Structure-property relationships in coextruded foam/film microlayers. *J. Cell. Plast.* 40 (6): 497–507.
**137** Sun, X. and Liang, W. (2016). Cellular structure control and sound absorption of polyolefin microlayer sheets. *Compos. Part B Eng.* 87: 21–26. a.

**138** Zhao, T., Wu, H., Guo, S. et al. (2015). Morphology and properties of polypropylene foaming sheet with alternating multilayered structure. *Polym. Plast. Technol. Eng.* 54 (9): 899–906.

**139** Zhao, T., Yang, M., Wu, H. et al. (2015). Preparation of a new foam/film structure poly (ethylene-co-octene) foam materials and its sound absorption properties. *Mater. Lett.* 139: 275–278.

**140** Zhao, T., Zhang, X., Wu, H. et al. (2015). Rheology, morphology, and mechanical properties of hmspp/poe blends and its alternate layered foam. *J. Appl. Polym. Sci.* 132 (4): 41339. (1–7).

**141** Leung, S.N., Park, C.B., and Li, H. (2006). Numerical simulation of polymeric foaming processes using modified nucleation theory. *Plast. Rubber Compos.* 35 (3): 93–100.

**142** Park, C.B., Baldwin, D.F., and Suh, N.P. (1995). Effect of the pressure drop rate on cell nucleation in continuous processing of microcellular polymers. *Polym. Eng. Sci.* 35 (5): 432–440.

**143** Leung, S.N., Wong, A., Wang, L.C., and Park, C.B. (2012). Mechanism of extensional stress-induced cell formation in polymeric foaming processes with the presence of nucleating agents. *J. Supercrit. Fluids* 63: 187–198.

**144** Leung, S.N., Park, C.B., Xu, D. et al. (2006). Computer simulation of bubble-growth phenomena in foaming. *Ind. Eng. Chem. Res.* 45 (23): 7823–7831.

**145** Wong, A., Chu, R.K.M., Leung, S.N. et al. (2011). A batch foaming visualization system with extensional stress-inducing ability. *Chem. Eng. Sci.* 66 (1): 55–63.

**146** Lee, P.C., Wang, J., and Park, C.B. (2005). Extruded open-cell foams using two semi-crystalline polymers with different crystallization temperatures. *Annu. Tech. Conf. - ANTEC, Conf. Proc.* 7: 177–181.

**147** Xu, X., Park, C.B., Xu, D., and Pop-Iliev, R. (2003). Effects of die geometry on cell nucleation of PS foams blown with $CO_2$. *Polym. Eng. Sci.* 43 (7): 1378–1390.

**148** Chen, Z., Faysal, A.A., Embabi, M. et al. (2021). A path to nano-cellular foams: constrained cell nucleation and growth in micro-/nano-layered structures. *Polymer (Guildf).* 235: 124272. (1–11).

**149** Wong, A. and Park, C.B. (2012). The effects of extensional stresses on the foamability of polystyrene-talc composites blown with carbon dioxide. *Chem. Eng. Sci.* 75: 49–62.

**150** Zhao, J., Wang, G., Zhang, L. et al. (2019). Lightweight and strong fibrillary PTFE reinforced polypropylene composite foams fabricated by foam injection molding. *Eur. Polym. J.* 119 (June): 22–31.

**151** Wang, G.G., Zhao, J., Wang, G.G. et al. (2020). Strong and Super thermally insulating in-situ nanofibrillar PLA/PET composite foam fabricated by high-pressure microcellular injection molding. *Chem. Eng. J.* 390 (November 2019): 124520/1–124520/14.

**152** Notario, B., Pinto, J., and Rodríguez-Pérez, M.A. (2015). Towards a new generation of polymeric foams: PMMA nanocellular foams with enhanced physical properties. *Polymer (Guildf).* 63: 116–126.

**153** Bethke, C., Goedderz, D., Weber, L. et al. (2020). Improving the flame-retardant property of bottle-grade PET foam made by reactive foam extrusion. *J. Appl. Polym. Sci.* 137 (35): 1–15.

**154** Sorrentino, L., Di Maio, E., and Iannace, S. (2010). Poly(ethylene terephthalate) foams: correlation between the polymer properties and the foaming process. *J. Appl. Polym. Sci.* 116 (5): 2658–2667.

**155** Li, D., Liu, T., Zhao, L., and Yuan, W. (2012). Controlling sandwich-structure of PET microcellular foams using coupling of $CO_2$ diffusion and induced crystallization. *AIChE J.* 58 (8): 2512–2523.

**156** Lee, S.T., Kareko, L., and Jun, J. (2008). Study of thermoplastic PLA foam extrusion. *J. Cell. Plast.* 44 (4): 293–305.

**157** Nofar, M. (2016). Effects of nano-/micro-sized additives and the corresponding induced crystallinity on the extrusion foaming behavior of PLA using supercritical $CO_2$. *Mater. Des.* 101: 24–34.

**158** Julien, J.M., Quantin, J.C., Bénézet, J.C. et al. (2015). Chemical foaming extrusion of poly(lactic acid) with chain-extenders: physical and morphological characterizations. *Eur. Polym. J.* 67: 40–49.

**159** Benning, C.J. (1967). Polyethylene foam III – "orientation in thermoplastic foams". *J. Cell. Plast.* 3 (4): 174–184.

**160** Langhe, D. and Ponting, M. (2016). Novel multilayered structures and applications. In: *Manufacturing and Novel Applications of Multilayer Polymer Films* (ed. D. Langhe and M. Ponting), 190–220. Elsevier.

# 7

# Heat-Setting of Polyester Films

*Jehuda Greener*[1] *and Andy H. Tsou*[2]

[1] *J. Greener Consulting, Rochester, NY, USA*
[2] *National Taiwan University, Houston, TX, USA*

## 7.1 Introduction

Heat-setting, or constrained high-temperature annealing, is an important step in the tenter-frame (tentering) process [1–3] commonly used for making biaxially oriented polyester films, with potentially significant impact on dimensional stability, mechanical properties, and other important physical characteristics of the stretched film. Heat-setting is usually applied immediately following the transverse direction (TD) stretch while the biaxially stretched film is constrained by the tenter-frame section of the filmmaking machine, although other process sequences are possible. The essential microstructural features of the film are established during the sequential stretching steps as the material undergoes molecular alignment and strain-induced crystallization. However, because of the low temperature and rapid crystallization in these steps (the kinetics of strain-induced crystallization is orders-of-magnitude faster than that of spontaneous thermal crystallization [4, 5]), the crystalline domains formed during stretching contain many lattice and conformational defects. These defects can be partially eliminated and replaced with a denser crystalline structure by suitable annealing at high temperatures. In turn, the thickening and perfecting of the crystalline domains and the attendant changes in morphology and orientation can significantly alter the physical performance and uniformity of the stretched film. The importance of heat-setting as a means to optimize the structure and performance of oriented films and fibers has been well recognized. The morphological changes underlying the heat-setting process step in poly(ethylene terephthalate) (PET),

*Polyester Films: Materials, Processes and Applications*, First Edition.
Edited by Miko Cakmak and Jehuda Greener.

poly(ethylene-2,6-naphthalene dicarboxylate) (PEN), and other semi-crystalline polymers have been studied extensively over the past six decades by means of X-ray diffraction, IR spectroscopy, calorimetry, rheology, and other characterization techniques [5–24]. Most studies conclude that the increase in the crystalline fraction and the size of the crystalline domains during heat-setting signify a qualitative change in the superstructure and morphology of the oriented matrix, but the exact nature of this microstructural change has been the subject of some debate in the open literature. In studies on biaxially stretched PET films and uniaxially drawn PET fibers, Schultz and coworkers [18,19] contend, in line with earlier observations by Fischer, Fakirov, and coworkers [7–9], Gupta et al. [13,14], and others [12], that the heat-setting process not only reduces the concentration of lattice defects but it also leads to the formation of lamellar crystals extending normal to the chain (or drawing) axis. In a study of the effects of heat-setting on the physical properties of biaxially stretched PET films, Gohil [20] has noted that the heat-setting process can be divided into two temperature regimes marked by different crystallization rates and different temperature dependencies of some key physical properties. In a wide-ranging study on the heat-setting of stretched polyester films, Greener et al. [21] compared the heat-set response of PET and PEN over a wide range of process conditions. They present a molecular mechanism similar to the one proposed by Schultz et al. [18,19] and previously by Fischer, Fakirov, and coworkers [7–9] to explain the myriad observations for both polyesters. In Section 7.2, drawing mostly from the study of Greener et al. [21], we illustrate the effects of a controlled heat-setting treatment on the morphology and physical performance of PET and PEN films and assess the similarity and differences between these materials stemming from their different molecular structures. The microstructural changes induced by heat-setting and the implications of these changes to some macroscopic properties of the stretched films are discussed in Section 7.3. Finally, in Section 7.4, we summarize the results and discuss practical implications to the manufacture of polyester films.

## 7.2 Some Observations on the Effects of Heat-Setting

In their phenomenological study on heat-setting, Greener et al. [21] applied controlled heat-setting treatment by annealing biaxially stretched PET and PEN films in a thermostated oil bath in a way that closely simulates actual heat-set conditions in a tenter-frame machine. Annealing times and temperatures (closely related to heat-set times and temperatures) varied over wide ranges, and the annealed/heat-set films were then characterized using a variety of analytical techniques, including calorimetry, densitometry, dynamic-mechanical analysis, thermo-mechanical analysis, X-ray diffraction, refractometry, and spectroscopy. It must be noted that the characterization of the various annealed samples was done offline, and some of the annealing conditions applied are not fully representative of real machine conditions. Yet, the results of this study can serve to illustrate general trends and highlight the key differences between two widely used polyester resins, PET and PEN. A more direct online methodology for measuring the rheo-optical properties of polyester films was developed by Cakmak et al. [25] and demonstrated for use in a tenter-frame process by Hassan and Cakmak [26, 27]. Some selected results of the study of Greener et al. [21] are presented below.

### 7.2.1 Densitometry

The evolution of density with anneal (soak) time ($t_{HS}$) at several heat-set temperatures ($T_{HS}$) is shown in Figures 7.1 and 7.2 for PET and PEN films, respectively. In both cases, the density increases asymptotically to a value that remains constant for a long time (~60 min) after saturation, but this asymptotic value is dependent on the applied heat-set temperature. The kinetics of the densification process appears to be independent of $T_{HS}$, but some differences in saturation time between the materials are noted. These differences are attributed to variations in film thickness, i.e. the kinetics of the heat-setting process at the experimental time scales is controlled, most likely, by thermal diffusivity. Indeed, in a study by Lee and Schultz [19], crystallization half times for the heat-setting process in PET

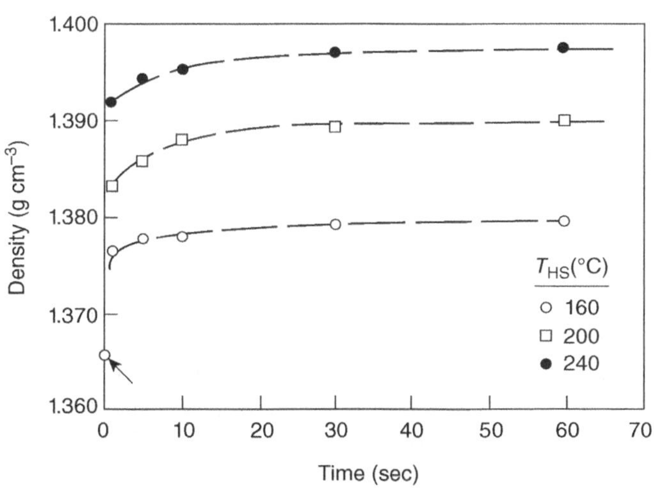


**Figure 7.1** Evolution of density with anneal time for PET films. Effect of heat-set temperature. Arrow indicates NHS film. *Source:* Greener et al. [21]/ with permission from John Wiley & Sons.

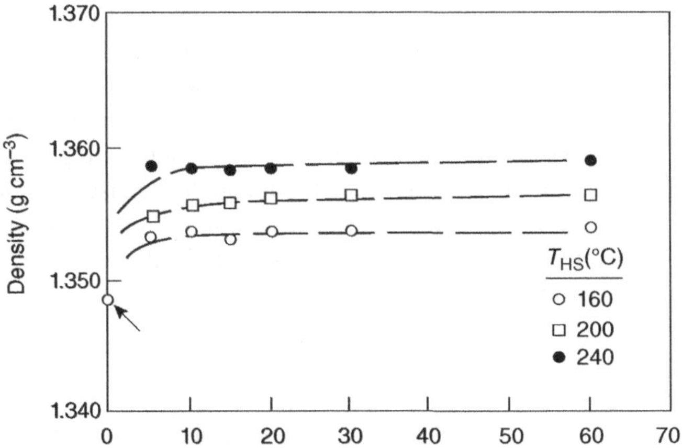


**Figure 7.2** Evolution of density with anneal time for PEN films. Effect of heat-set temperature. Arrow indicates NHS film. *Source:* Greener et al. [21]/ with permission from John Wiley & Sons.

fibers were found to be of the order of 10–100 msec, much less than the annealing times covered in the study of Greener et al., i.e. the crystallization process may be considered to be instantaneous. In both cases, saturation is attained at times <20 s, with the actual equilibration time depending primarily on the thickness of the sample. In all cases, the heat-set (annealed) samples are benchmarked against corresponding biaxially stretched samples that were unannealed (designated NHS).

Figure 7.3 shows the effect of $T_{HS}$ on density at saturation ($t_{HS} = 30$ s). A simple, monotone relationship is found for both materials, indicating a steady increase in the crystalline fraction with heat-set temperature, with PEN exhibiting lower densities overall and a lower rate of change of density with $T_{HS}$. Assuming a two-phase model, the corresponding changes in the crystalline volume fraction ($X_c$) can be estimated from:

$$X_c \approx \frac{\rho - \rho_a}{\rho_c - \rho_a} \tag{7.1}$$

where $\rho_c$ and $\rho_a$ are the densities of the crystalline and amorphous domains. The values of $\rho_a$ and $\rho_c$ are 1.335 and 1.455 g cm$^{-3}$ for PET [28] and 1.325 and 1.407 g cm$^{-3}$ for PEN [29], respectively. Although the use of Eq. (7.1) has been called into question on the grounds that a two-phase model is not strictly valid for highly oriented semi-crystalline polymers [8], it does provide a reasonable estimate of the degree of crystallinity. A correlation between $X_c$, and $T_{HS}$ for PET and PEN is given in Figure 7.4.

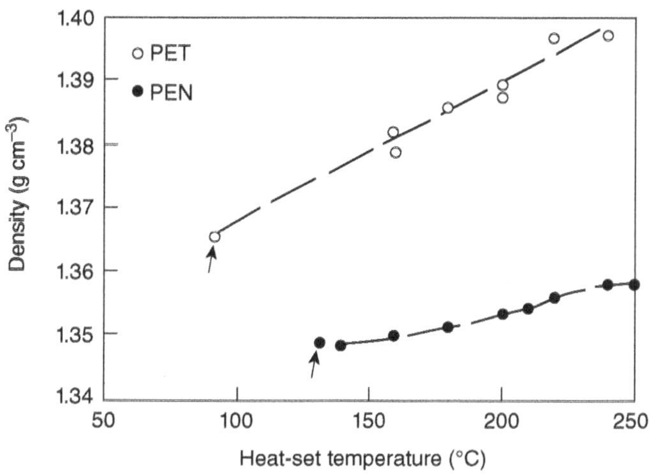


**Figure 7.3** Density vs. heat-set temperature for $t_{HS} = 30$ s. Arrows indicate NHS films. *Source:* Greener et al. [21]/ with permission from John Wiley & Sons.

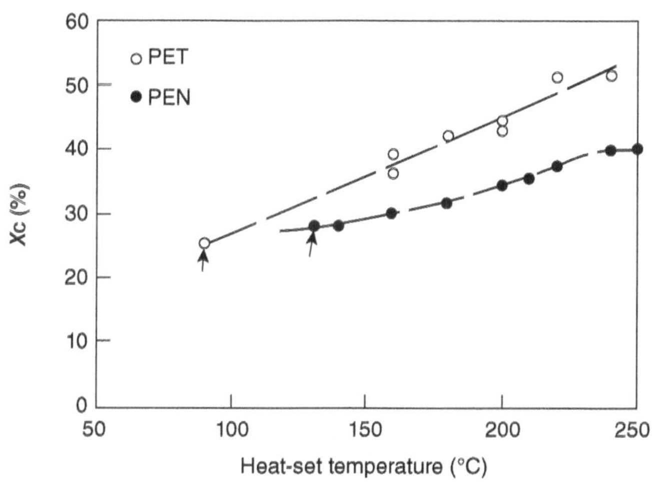


**Figure 7.4** Degree of crystallinity (volume fraction) vs. heat-set temperature ($t_{HS} = 30$ s). Arrows indicate NHS films. *Source:* Greener et al. [21]/ with permission from John Wiley & Sons.

We note that, over the wide range of temperatures covered, the change in crystalline fraction is substantial for both materials but especially for PET (25 → 52% vs. 29 → 41% for PEN), and $X_c$ increases approximately linearly with $T_{HS}$. (Buchner et al. [30] have proposed a higher value for $\rho_a$ for PEN, 1.340 g cm$^{-3}$, which will tend to lower the estimated crystalline fraction by >30%, so the results listed above may be considered a "high" estimate for PEN.) This result indicates that the heat-setting process brings about a significant transformation in the material, the extent of which depends strictly on the heat-set temperature once thermal equilibration has been attained. Also, the "efficiency" of the process in terms of its ability to crystallize amorphous matter is somewhat lower for PEN than for PET. This may be due to lower inherent crystallizability of the PEN matrix or simply a result of some uncertainty in the values of the amorphous and crystalline densities of this material (see above comment).

### 7.2.2 Thermal Analysis

Differential scanning calorimetry (DSC) of the heat-treated films also reveals some interesting features of the heat-setting process. Figures 7.5 and 7.6 show DSC scans of film samples heat-set at various temperatures for a fixed soak (heat-set) time of 30 s.

The thermograms for both materials feature a low-temperature "secondary" melting peak ($T'_m$) at temperatures that nearly coincide with the actual heat-set temperature applied to prepare the sample. The enthalpy associated with the secondary endotherm is considerably smaller than that of the primary peak, but even if heat-setting takes place within the primary melting range of the material (>240 °C), the secondary peak retains its identity and appears to simply superimpose on top of the primary melting endotherm; The two melting events, which can be readily deconvoluted, represent, most likely, two distinct morphologies, one

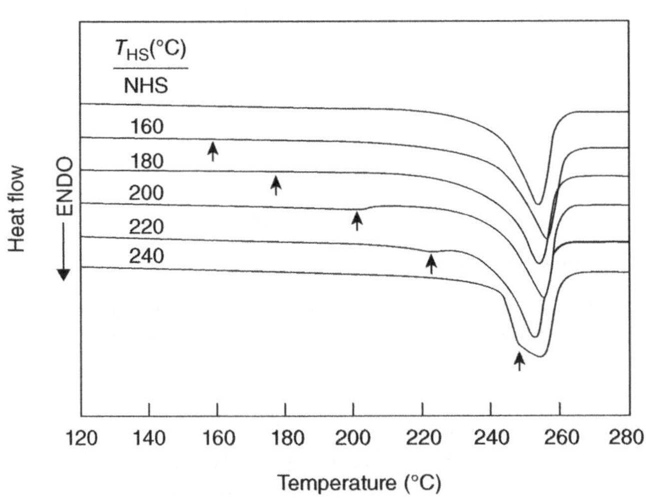


**Figure 7.5** DSC thermograms for PET films. Effect of heat-set temperature ($t_{HS}$ = 30 s). *Source:* Greener et al. [21]/ with permission from John Wiley & Sons.

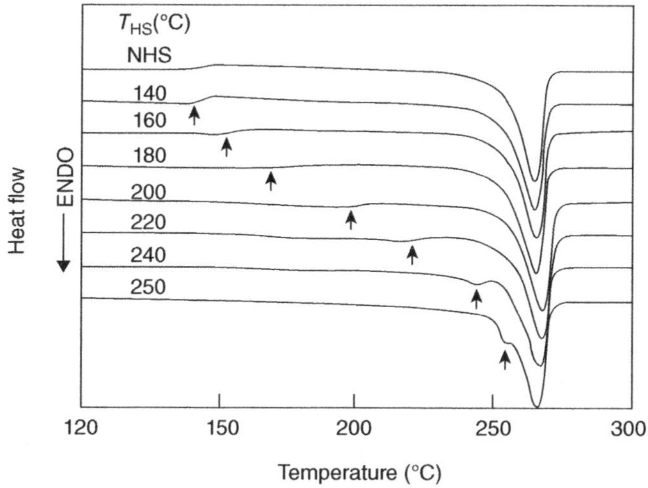


**Figure 7.6** DSC thermograms for PEN films. Effect of heat-set temperature ($t_{HS}$ = 30 s). *Source:* Greener et al. [21]/ with permission from John Wiley & Sons.

(represented by $T'_m$) intimately associated with the heat-setting process and the other ($T_m$) practically independent of the heat-setting treatment. A plot of the secondary melt temperature vs. heat-set temperature, Figure 7.7, shows a near identity between these temperatures, suggesting that some of the crystalline domains formed during the heat-setting step are relatively unstable and thus can be melted out at temperatures only slightly higher than their formation temperature.

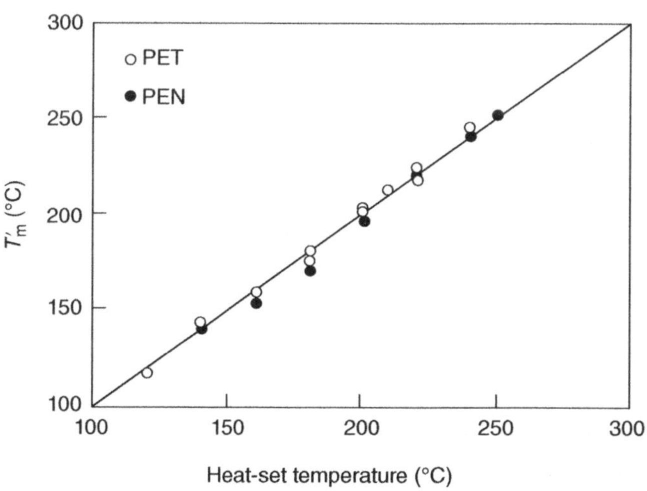


**Figure 7.7** Position of secondary melting peak ($T'_m$) vs. heat-set temperature. Line represents identity. *Source:* Greener et al. [21]/ with permission from John Wiley & Sons.

Studies on the dual-crystallization behavior of PET and other polymers were reported by Fakirov et al. [9], Holdsworth and Turner-Jones [31], Hassan and Cakmak [26,27], and others [20,32–35], all of whom detected secondary melting peaks at temperatures close to or above the annealing temperature. However, Fakirov et al. observed a dual-melting behavior, similar to that represented in Figures 7.5 and 7.6 [21], only for undrawn amorphous PET, while the secondary melting temperatures for highly drawn (5×) fibers were generally much higher than the treatment temperatures and strongly overlapped with the primary peaks. A similar result for highly drawn PET fibers was noted by Buckley and Salem [15]. Oswald et al. [35], who studied the annealing of PET fibers without constraint (not strictly heat-setting!), show similar results to the ones reported by Greener et al. [21] except that their secondary melting peaks appear to be weakly dependent on the annealing time beyond thermal equilibration, contrary to the observations reported above. The evolution of the secondary peak with time for PET film is illustrated in Figure 7.8. The peaks for short annealing times (i.e. before equilibration) appear at lower temperatures, i.e. the effective temperature "seen" by the material prior to full equilibration is lower due to thermal lag in the film. These results as well as similar results for PEN film, together with the transient density data in Figures 7.1 and 7.2, validate the conclusion that, under the experimental conditions for the data reported here, the kinetics of the heat-setting process is controlled by thermal diffusivity.

Hassan and Cakmak [24] also presented DSC scans for simultaneously and sequentially biaxially stretched PET film samples heat-set at 180 °C. The

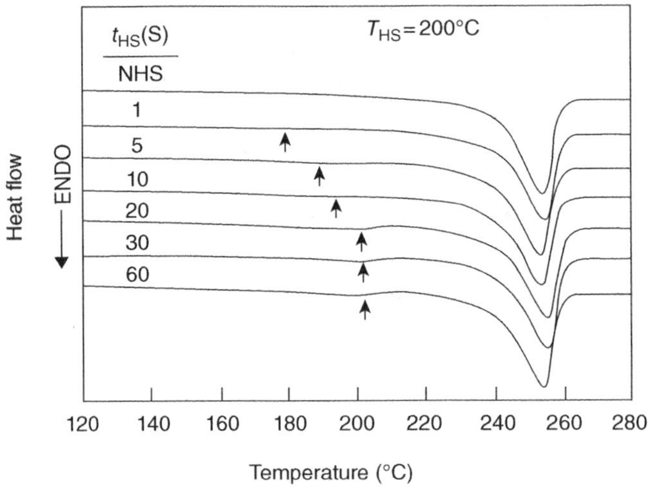


**Figure 7.8** DSC thermograms for PET films at various anneal times ($T_{HS}$ = 200 °C). *Source:* Greener et al. [21]/ with permission from John Wiley & Sons.

as-stretched samples exhibited a cold crystallization exotherm at 135–140 °C and a melting endotherm at approximately 250 °C. The position of the primary melting peak did not change with stretch ratio and stretching mode. Heat-setting at 180 °C for 1800 s caused the disappearance of the cold crystallization exotherm for all stretch ratios and modes and the appearance of a secondary melting peak similar to the one observed by Greener et al. [21]. This endotherm appeared to move closer to the primary melting endotherm with increase in the area stretch ratio.

### 7.2.3 Thermomechanical Analysis

Another characterization method that can be used to probe the microstructure of the heat-treated films is thermomechanical analysis (TMA). The TMA test traces the dimensional change (shrinkage or expansion) of the sample by scanning over a set temperature range at a given rate. TMA scans for the test materials for various heat-set temperatures ($t_{HS}$ = 30 s) are shown in Figures 7.9 and 7.10.

The most striking feature in these traces is the appearance of a distinct shrinkage signature in the film above some critical temperature, which seems to roughly coincide with the corresponding heat-set temperature. The results for PET and PEN suggest that the secondary crystalline structure formed during heat-setting has an "interlocking" effect on the sample, which prevents

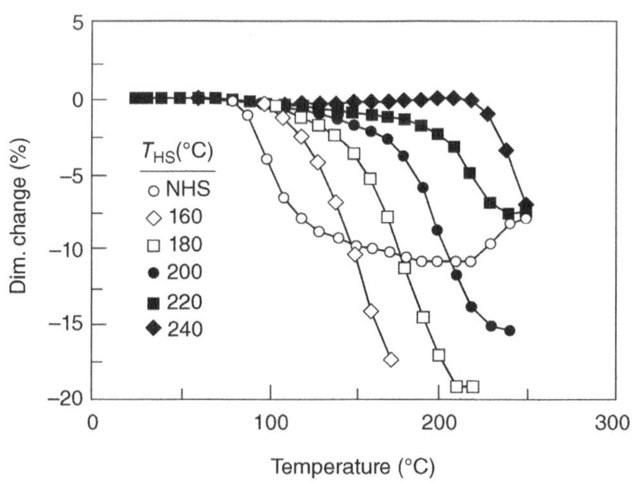


**Figure 7.9** TMA scans for PET film: % change in sample dimension (along MD) vs. temperature. Effect of heat-set temperature ($t_{HS}$ = 30 s). *Source:* Greener et al. [21]/ John Wiley & Sons.

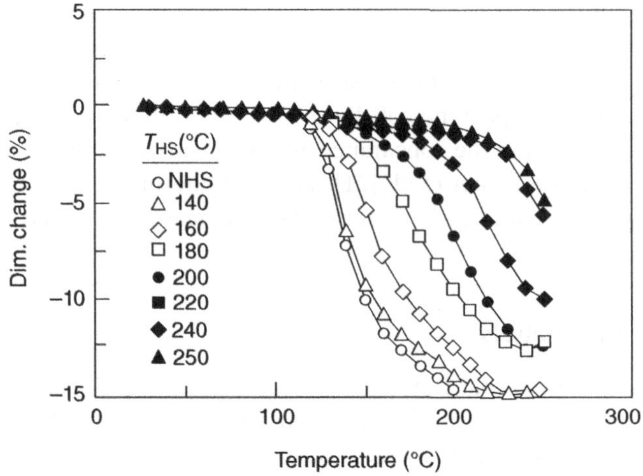


**Figure 7.10** TMA scans for PEN film: % change in sample dimension (along MD) vs. temperature. Effect of heat-set temperature ($t_{HS}$ = 30 s). *Source:* Greener et al. [21]/ with permission from John Wiley & Sons.

shrinkage or disorientation of the taut amorphous phase at temperatures below the melting of the "secondary" crystallites. However, once these crystallites are melted out, at $T > {\sim}T_{HS}$, the highly strained amorphous phase is free to disorient (shrink) and thereby lower its free energy. By contrast, it appears that the "primary" crystalline phase formed during the stretching steps (and partly during the DSC scan [25]) does not have the "interlocking" structure needed to prevent slippage and retraction of the oriented amorphous domains above the glass transition temperature. It is interesting to note, however, that even at high $T_{HS}$, both film types exhibit a mild inflection in the TMA trace (slight shrinkage) at a temperature that is nearly coincident with the glass transition temperature of the material (~94 and 120 °C for PET and PEN, respectively). This indicates that a small portion of the amorphous phase is not fully "locked" by the crystalline network formed during heat-setting, and it thus can partially retract once the amorphous domains gain sufficient mobility at $T_g$. Eliminating this "subtle" shrinkage may require additional annealing steps, which would have to be applied offline under low conveyance tension. The evolution in the shrinkage response of PET film with anneal time for $T_{HS}$ = 200 °C is shown in Figure 7.11.

The onset of shrinkage in the TMA traces appears to be shifted to lower temperatures for short exposure times (heat-set times), in line with the observation that incomplete heating due to thermal lag corresponds to exposure to lower $T_{HS}$ (cf. Figure 7.9).

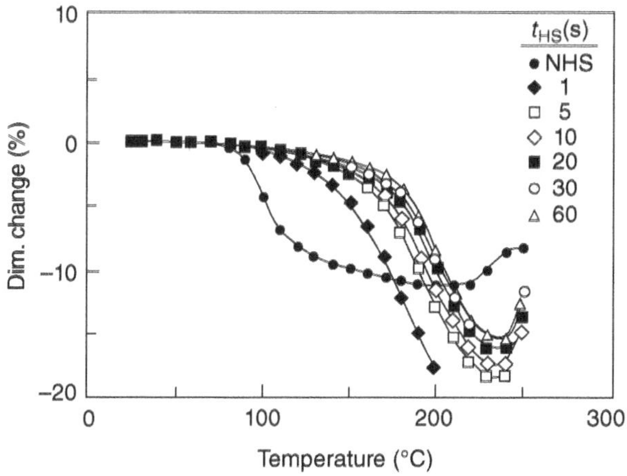


**Figure 7.11** TMA scans for PET film: % change in sample dimension (along MD) vs. temperature. Effect of heat-set time ($T_{HS}$ = 200 °C). *Source:* Greener et al. [21]/ with permission from John Wiley & Sons.

### 7.2.4 Dynamic-Mechanical Analysis

The attendant morphological changes during heat-setting also have a pronounced effect on the dynamic-mechanical response of the films. Figures 7.12 and 7.13 show isochronous plots of loss tangent vs. temperature for the test materials at several heat-set temperatures. Two outstanding features are noted for both

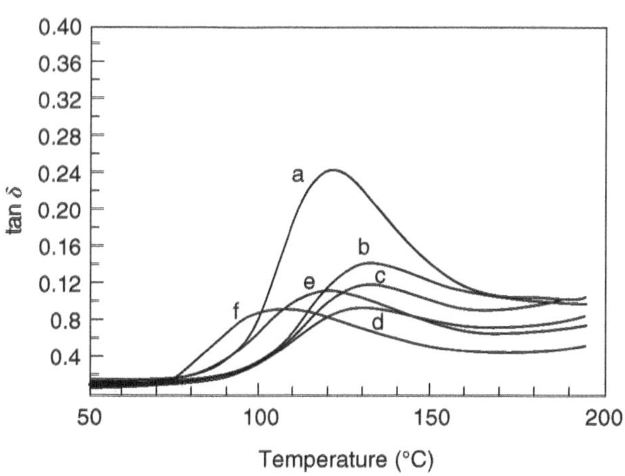


**Figure 7.12** Isochronous dynamic-mechanical response of PET films. Loss tangent vs. temperature at 1 Hz. Effect of heat-set temperature: (a) NHS film, (b) 160, (c) 180, (d) 200, (e) 220, and (f) 240 °C, ($t_{HS}$ = 30 s). *Source:* Greener et al. [21]/ with permission from John Wiley & Sons.

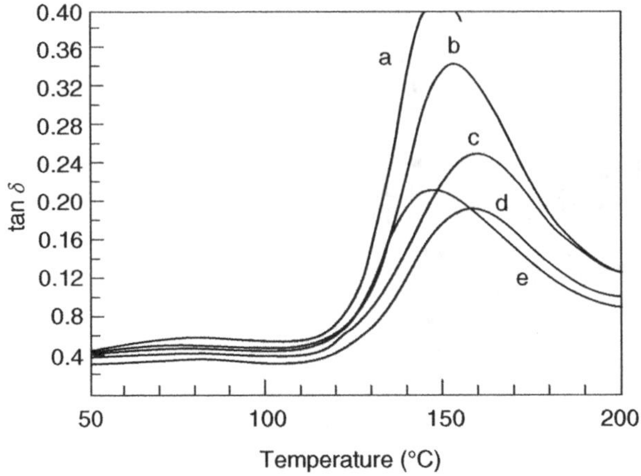


**Figure 7.13** Isochronous dynamic-mechanical response of PEN films. Loss tangent vs. temperature at 1 Hz. Effect of heat-set temperature: (a) NHS film, (b) 140, (c) 180, (d) 220, (e) 240 °C, ($t_{HS}$ = 30 s). *Source:* Greener et al. [21]/ with permission from John Wiley & Sons.

materials: (i) the primary α peak, representing the effective glass transition, is strongly suppressed by heat-setting, and (ii) the position of the α peak ($T_\alpha$) first increases and then decreases with $T_{HS}$. These effects are illustrated in Figures 7.14 and 7.15. In both cases, the peak height drops with temperature to an asymptotic value at ~200 °C, with the absolute change in peak height being generally greater for the PEN film. Similarly, the peak temperature (closely related to $T_g$) appears to go through a shallow maximum (at ~180 °C for PET and 200 °C for PEN) and then

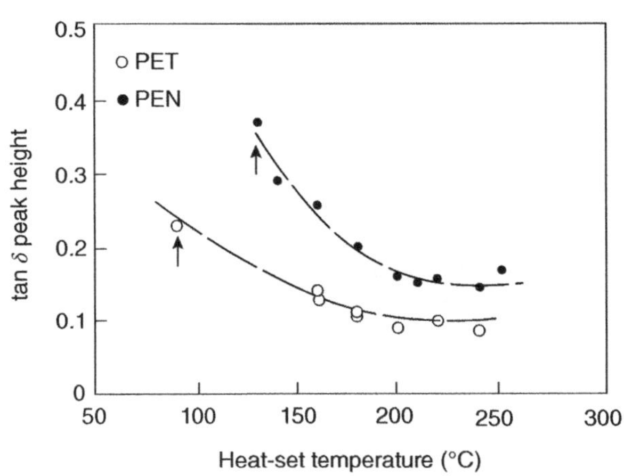


**Figure 7.14** Loss tangent peak height (at $T_\alpha$) vs. heat-set temperature ($t_{HS}$ = 30 s). Arrows indicate NHS films. *Source:* Greener et al. [21]/ with permission from John Wiley & Sons.

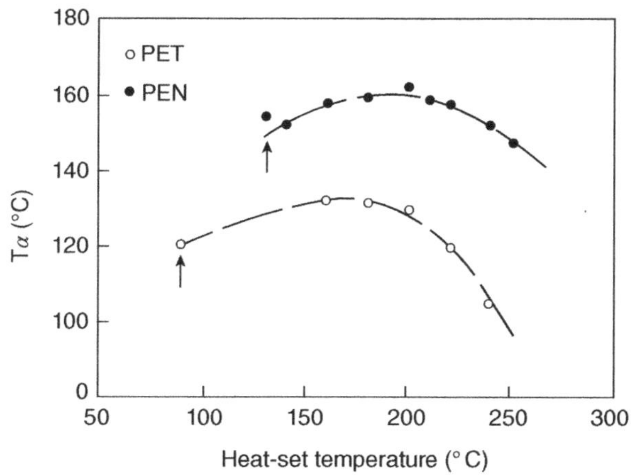


**Figure 7.15** α Transition temperature ($T_\alpha$) vs. heat-set temperature ($t_{HS}$ = 30 s). Arrows indicate NHS films. *Source:* Greener et al. [21]/ with permission from John Wiley & Sons.

drop substantially with $T_{HS}$ (Figure 7.15). The suppression of the tan $\delta$ peak can be attributed to the decrease in the amorphous fraction, consistent with the results in Figure 7.4, but the decrease in the position of the peak at high heat-set temperatures suggests the possibility of some microstructural transformation beyond a certain temperature threshold ($T_{HS}$*).

This transformation apparently removes some constraints from the interlocked amorphous domains, thus increasing their segmental mobility and thereby lowering the glass transition temperature of the film. A similar conclusion was arrived at by Illers and Breuer [36] based on similar results on the effect of crystallinity on the α transition temperature of PET.

### 7.2.5 Optical Properties

The optical properties of the stretched films are also strongly affected by the heat-setting process. Generally, in line with the Lorentz-Lorenz equation, we observe a nearly linear dependence of the average refractive index on density. But, beyond this change, the molecular orientation (or optical anisotropy) in the polymer matrix, initially formed during the stretching steps of the film process, is also modified by the heat-setting treatment. The effect of $T_{HS}$ on the planar and in-plane birefringence components (PBR and BR, respectively) is shown in Figures 7.16 and 7.17 where the average refractive index and two birefringence components are defined as:

$$n_{av} \equiv \frac{n_1 + n_2 + n_3}{3}, \text{ BR} \equiv n_1 - n_2, \text{and PBR} \equiv \frac{n_1 + n_2 - 2n_3}{2} \tag{7.2}$$

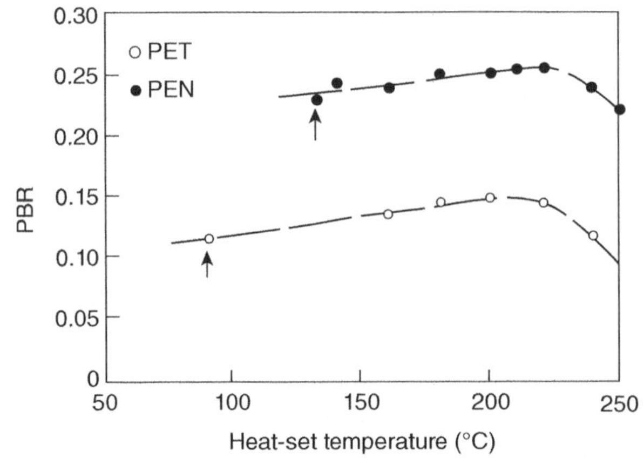


**Figure 7.16** Planar birefringence vs. heat-set temperature ($t_{HS}$ = 30 s). Arrows indicate NHS films. *Source:* Greener et al. [21]/ with permission from John Wiley & Sons.

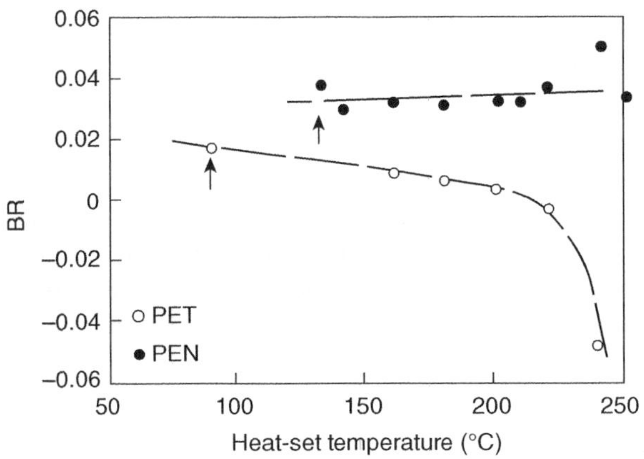


**Figure 7.17** In-plane birefringence vs. heat-set temperature ($t_{HS}$ = 30 s). Arrows indicate NHS films. *Source:* Greener et al. [21]/John Wiley & Sons. α transition temperature ($T_\alpha$) vs. heat-set temperature ($t_{HS}$ = 30 s). Arrows indicate NHS films. *Source:* Greener et al. [21]/ with permission from John Wiley & Sons

and $n_1$, $n_2$, and $n_3$ are the refractive indices along the machine direction [1], the transverse direction [2], and the thickness direction [3], respectively.

The value of the planar birefringence (PBR) is considerably higher for PEN owing to a higher planar alignment of the bulky and "flat" naphthyl moiety in this polymer chain. But in both cases, PBR shows a similar dependence on heat-set temperature; it initially increases then drops with $T_{HS}$, as in the case of $T_\alpha$ (Figure 7.15). The shallow peak in PBR occurs at a slightly higher temperature than the peak in $T_\alpha$, and the corresponding value for PEN (~220 °C) is ~20 °C higher than for PET. This behavior, again, underlies some morphological changes taking place at some high heat-set temperature, $T_{HS}$*, near the onset of the primary melting range of the polymer. Essentially, the planar alignment is enhanced by heat-setting at low temperatures but is clearly suppressed at temperatures approaching the primary melting range of the material. The in-plane birefringence (BR) for PET also shows a systematic change with $T_{HS}$, see Figure 7.17, with the material becoming more aligned toward the TD (negative BR) at higher temperatures ($T_{HS} > 220\,°C$). In line with the PBR data, the in-plane orientation of PEN is nearly independent of heat-set temperature, indicating that the relatively high molecular alignment attained during the stretching steps is not modified to any measurable extent by the heat-setting treatment of these films.

### 7.2.6 X-Ray Diffraction

Wide-angle X-ray (WAXS) diffraction analysis was used to probe the structure of the crystalline phase [21, 37]. Reflection mode $\theta/2\theta$ diffraction scans for PET and PEN films annealed at 200 °C for 30 s are shown in Figure 7.18. The PET pattern is characterized by a strong (100) diffraction peak at $2\theta = 25.5°$. The crystal structure of PEN is different from that of PET [29, 38]. The peak at $2\theta = 26°$ in Figure 7.19 (for PEN) is assigned to the (110) reflection. Transmission mode $\theta/2\theta$ diffraction patterns are shown in Figure 7.19. The PET diffraction pattern features an intense peak at $2\theta = 43°$, which is assigned to the ($\bar{1}05$) plane, while the PEN diffraction pattern is dominated by an intense (010) peak at $2\theta = 14.5°$. Since the ($\bar{1}05$) (for PET) and (010) (for PEN) planes are closely aligned with the chain axes of the respective polymers, the corresponding transmission peaks were used to generate complete pole figures for both film types. Figure 7.20 shows the effect of heat-set temperature ($t_{HS} = 30\,s$) on the crystallite size as estimated from the breadth of a characteristic reflection ($\bar{1}05$) for PET and (010) for PEN, based on the Scherrer equation [37]. Despite significant differences in crystal unit cell structure, both materials exhibit similar levels of crystallite size and similar dependence of crystallite size on $T_{HS}$; over the range of temperatures covered, the crystallite size grows steadily with $T_{HS}$ with a relatively steep upturn at the highest temperatures for both materials. These results are in line with observations of other investigators [9, 12, 13, 20] and may suggest, again, the existence of some morphological transformation at temperatures near the foot of the primary melting range of the material.

The effect of heat-set temperature on the planar and in-plane crystalline orientation functions, as expressed by *HOFc* (Herman Orientation Function – crystalline) and *HOFa* (Herman Orientation Function – amorphous) [21], is shown in Figures 7.21 and 7.22. Those HOF orientation values were calculated

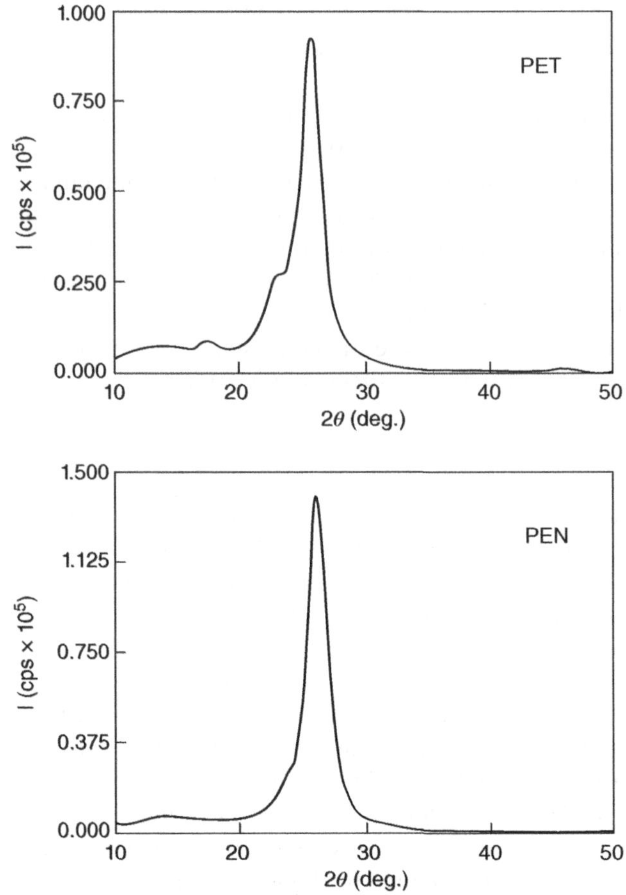


**Figure 7.18** Reflection mode $\theta/2\theta$ diffraction patterns for biaxially oriented PET and PEN films heat-set at 200 °C for 30 s. *Source:* Greener et al. [21]/ with permission from John Wiley & Sons.

from the pole figure results. The planar orientation (*HOFc*) of the PET films follows closely the results for the PBR (Figure 7.16), with the orientation peaking at ~220 °C. However, the planar orientation of the PEN films is considerably lower and appears to be independent of $T_{HS}$. Since the BR represents the combined contributions of the amorphous and crystalline phases to the molecular orientation, the data in Figure 7.21 imply that the main contribution to the high PBR of the PEN films does not come from the crystalline phase but results from the highly oriented amorphous phase. The in-plane orientation function (Figure 7.22) follows a similar pattern as the in-plane BR (Figure 7.17). The orientation of the PEN films is relatively unaffected by the heat-set conditions, while the PET crystals

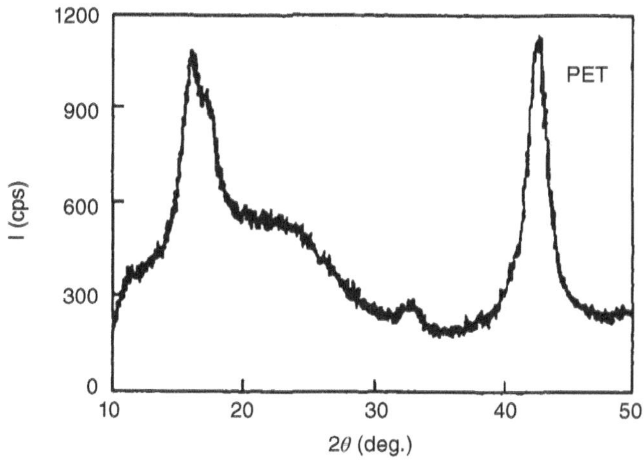


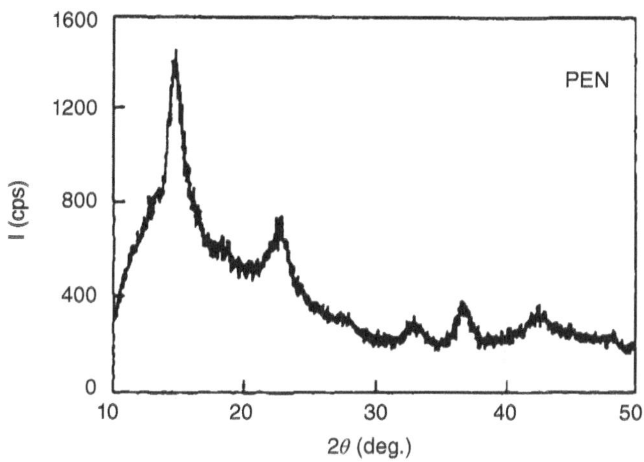


**Figure 7.19** Transmission mode $\theta/2\theta$ diffraction patterns for biaxially oriented PET and PEN films heat-set at 200 °C for 30 s. *Source:* Greener et al. [21]/ with permission from John Wiley & Sons.

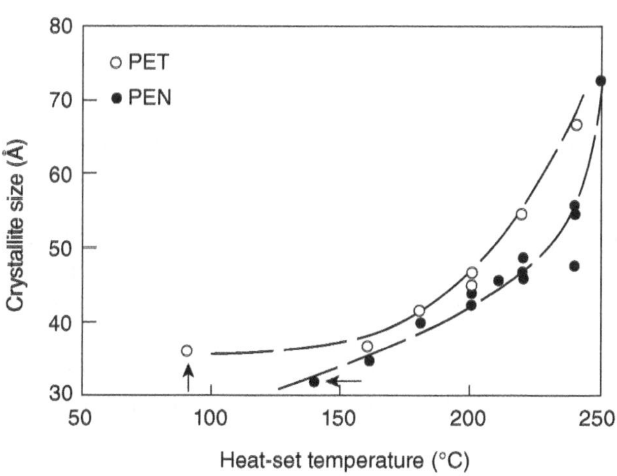


**Figure 7.20** Crystallite size vs. heat-set temperature ($t_{HS}$ = 30 s). Arrows indicate NHS films. *Source:* Greener et al. [21]/ with permission from John Wiley & Sons.

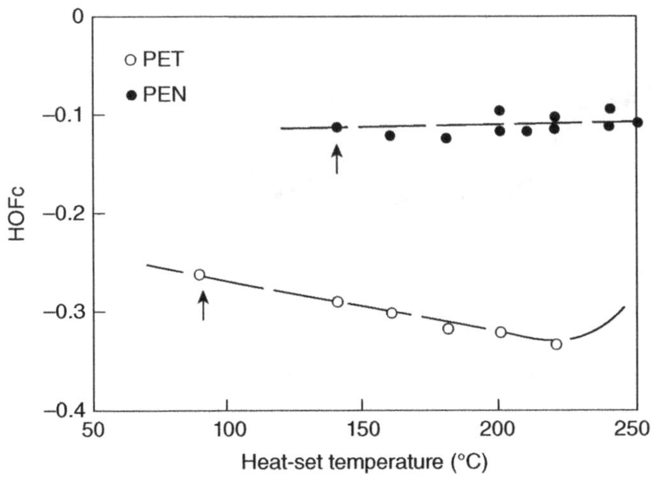


**Figure 7.21** Planar crystalline orientation ($HOF_C$) vs. heat-set temperature ($t_{HS}$ = 30 s). Arrows indicate NHS films. *Source:* Greener et al. [21]/ with permission from John Wiley & Sons.

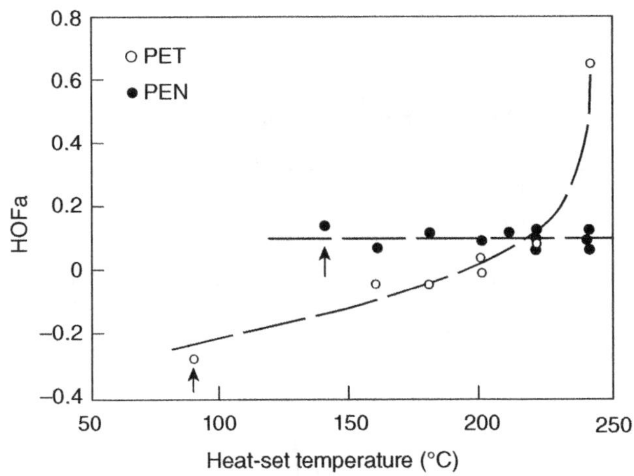


**Figure 7.22** In-plane crystalline orientation ($HOF_a$) vs. heat-set temperature ($t_{HS}$ = 30 s). Arrows indicate NHS films. *Source:* Greener et al. [21]/ with permission from John Wiley & Sons.

seem to realign toward the TD with increase in $T_{HS}$. The PET crystallites are initially aligned in the machine direction but, with increasing heat-set temperature, they become increasingly aligned in the TD. The overall orientation patterns as extracted from the X-ray and BR data are generally consistent, and they indicate that the heat-setting process is capable of imparting significant molecular realignment in PET films but only subtle changes in the orientation of the PEN films are observed. (Note that the reference direction in the X-ray analysis is the TD, while for the BR measurements the orientation is referenced to the machine direction.)

Hassan and Cakmak [27] carried out a WAXS study on the heat-setting of simultaneously and sequentially biaxially stretched PET film samples. They generally observe that heat-setting leads to intensification and sharpening of the diffraction peaks as compared to those of the stretched non-heatset films, in line with the observations of Greener et al. [21]. Heat-setting of low stretch ratio samples resulted in thermal crystallization manifested by crystallization rings indicating that the crystalline domains have no preferred orientation. This was observed for the 1.5×1.5 simultaneously-stretched and 1.5×1 and 1.5×1.5 sequentially stretched samples in their study. During heat-setting, the (100) crystal plane preferentially orients in the film plane, and the percent crystallinity as well as the lateral dimensions of crystallites also increase, which is manifested for the intermediate and high stretch ratio samples as sharpening of the diffraction peaks. Also, the plane (010), which was not observed in the as-stretched samples, is clearly visible in the heat-set samples as a result of the higher alignment of the phenyl rings in the plane of the film.

## 7.3 Morphological and Microstructural Implications

Data presented in Section 7.2 illustrate the pronounced effect of heat-setting treatment on the microstructure and macroscopic properties of two polyester films – PEN and PET; these effects are manifested by notable changes in crystallinity, crystallite size, density, orientation, dimensional stability, and other related properties. Despite differences in molecular structure, crystal unit cell, and crystallization kinetics, both polyesters appear to undergo similar morphological changes in response to the controlled heat-setting treatment applied. Aside from differences in temperature dependence, which scale with the corresponding glass transition temperatures, the main difference between the PET and PEN films concerns the effect of heat-setting on molecular orientation. Based on BR and X-ray data, the PET films exhibit significant molecular realignment during heat-setting, while the apparent orientation of the PEN films is only weakly affected by this heat treatment. These differences notwithstanding, the general morphological transformation imparted by heat-setting and the attendant changes in physical response appear to be similar for both materials.

The morphological models of Fischer and Fakirov [8,9] and Schultz and coworkers [18,19] provide a useful framework for analyzing the results presented in Section 7.2. Both models deduce from small-angle X-ray scattering (SAXS) data on oriented PET fibers that constrained annealing gives rise to two distinct changes in the crystal morphology of the oriented matrix: (i) sharpening of diffraction nodes upon annealing, which represents a reduction in lattice and conformational defects, and (ii) formation of a "secondary" crystalline morphology in the form of folded chain lamellae, which grow perpendicular to the chain axis or drawing direction. These lamellae are superimposed on top of the "primary" fibrillar structure generated during the stretching steps by a strain-induced crystallization process, thus forming a paracrystalline "mosaic" network. It is very likely that the lamellar crystals nucleate and grow epitaxially on the existing microfibrils and are organized in a "shish-kebab"-like superstructure. These morphological changes are represented schematically in Figure 7.23.

The lamellae grow in size with increase in $T_{HS}$ because of the increased fraction of taut-tie amorphous chains with sufficient mobility to partake in the secondary crystallization process that involves chain folding. Consequently, the density (or the crystalline fraction), the average refractive index, and the crystallite size grow with annealing temperature. However, since the secondary crystals are attached to highly

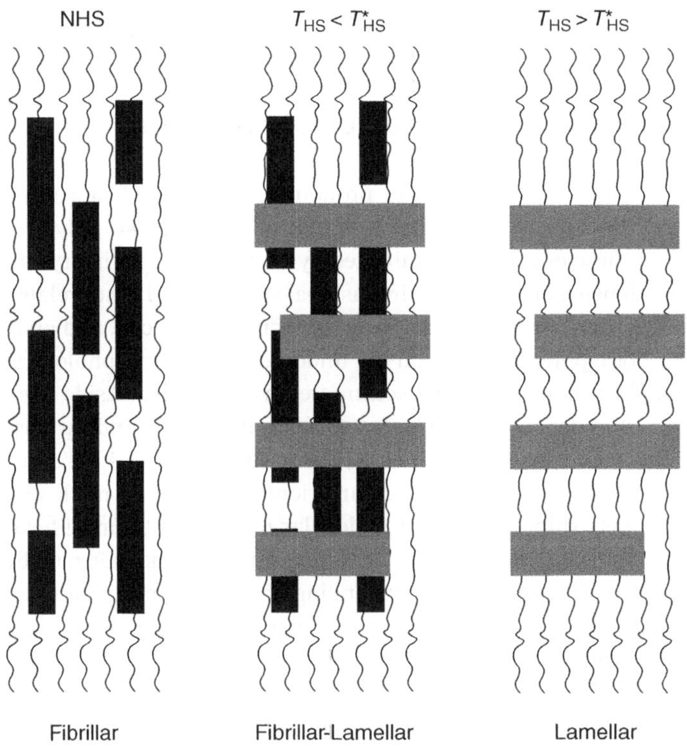


**Figure 7.23** Schematic morphological model of fibrillar-to-lamellar transition (*Source:* Chang et al. [18]; Lee and Schultz [19]) showing the changes in the crystalline network with increase in heat-set temperature. Scheme shown here for uniaxially drawn fiber can be generalized for biaxially drawn film.

oriented amorphous ("taut-tie") chains, they are relatively unstable and have, therefore, a suppressed melting point ($T'_{\mathrm{m}}$). Indeed, the corresponding melting point, based on DSC data, is nearly coincident with the formation temperature of the crystal (Figure 7.7), indicating that once the taut-tie chains regain sufficient mobility, the secondary structures melt out due to the potential gain in free energy resulting from an increase in melting entropy of the surrounding amorphous domains, i.e.

$$T'_{\mathrm{m}} = \frac{\Delta H'_{\mathrm{m}}}{\Delta S'_{\mathrm{m}}} \tag{7.3}$$

where $\Delta H'_{\mathrm{m}}$ and $\Delta S'_{\mathrm{m}}$ are the changes in enthalpy and entropy for the secondary melting process. The depression in the secondary melting point can also be attributed to the observed changes in lamellar thickness through the Hoffman equation [39]:

$$T_{\mathrm{m}}(l) = T_{\mathrm{m}}^{0}\left[1 - \frac{2\sigma_{\mathrm{c}}}{\Delta H_{\mathrm{f}} \cdot l}\right] \tag{7.4}$$

where $T_m^0$ and $\Delta H_f$ are the melting point and latent heat of fusion of a perfect crystal (crystal with infinite dimensions), $l$ is the average thickness of the lamellar crystal, and $\sigma_c$ is the surface free energy of the crystal. By adjusting the parameters in this equation, it is possible to show that the observed changes in crystallite size, $l$, (Figure 7.20) can be directly correlated with the changes in the secondary melting temperature.

The apparent enthalpy associated with the secondary melting endotherm, based on the DSC data (Figures 7.5 and 7.6), is relatively small compared to the total amount of crystalline matter formed during the heat-setting process (cf. Figure 7.4). This implies that upon melting (at $T'_{\mathrm{m}}$) the secondary crystallites immediately recrystallize into more stable-folded chain crystals that ultimately melt at $T_{\mathrm{m}}$, so that the secondary melting peak represents only a fraction of the crystalline matter formed during heat-setting. Also, as noted above, some of the crystallinity gain is associated with the perfection of the existing fibrillar crystallites (rather than buildup of secondary lamellae) and is, therefore, not represented by the secondary melting peak. It should be noted that the dual-melting behavior of PET and other polymers has been the subject of some debate in the literature [9, 20, 21, 31–35], and the general picture painted above is not universally accepted. It is generally agreed, however, that the two melting endotherms represent two distinct crystalline morphologies that depend on the thermal history and orientation of the polymer matrix and annealing mode (constrained vs. free) of the film sample. One form, corresponding to the secondary melt temperature, $T'_{\mathrm{m}}$, is closely associated with the heat-setting process, while the second form, corresponding to $T_{\mathrm{m}}$, is practically independent of the heat-set conditions. Holdsworth and Turner-Jones [31] have proposed, in fact, that the primary melting peak, $T_m$, is the melting point of stable-folded chain crystals that are formed throughout the DSC scan by a process of continuous melting and recrystallization, while the secondary peaks are the melting transitions of the thin crystalline lamellae formed during heat treatment. Oswald et al. [35] have concluded, from studies on annealing of unconstrained PET fibers, that the secondary melting peak is a result of lamellar crystals ("nuclei") formed during heat treatment on the present fibrils, with the melt temperature being related to the size of the crystals. They observed, however, that the secondary melting temperature increased somewhat with anneal time, unlike the result of Greener et al. [21] in which the asymptotic value of the secondary melt temperature is constant (annealing time of ca. 60 minutes gave essentially identical $T'_m$ as 30 s). This difference can be attributed to the fact that the heat-set samples were annealed under lateral constraint. Oswald et al.'s interpretation of the effects of heat treatment on the nature of the secondary melting process is, however, similar to the mechanism given above.

Although the general morphological changes during heat-setting observed by Greener et al. [21] and other researchers are consistent with the phenomenological models of Fischer and Fakirov [8,9] and Schultz and coworkers [18,19], neither model pays particular attention to the apparent morphological transition at high heat-set temperatures ($>T_{HS}$*). This transition is marked by distinct maxima in PBR and α transition temperature ($T_\alpha$) and abrupt change in crystallite size and in-plane orientation at temperatures close to the primary melting range of the material. Gohil [20] first noted this "transition," which he attributes to a peak in crystallization kinetics, and assigned it as the boundary between two temperature regimes for heat-setting. This transition can also be interpreted by the phenomenological model of Schultz et al. as the onset of the primary melting range of the material, where the fibrillar core structure starts to melt out and recrystallize into a more stable lamellar form such that the general population of fibrillar crystals is depleted and replaced with lamellar crystals. In support of this idea, Statton et al. [10] and Wilson [11] observed that annealing of drawn PET at high temperatures (>200 °C) gives rise to a significant increase in the number of chain folds, which is characteristic of lamellar crystals.

The tie chains that connect two crystals are the critical amorphous topological component controlling the toughness of a semi-crystalline polymer. There are generally three amorphous topological components, namely folds, tails, and tie chains. The statistic tie chain theory [40] allows the calculation of the tie chain probability from a polymer based on its coil dimension and crystal size. This theory is derived from the fundamental assumption of coil dimension preservation during crystallization of a polymer, which was proven experimentally in 1978 by neutron scattering. The tie chain concentration increases with increasing crystallinity, but it would be lowered with an increase in crystallite size at a given crystallinity. The apparent morphological transition at $T_{HS}$* may be associated with the tie chain concentration maximum considering that both crystallinity and crystal size are raised with the heat-set temperature. At $T_{HS}$*, the increase in the number of tie chains with crystallinity is contrasted by the decrease in tie chains with crystal size. Although tie chains are difficult to measure, they are known to control the material's toughness. As shown in Figure 7.24, a PET film toughness attains a maximum at $T_{HS}$*. With the reduction in tie chain concentration for PET beyond $T_{HS}$*, a reduction in its PBR is indicated. It should be noted that a reduction in tie chains corresponds to an increase in chain folds.

Through their interlocking (crosslinking) effect, the secondary crystallites restrict the mobility of the oriented amorphous chains, thereby enhancing the dimensional stability and stiffness (modulus) of the matrix. As soon as these crystallites are melted out (at $T'_m$), the oriented amorphous (taut-tie) chains are free to disorient and the film exhibits significant shrinkage (Figures 7.9–7.11). The large distortion beyond $T'_m$ suggests that once the secondary crystals are melted, there is nothing to prevent the highly strained amorphous domains from collapsing and moving the system into a lower free energy state. These secondary crystallites formed during heat-setting are qualitatively different and apparently much less stable than those formed during stretching. The prevailing morphology of the strain-induced (primary) crystallites is that of a "shish kebab" or a bundle-like microfibril [42], see Figure 7.23, which contains a relatively small number of chain

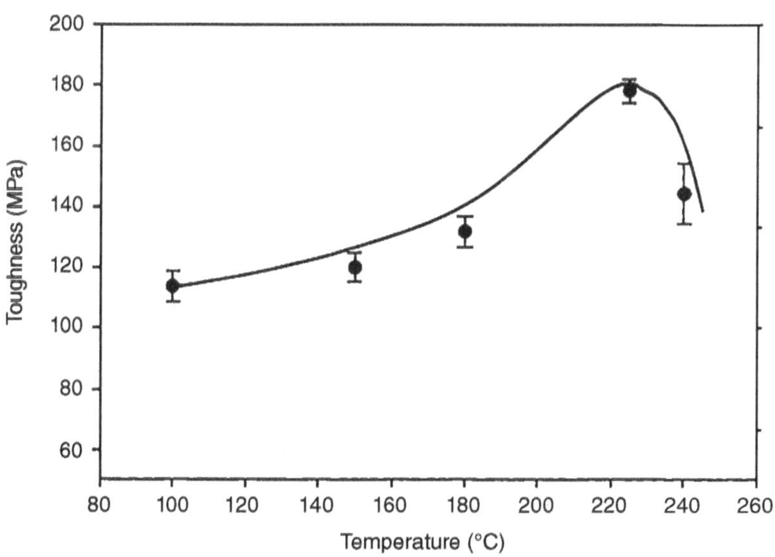


**Figure 7.24** Toughness of biaxially oriented PET film as a function of heat-set temperature. *Source:* Adapted from [42].

folds in its fibrillar core ("shish") and small lamellae growing perpendicular to the draw direction ("kebab"). Unlike the secondary lamellae, these crystals apparently cannot prevent the amorphous chains, interspersed between the microfibrils, from disorienting (shrinking) at temperatures above the glass transition, although they contribute to the overall mosaic structure of the crystalline network and thus enhance the stiffness and strength of the matrix (at least along the draw direction). Once the melting range of the fibrillar crystals (or their stable recrystallized form) is approached, these crystals start to melt and transform into more stable-folded chain lamellae. This transformation gives rise to the decrease in PBR and the glass transition temperature, $T_{\alpha}$, above $T_{HS}$*. The decrease in $T_{\alpha}$ suggests also that the transformation at $T_{HS}$* "loosens up" the amorphous domains and imparts higher segmental mobility to the non-crystallized chains, as suggested by Illers and Breuer [36] in a study on the effect of crystallinity on the dynamic-mechanical properties of PET. Elenga et al. [34] also studied high-temperature annealing of oriented PET films, but they argue that the observed changes in mechanical properties after long annealing at high temperatures are due to a chemical transformation caused by a trans-esterification reaction. This, of course, cannot be ruled out and it may be happening in conjunction with the melting of the primary fibrillar crystals (or recrystallized folded chain crystals) over longer time scales. Whatever the origin of the morphological transformation, it must be recognized that there exists a critical heat-set temperature, $T_{HS}$*, beyond which the mechanical and physical performance of the film is qualitatively altered.

## 7.4 Summary

Heat-setting is a critical step in the tenter-frame process used for the manufacture of biaxially oriented polyester films, with potentially significant impact on the microstructure and attendant physical properties of the film. Although most past studies of heat-setting focused on PET, the most common polyester resin, available data on PEN and other polyesters [21] show similar trends such that the overall effects of heat-setting can be generalized to other film-grade polyesters. The salient effects of heat-setting can be summarized as follows:

1) Heat-setting can substantially increase the degree of crystallinity and crystallite size of biaxially stretched films. The crystalline fraction grows nearly linearly with heat-set temperature and can be nearly doubled over a wide temperature range. The changes in the crystalline fraction correspond to similar changes in density and average refractive index.
2) A secondary melting endotherm can be detected in a DSC thermogram for a heat-set film at a temperature that nearly coincides with the applied heat-set

temperature. The secondary melting temperature does not change with the applied heat-set (anneal) time once thermal equilibration is reached, and it involves considerably lower enthalpy change compared to the primary melting transition of the polymer. This characteristic "signature" can be used effectively for process and product diagnostics. It must be noted, though, that in some high line-speed operations when thermal equilibrium is not fully attained in the heat-set section of the tenter-frame machine, the effective heat-set temperature would be lower than the set machine temperature.

3) The heat-set step has a pronounced effect on the dimensional stability of the oriented film. Both PET and PEN films exhibit significant shrinkage at temperatures near and above the corresponding heat-set temperature. Thus, by raising the heat-set temperature, it should be possible to improve the overall high-temperature dimensional stability of the stretched film and make it better suited for applications requiring high-temperature fabrication steps and tight registration tolerances. Subtle dimensional changes may be observed above the glass transition temperature even for a fully heat-set film. Such changes may require additional high-temperature heat relaxation steps in order to minimize shrinkage in the polyester film.
4) X-ray and BR data indicate that the changes in crystalline morphology are accompanied by some molecular realignment in the polymer matrix during heat-setting. The changes in orientation in PEN films are small compared to those in PET films. In both cases, however, the planar orientation goes through a broad maximum at temperatures approaching the onset of the primary melting range of the corresponding polyester. Also, the crystallite size grows steadily with heat-set temperature, with some indication of a sharp upturn near the onset of the primary melting range.
5) Heat-setting leads to a strong suppression and a shift in the position of the loss tangent peak corresponding to the $\alpha$ transition (related to $T_g$). Like the PBR, the position of the $\alpha$ transition (position of tan $\delta$ peak) first increases and then decreases with $T_{HS}$. The maximum in the $\alpha$ transition temperature occurs near the onset of the primary melting range of the material.
6) The maxima in PBR and $T_g$ and, in the case of PET, the steep increase in in-plane orientation at temperatures close to the onset of the primary melting range of the polyester, suggest the presence of a morphological transformation at some critical heat-set temperature, $T_{HS}^*$. This transformation is associated with changes in the physical and mechanical properties of the film, in line with observations made by Gohil [20] and Greener et al. [21]. In addition to increasing the effective softening point of the film, operating at high heat-set temperatures, $>T_{HS}^*$, can improve its mechanical properties and, in particular, its finishing and cutting characteristics corresponding to a reduction in toughness. This reduction in toughness can be explained also in terms of the decrease in the number of tie chains.

7) The morphological model discussed in Section 7.3, based on suggestions by Fischer and Fakirov [8,6], Schultz et al. [18,19], and Greener et al. [21], provides a useful framework for explaining many experimental observations of the effects of heat-setting. The basic premise of the model is that the heat-setting process induces the formation of chain folded lamellae, which nucleate epitaxially on the fibrillar structures formed during the sequential stretching steps. The lamellae grow perpendicular to the stretching direction, with the lamellar size being directly dependent on the heat-set temperature, thus producing an "interlocking" paracrystalline network that enhances the dimensional stability of the oriented matrix. The transition corresponding to $T_{HS}$* may be due to partial melting and recrystallizing of the fibrillar core crystals in the vicinity of the onset of the primary melting range of the polymer.

## References

1 Tobita, K., Miki, T., and Takeuchi, N. (1999). Biaxially oriented film. In: *Film Processing* (ed. T. Kanai and G.M. Campbell), 244–280. Munich: Hanser Publishers.
2 Tsunashima, K., Toyoda, K., and Yoshii, T. (1999). Stretching conditions, orientation and physical properties of biaxially oriented film. In: *Film Processing* (ed. T. Kanai and G.M. Campbell), 320–352. Munich: Hanser Publishers.
3 Yamada, T., Nonomura, C., and Kase, S. (1999). Theoretical analysis of the tentering process. In: *Film Processing* (ed. T. Kanai and G.M. Campbell), 353–385. Munich: Hanser Publishers.
4 J. M. Schultz, Structure development in polyesters in *Solid State Behavior of Linear Polyesters and Polyamides*, pp. 97–103, J. M. Schultz and S. Fakirov, eds., Prentice-Hall, New York (1991).
5 Smith, F.S. and Steward, R.D. (1974). The crystallization of oriented poly (ethylene terephthalate). *Polymer* 15 (283).
6 Prevorsek, D.S., Tirpak, G.A., Harget, P.J., and Reimchuessel, A.C. (1974). Effects of thermal contraction on structure and properties of PET fibers. *J. Macromol. Sci. Phys.* B9: 733.
7 Fakirov, S., Fischer, E.W., and Schmidt, G.F. (1975). Unit cell dimensions of poly (ethylene terephthalate). *Macromol. Chem.* 176: 2459.
8 Fischer, E.W. and Fakirov, S. (1976). Structure and properties of polyethylene-terephthalate crystallized by annealing in the highly oriented state. *J. Mater. Sci.* 11: 1041.
9 Fakirov, S., Fischer, E.W., Hoffmann, R., and Schmidt, G.F. (1977). Structure and properties of poly(ethylene terephthalate) crystallized by annealing in the highly oriented state: 2. Melting behaviour and the mosaic block structure of the crystalline layers. *Polymer* 18: 1121.

**10** Statton, W.O., Koenig, J.L., and Hannon, M. (1975). Characterization of chain folding in poly (ethylene terephthalate) fibers. *J. Appl. Phys.* 41: 4290.
**11** Wilson, M.P. (1974). Shrinkage and chain folding in drawn poly (ethylene terephthalate) fibres. *Polymer* 15: 277.
**12** Hindeleh, A.M. and Johnson, D.J. (1978). Crystallinity and crystallite size measurement in polyamide and polyester fibres. *Polymer* 19: 27.
**13** Gupta, V.B. and Kumar, S. (1981). The effect of heat setting on the structure and mechanical properties of poly (ethylene terephthalate) fiber. I. Structural changes. *J. Appl. Polym. Sci.* 26: 1865.
**14** Gupta, V.B., Ramesh, C., and Gupta, A.K. (1984). Structure–property relationship in heat-set poly (ethylene terephthalate) fibers. I. Structure and morphology. *J. Appl. Polym. Sci.* 29: 3115.
**15** Buckley, C.P. and Salem, D.R. (1987). High-temperature viscoelasticity and heat-setting of poly (ethylene terephthalate). *Polymer* 28: 69.
**16** Buckley, C.P. and Salem, D.R. (1990). Heat setting of oriented poly (ethylene terephthalate): effects of deformation mode, strain level, and heating medium. *J. Appl. Polym. Sci.* 41: 1707.
**17** Bechev, C.D. and Mishinev, J.T. (1992). Structural changes in draw-textured poly (ethylene terephthalate) filaments during heat-setting. *J. Appl. Polym. Sci.* 45: 29.
**18** Chang, H., Schultz, J.M., and Gohil, R.M. (1993). Morphology of biaxially stretched poly (ethylene terephthalate) films. *J. Macromol. Sci. Phys.* B32: 99.
**19** K.-G. Lee and J. M. Schultz, The development of structure and mechanical properties in poly (ethylene terephthalate) fibres during heat treatment under stress *Polymer*, 34, 4455 (1993);see also H. Chang, K. G. Lee and J. M. Schultz, Structure development of polyethylene terephthalate fibers during postspinning annealing *J. Macromol. Sci. Phys.*, B33, 105 (1994).
**20** Gohil, R.M. (1994). Morphology–property relationship in oriented PET films: microstructural reorganization during heat treatment. *J. Appl. Polym. Sci.* 52: 925.
**21** Greener, J., Tsou, A.H., and Blanton, T.N. (1999). Physical and microstructural effects of heat setting in polyester films. *Polym. Eng. Sci.* 39 (12): 2403.
**22** Maruhashi, Y. and Asada, T. (1992). Structure and physical properties of biaxially stretched polyethylene terephthalate sheet after heat-setting. *Polym. Eng. Sci.* 32 (7): 481.
**23** Maruhashi, Y. (2001). Structure and physical properties of biaxially stretched polyethylene terephthalate sheets under different heat-set and stretch conditions. *Polym. Eng. Sci.* 41 (12): 2194.
**24** Schoukens, G., Samyn, P., Maddens, S., and van Audenaerde, T. (2003). Shrinkage behavior after the heat setting of biaxially stretched poly (ethylene 2, 6-naphthalate) films and bottles. *J. Appl. Polym. Sci.* 87: 1462.

**25** Cakmak, M., Hassan, M., Unsal, E., and Martins, C. (2012). A fast real time measurement system to track in and out of plane optical retardation/birefringence, true stress, and true strain during biaxial stretching of polymer films. *Rev. Sci. Inst.* 12 (83): 123901/1–123901/10.

**26** Hassan, M. and Cakmak, M. (2013). Mechano optical behavior of polyethylene terephthalate films during simultaneous biaxial stretching: real time measurements with an instrumented system. *Polymer* 54: 6463.

**27** Hassan, M. and Cakmak, M. (2014). Mechanisms of structural organizational processes as revealed by real time mechano optical behavior of PET film during sequential biaxial stretching. *Polymer* 55: 5245.

**28** de Daubeny, R.P., Bunn, C.W., and Brown, C.J. (1954). The crystal structure of polyethylene terephthalate. *Proc. Roy. Soc. London., Ser. A* 226: 531.

**29** Mencik, Z. (1967). Crystal structure of polycondensate of naphthalene-2, 6-dicarboxylic acid with ethylene glycol. *Chem. Prim.* 17: 78.

**30** Buchner, S., Wiswe, D., and Zachmann, H.G. (1989). Kinetics of crystallization and melting behaviour of poly (ethylene naphthalene-2, 6-dicarboxylate). *Polymer* 30 (480).

**31** Holdsworth, P.J. and Turner-Jones, A. (1971). The melting behaviour of heat crystallized poly (ethylene terephthalate). *Polymer* 12: 195.

**32** J. P. Bell and J. H. Dumbleton, Relation between melting behavior and physical structure in polymers *J. Polym. Sci., Polym. Phys. Ed.*, 7, 1033 (1969); J. P. Bell and T. Murayama, Relations between dynamic mechanical properties and melting behavior of nylon 66 and poly (ethylene terephthalate) Ibid., p. 1059.

**33** Nealy, D.L., Davis, T.G., and Kibler, C.J. (1970). Thermal history and melting behavior of poly (ethylene terephthalate). *J. Polym. Sci., Polym. Phys. Ed.* 8: 2141.

**34** Elenga, R., Seguela, R., and Rietsch, F. (1991). Thermal and mechanical behaviour of crystalline poly (ethylene terephthalate): effects of high temperature annealing and tensile drawing. *Polymer* 32: 1975.

**35** Oswald, H.J., Turi, E.A., Harget, P.J., and Khanna, Y.P. (1977). Development of a middle endotherm in DSC thermograms of thermally treated drawn pet yarns and its structural and mechanistic interpretation. *J. Macromol. Sci. Phys.* B13: 231.

**36** Illers, K.H. and Breuer, H. (1963). Molecular motions in polyethylene terephthalate. *J. Colloid Sci.* 18: 1.

**37** L. E. Alexander, *X-Ray Diffraction Methods in Polymer Science*, Wiley Interscience, New York (1969);see also, B. D. Cullity, Elements of X-Ray Diffraction, 2, Addison-Wesley, Reading, Mass. (1975).

**38** Cakmak, M., Wang, Y.D., and Simhambhatla, M. (1990). Processing characteristics, structure development, and properties of uni and biaxially stretched poly (ethylene 2, 6 naphthalate)(PEN) films. *Polym. Eng. Sci.* 30: 721.

**39** Wunderlich, B. (1980). Vol. 3. Crystal melting. In: *Macromolecular Phys.*, 30–33. New York: Academic Press.

**40** Magill, J.H. (1977). Treatise on materials science and technology. In: *Vol. 10, Properties of Solid Polymeric Materials (Part A)* (ed. J.M. Schultz), 89–122. New York: Academic Press.

**41** Huang, Y.L. and Brown, N. (1991). Dependence of slow crack growth in polyethylene on butyl branch density: morphology and theory. *J. Polym. Sci., B* 29: 129.

**42** Rao, Y., Greener, J., Avila-Orta, C.A. et al. (2008). The relationship between microstructure and toughness of biaxially oriented semicrystalline polyester films. *Polymer* 49 (10): 2507.

# 8

# Polyester Materials in Biomedical Applications

*Ebrahim Jalali Dil[1], Abdellah Ajji[2], and Amir Saffar[3]*

[1] *Polyexpert Inc., Laval, Quebec, Canada*
[2] *Chemical Engineering Department, Polytechnique Montreal, Montreal, Quebec, Canada*
[3] *ProAmpac Inc., Terrebonne, Quebec, Canada*

## 8.1 Introduction

Market value for medical devices reached more than US$ 360 billion in 2017 [1] and is expected to grow at 4.1% from 2020 and reach US$ 432 billion by 2025 [2]. Polymer-based medical devices including medical disposals, prosthetics, medical tools, and drug delivery systems had a market value of more than US$ 21 billion by the end of 2018 and are expected to grow at 6.8% and reach about US$ 32 billion at the end of 2024 [3]. This can be attributed to the significant potential of polymer materials in biomedical applications due to their low processing cost and design flexibility. Among different polymer materials, polyesters and their copolymers have attracted considerable attention due to their tunable mechanical properties, biocompatibility, and biodegradability [4–6]. These materials can now be found in many commercial biomedical products including interference screws, fracture repair kits, fixation devices, suture as well as highly complex drug delivery systems for anticancer drugs. Due to the wide variety of polyester materials in biomedical application, this chapter will only focus on the most important polyester materials in this field. In the first section, the general requirements of polymer materials in biomedical applications will be presented. Then, previous studies in the literature on polyesters materials will be categorized in four main groups: (i) tissue engineering applications, (ii) drug delivery and controlled release systems, (iii) orthopedic applications, and (iv) medical devices. In each category, previous studies in the literature will be reviewed, and commercially available products in each category will be presented.

*Polyester Films: Materials, Processes and Applications*, First Edition.
Edited by Miko Cakmak and Jehuda Greener.

## 8.2 Perquisites for Biomedical Applications

Before discussing biomedical applications of polyesters, it is necessary to present an overview of the selection criteria for polyester materials used in biomedical applications. Angelova and Hunkeler [7] presented a flow chart for the selection of polymers for biomedical applications. General requirements for this field can be categorized as biocompatibility, non-antigenic, non-carcinogenic, and non-toxic. In addition, biodegradability of polymers is another important factor that needs to be considered, especially in tissue engineering and orthopedic applications.

### 8.2.1 Biocompatibility

Considering the interactions of cells and biofluids with polymers is critical in biomedical applications, it has been shown that polyester materials exhibit negative charges on their surface [8]. Considering the negative charges of cell membranes, the surface charge of polyesters inhibits cell adhesion and proliferation [9]. In addition, polyesters have the tendency to form blood clots due to their thrombogenicity. Two possible scenarios when an incompatible foreign surface is introduced into the bloodstream are [10] (i) protein adsorption followed by platelet aggregation and formation of white thrombus, and (ii) protein adsorption followed by fibrin formation, platelet aggregation and trapping red cell resulting in the formation of red thrombus. Both scenarios will cause embolization. Therefore, it is necessary to modify the surface of polyester materials for biomedical applications [11]. Different surface modification approaches have been employed to enhance interaction of polyester materials with biofluids and cells including chemical surface modification [12–17], layer-by-layer (LbL) assembly [9, 18–21], surface polymerization and immobilization [22–24], irradiation (UV or plasma) [25–32], surface micro- and nanopatterning [33], and blending with other polymers [34, 35]. For example, Ghavidel et al. [9, 18]. used LbL deposition of polycations followed by deposition of chitosan on the surface of a porous PCL scaffold and found that cell adhesion and penetration in the scaffold was enhanced. In another example, Kim et al. [13]. showed that chemical surface modification of PET blood tubes could enhance interactions of serum with the tube wall and reduce blood clots formation. Selection criteria for the modification approach depend on final intended applications and structure of the material. For example, UV irradiation technique are appropriate for films and 2D surfaces, while LbL deposition and chemical modification are more suitable options for 3D structures. More detail for surface modification techniques for biomedical applications can be found in the review paper of Goddard et al. [36].

Stability of modified surfaces is another important criterion that needs to be considered. Jalali Dil et al. [16]. examined the surface tension of chemically

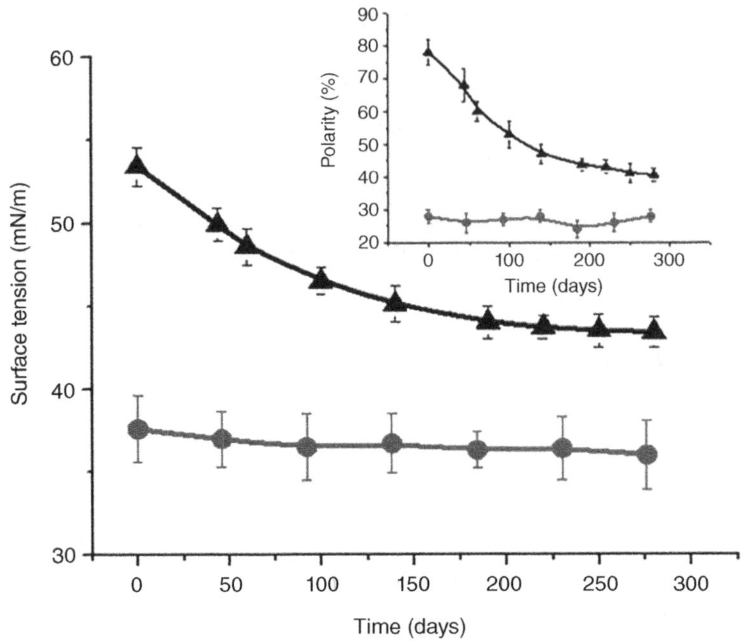


**Figure 8.1** Variation of surface tension for PET (●) and chemically modified PET (▲). The inset shows variations of surface polarity of the same samples. *Source:* Jalali Dil et al. [16]/with permission from Elsevier.

modified PET over 280 days and showed that both surface tension and polarity of the modified PET surface decreased considerably in the studied period. This is very important aspect to consider in applications where surface stability is required for long times (Figure 8.1).

### 8.2.2 Biodegradability

Biodegradability is an important factor for in-vivo applications where degradation of polymer implanted structure or device is required. Polymers with the ability of in-vivo degradation have been referred to using different terms such as biodegradable, bioabsorbable, and bioresorbable. Li and Vert [37] tried to clarify these terms and provided definitions for each term to avoid confusion. They mentioned that biodegradable polymers degrade at least partially by mediation of a biological system while bioabsorbable polymers are digested by bio-systems. According to their definitions, the term bioresorbable can be used when the by-products of degradation of polymer can be eliminated through natural pathways. The biodegradation mechanism of biodegradable polyesters occurs by random hydrolysis of ester linkages through the polymer chain, which reduces the molecular weight and mechanical properties of a polyester [38]. In addition, the resulting substances produced in their biodegradation process can be eliminated through natural pathways [39]. Therefore, biodegradable polyesters have been referred to as bioresorbable polymers. The degradation process can occur in two different ways [40]: (i) surface erosion, and (ii) bulk erosion or bulk hydrolysis. In the first mechanism, the degradation occurs at the surface of polymer and the polymer volume decreases by time. As erosion begins at the surface, surface characteristics such as surface hydrophobicity and surface area are also important factors in biodegradation rate [41]. Bulk erosion involves two steps: in the first step, water penetrates polymer and breaks hydrolytically unstable chemical bonds in the amorphous phase; thus, breaking down polymer chains, and in the second step enzymes attack to the water-soluble fragments resulting in erosion to occur through the bulk. The mechanism of degradation is an important factor in designing polyester devices for biomedical applications. For example, it has been shown that PLA, PGA, and PLGA undergo bulk hydrolysis [37,42] while polyorthoesters degradation occurs by surface erosion [40]. This indicates that polyorthoester devices will maintain their integrity longer than devices based on PLA, PGA, or PLGA.

Biodegradation rate of polymer materials is another important factor for biomedical applications and can be affected by polymer chemical structure [38,43], stereo isomerism [44], molecular weight and molecular weight distribution [45], end group chemistry [46], crystallinity and crystal structure [41]. For example, degradation of polyorthoesters are in order of hours or days while it takes months or years for aliphatic polyesters to degrade [43,47]. Accessibility of hydrolyzable/oxidable functional groups can also significantly affect biodegradation rate. For example, the repeat units of PLA and PGA differ only in an extra $CH_3$ in PLA; however, this increases the hydrophobicity and accessibility to the ester linkage by increasing steric hinderance and reduces significantly the biodegradation rate of PLA [48]. Similarly, in semi-crystalline polymers, the crystalline region degrades after the amorphous region due to the highly packed structure of the crystalline phase that reduces accessibility of hydrolysable groups.

## 8.3 Biomedical Applications of Polyesters

### 8.3.1 Tissue Engineering

A tissue can be defined as a group of cells with similar structure that act together to perform a specific task in a living body [49]. Replacing damaged, diseased, aged, or genetically deficient tissues of human body using his own tissue was a revolutionary step in traditional medicine. Langer and Vacanti defined tissue engineering as "the principles of engineering and the life sciences toward the development of biological substitutes that restore, maintain, or improve tissue function" [50]. A common strategy to achieve this objective consists of the following steps [51]: (i) fabrication of a bioabsorbable 3D porous scaffold with micro-structure and mechanical properties mimicking the natural cell medium, (ii) cell seeding onto the fabricated scaffold, (iii) growth of premature tissue in a dynamic environment, (iv) Growth of the mature tissue in a bioreactor, and (v) surgical transplantation. Polymers have been widely used for scaffold fabrication in tissue engineering due to their design flexibility that allows altering their properties by controlling their composition/structure. Among different polymers, polyesters and especially degradable polyesters such as polyglycolide (PGA), poly(lactic acid) or PLA, poly(lactic acid-co-glycolide) or PLGA, polycaprolactone (PCL), and polydioxanone (PDS) have been studied more extensively in tissue engineering due to their biocompatibility and biodegradable nature. The porous scaffolds or engineering applications can be produced by different routes. Here we review the most common approaches to fabricate these scaffolds.

1) Solution casting [52–54]: This method is interesting for applications where high mechanical strength is not required. In this method, solution of polymer is precipitated in a non-solvent, and the gummy-formed precipitant is then molded, sterilized, and implanted. Schmitz and Hollinger [52] used this approach to prepare scaffolds for calvarial defects by dissolving PLGA (50 : 50) copolymer in chloroform, followed by precipitation in ethanol and then mixed with demineralized freeze-dried bone (DFDB). The formed dough was then heat cured for 24 hours at 45–48 °C and sterilized with ethylene oxide and then passed through a lyophilization step (to remove ethylene oxide). Their results showed significantly greater volume of trabecular bone formation in the defect with implant compared to a control defect.
2) Solvent casting/particulate leaching [55–62]: This method has been widely used to prepare porous scaffolds due to its simplicity and lower costs compared to other methods. Here, polymer is dissolved in a solvent and then solid particles (salts or in some cases sugar) are added to the polymer solution. The solution is then cast into a mold followed by solvent removal to obtain a solid substrate. In the final step, the porous structure is obtained by leaching out solid particles using a selective solvent that dissolves only the particles. Liao et al. [62]. used this approach to prepare PLGA (85/15) copolymer scaffolds and showed that increasing copolymer content from 5 wt.% to 12.5 wt.% reduced the scaffold porosity from 95% to almost 80%. Figure 8.2 shows scanning electron microscopy (SEM) of their scaffold at 10 wt.% of PLGA.

Freed et al. [63]. prepared porous 3D PLLA scaffold for cartilage regeneration using this technique and found that the scaffold could keep its 3D structure for six

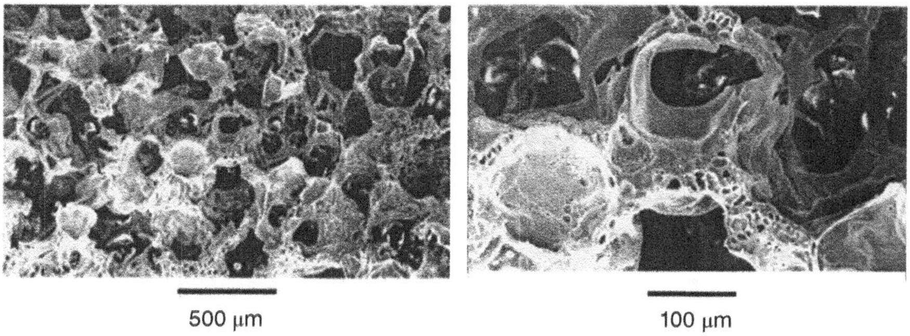


**Figure 8.2** SEM image of porous PLGA scaffold after leaching of salt with deionized water. The sample had initially 90% of salt. *Source:* Liao et al. [62]/ with permission from John Wiley & Sons.

months and observed chondrocytes growth on scaffold, which also contained sulfated glycosaminoglycans (S-GAG) and type II collagen, and closely resembled cartilage histologically.

3) Phase separation technique: This method is based on thermally induced phase separation or as sometimes referred to TIPS [64–69]. A solvent is selected in which polymer/solvent shows an upper critical solution temperature (UCST) phase behavior similar to the one shown in Figure 8.3. Mixing of polymers with the solvent is done at temperatures above the phase separation curve where a miscible solution is formed. Then the solution is quenched to a temperature in the two-phase region (below the phase separation curve) which induces phase separation. Increasing annealing time at quenching temperature increases the phase separated domain size and consequently, the pore size of the final scaffold. After achieving the desired pore size, solvent is removed by freeze drying to maintain the porous structure.

Hua et al. [65, 70]. used PLA and combination of water and diethyl dioxane for the solvent and, as can be seen in Figure 8.4, showed that a small change in the solvent composition could significantly affect the final pore sizes especially at longer annealing times. PLA sponges prepared by phase separation has shown promising performance as scaffold in some cartilage regeneration studies [71, 72].

4) Polymer particle sintering [73, 74]: In this method, a bed of polymer particles is heated to partially melt the particles and fuse their surfaces together resulting in the formation of a porous structure. Heating could be done using different sources and approaches. For example, Simpson et al. [74]. used

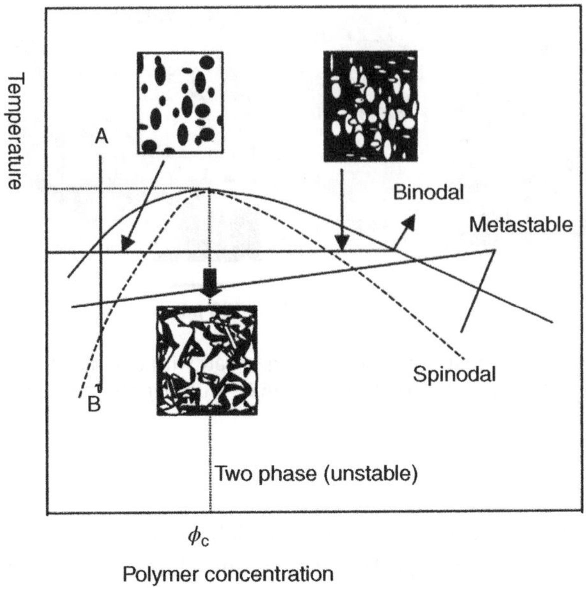


**Figure 8.3** Different regions in the phase separation curve in a polymer/solvent system. *Source:* Hua et al. [70]/with permission from John Wiley & Sons.

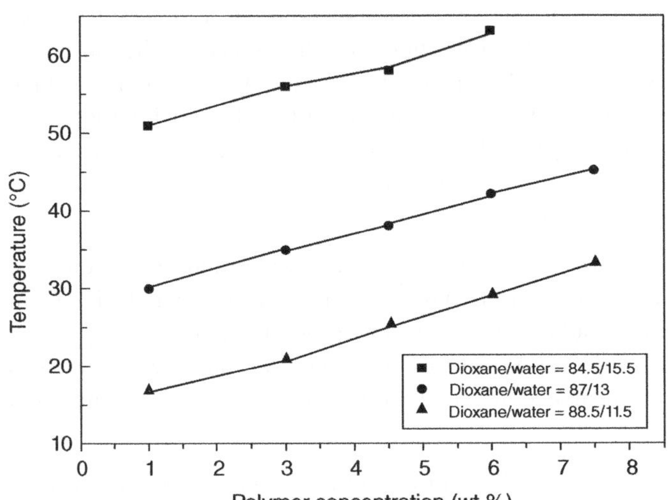


**Figure 8.4** Effect of solvent composition on the cloud point of PLA/water/Dioxane ternary system. *Source:* Hua et al. [70]/with permission from John Wiley & Sons.

selective laser sintering to produce porous PLGA/hydroxylapatite (HA) scaffold for trabecular bone.

5) Polymer blends [9, 75, 76]: In this approach, the porous scaffold is produced by selective extraction of a polymer component from a co-continuous binary blend of two polymers. Ghavidel et al. [18, 75]. blended polyethylene oxide (PEO) with PCL and then extracted PEO with water to produce a porous 3D PCL scaffolds for human bone marrow mesenchymal stem cells (MSCs) growth. They showed that thermal annealing of the blend before extraction of PEO could be used as an effective step in controlling the final pore sizes. Figure 8.5 shows μ-CT scans and SEM images of the scaffolds prepared at different thermal annealing times with average pore sizes of 84 and 141 μm.

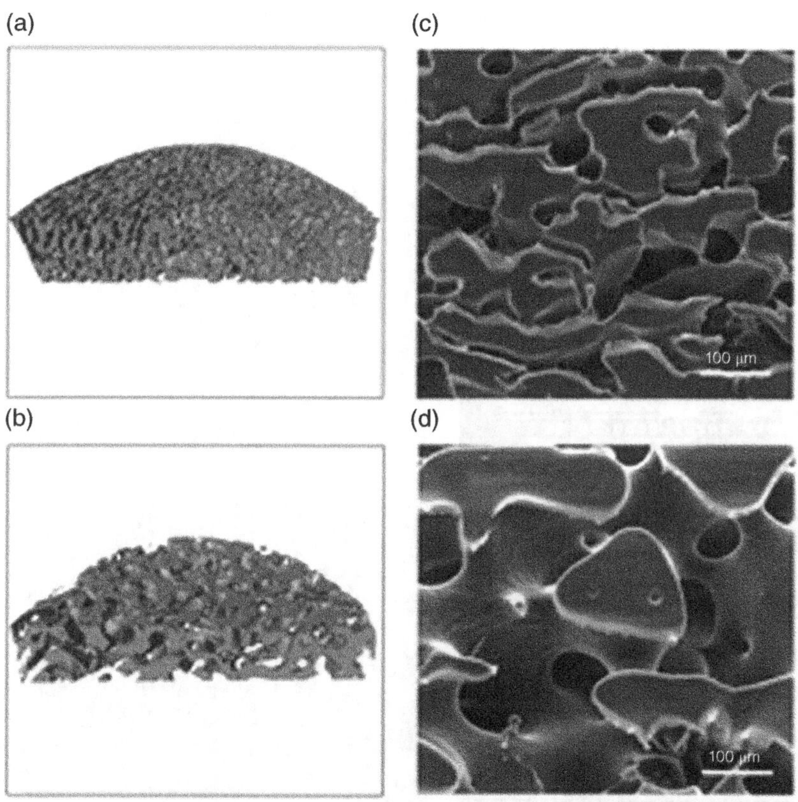


**Figure 8.5** Structure of PCL porous scaffolds prepared from PCL/PEO: (a and c) PCL/PEO blend was annealed at 160 °C for one hour, (b and d) PCL/PEO blend was annealed at 160 °C for two hours. (a) and (b) show μ-CT images while (c) and (d) show SEM images. PEO phase was extracted with water in all samples. *Source:* Mehr et al. [76]/with permission from John Wiley & Sons.

Gas Foaming: In this method, a polymer solution containing a gas-foaming agent is cast into a mold. Increasing temperature results in simultaneous evaporation of solvent and formation of gas by the gas foaming agent. This results in a porous 3D polymer scaffold. This method was initially introduced by Nam et al. [77], who prepared a porous PLLA structure by mixing ammonium bicarbonate with polymer solution followed by heating in hot water bath. The morphology of the final sample is shown in Figure 8.6. The pore size can be controlled by changing polymer/gas foaming agent ratio. Since then, this technique has been used by many authors to produce porous polyester scaffolds [78–80].

The presence of traces of solvent in the final scaffold in methods mentioned so far could deteriorate the scaffold performance. Mooney et al. [81]. prepared porous PLGA scaffolds by exposing disk of polymer to high-pressure $CO_2$ for

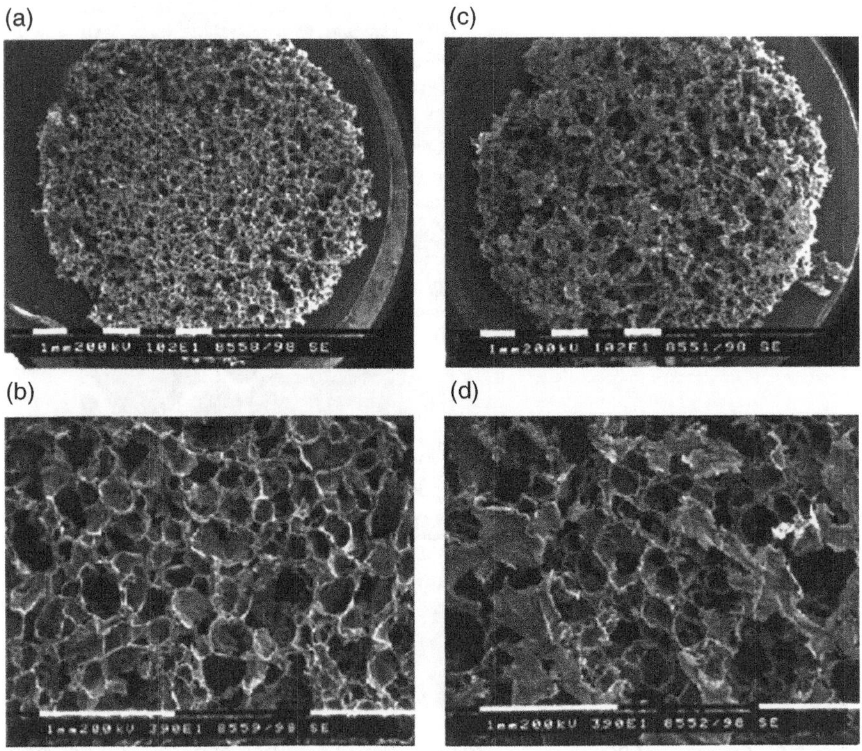


**Figure 8.6** SEM images of porous PLLA scaffolds prepared by gas foaming method: (a) ammonium bicarbonate/PLLA weight ratio = 6 : 1, (c) ammonium bicarbonate/PLLA weight ratio = 10 : 1; (b) and (d) show higher magnifications of (a) and (c). *Source:* Nam et al. [77]/ with permission from John Wiley & Sons.

72 hours. $CO_2$ pressure was then reduced, which led to thermodynamic instability and nucleation of $CO_2$ dissolved in the previous step in the polymer matrix. They reported that using this approach, scaffolds with large pore sized of 100 μm and porosity of 93% could be obtained.

6) Additive manufacturing techniques [82–85]: additive manufacturing techniques such as 3D printing and fused deposition modeling (FDM) have also been used for scaffold fabrication. FDM is an additive technique in which a molten polymer filament is deposited LbL to form 3D structures such as for porous scaffold. Figure 8.7 schematically shows the FDM process.

FDM techniques allow very high control on the micro-structure of the scaffold. An example of a porous scaffold produced from PCL filaments by FDM technique is shown in Figure 8.8.

For example, Yen et al. [86]. prepared a PLGA scaffold with different filament sizes and spacing using FDM technique and filled the scaffold with type II collagen. They then seeded the scaffold with chondrocytes isolated from the articular cartilage of femoral condyles of pigs. After 28 days, they found that the scaffold with the largest fiber spacing (1.15 mm) showed good distribution of

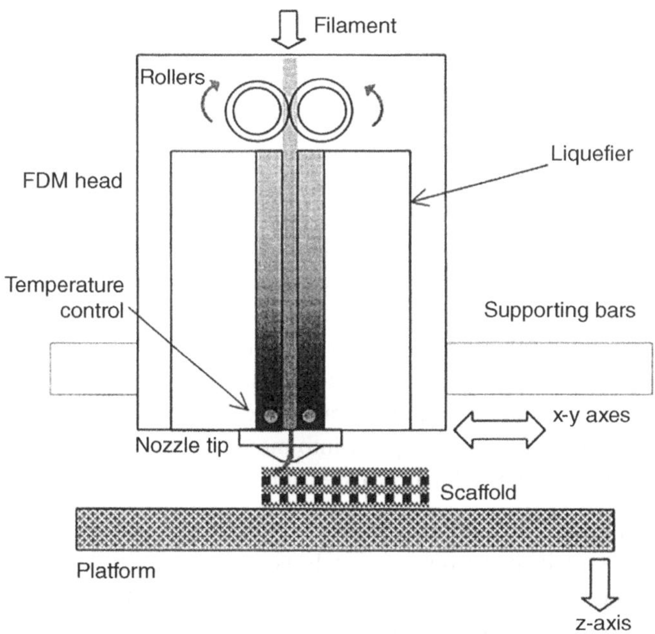


**Figure 8.7** Schematic diagram showing fused deposition modeling (FDM). *Source:* Hutmacher et al. [82]/with permission from John Wiley & Sons.

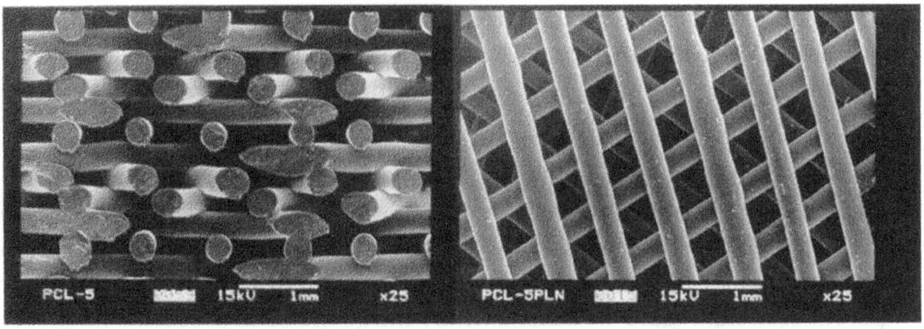


**Figure 8.8** SEM images of microstructure of PCL scaffolds prepared by FDM. (Left) side view and (right) top view. *Source:* Reproduced from [82]. Hutmacher et al. [82]/ with permission from John Wiley & Sons.

chondrocytes and formation of neocartilage around the scaffolds. The main limitation of FDM technique is its resolution which does not allow preparing nanostructured scaffolds such as the one needed for nerve regeneration.

7) Electrospinning: Electrospinning has been widely used to prepare nanofibrous porous scaffolds of polyesters for tissue engineering [87–93]. In this process, a solution of polymer and solvent is injected through a needle at low flow rates. A high electrical voltage is applied between the needle and a conductive target surface (which is called collector) at a distance from the needle. This results in the formation of very thin fibers from the ejected liquid. The evaporation of solvent results in the formation of solid nanofibers that are collected on the target surface. Li et al. [92]. used this approach to prepare electrospun scaffold of PLGA for MSC growth and observed guided growth of cells along with nanofiber orientation. Binan et al. [94]. used co-axial electrospinning to prepare nanofibers with core-shell structure where PLLA was the core surrounded by a shell of gelatin containing an active ingredient. They cross-linked gelatin using glyceraldehyde to control degradation rate of the shell layer and controlled release of the active ingredient. Figure 8.9 shows the microstructure of their co-electrospun PLLA/gelatine scaffold. They showed that seeded neural stem cells (NSCs) on the scaffold could proliferate and differentiate rapidly into motor neurons on the scaffold. The active ingredient enhanced neurite extension.

Vascular grafts are an interesting application for electrospun polyester scaffolds that are critically important in heart diseases such as congenital heart disease (CHD) [95,96]. The main characteristics of an ideal graft can be summarized as adequate mechanical strength for surgical implantation and withstanding blood flow, low rates of thromboembolic events, resistance to infection, and cells growth potential [97]. The graft materials that are commonly used in this application can

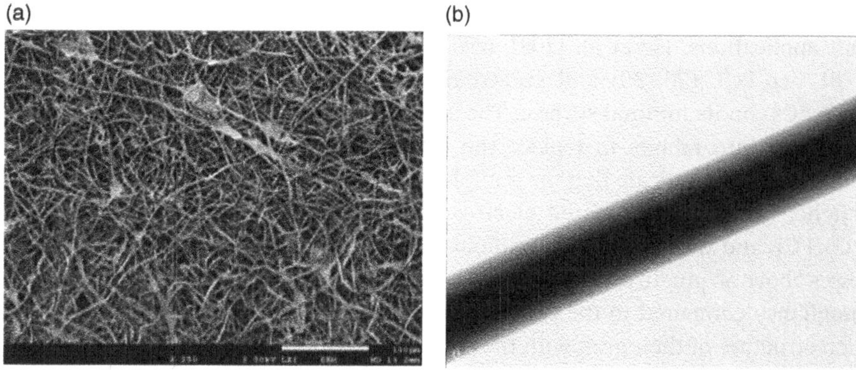


**Figure 8.9** (a) SEM images of co-electrospun fibers with PLLA core and collagen shell, (b) TEM images showing core and shell region in a fiber. *Source:* Binan et al. [94]/ with permission from Elsevier.

be categorized as [98]: (i) artificial grafts, (ii) autologous tissues, such as pericardium and saphenous vein, (iii) allografts, and (iv) xenografts. The main commercial artificial grafts are made from expanded polytetrafluorethylene (ePTFE, Gore-Tex), PTFE (Teflon, Dupont), and PET (Dacron). These materials are associated with postoperative complications including stenosis, thromboembolisation, calcium deposition, and infection [99,100] especially in small diameter vessels [101]. Despite a lower rate of thromboembolic events in biologic grafts like autologous tissues, allografts, and xenografts, they have shorter durability compare with artificial grafts, which requires reoperation and increases significantly rates of mortality [98]. The pioneer work of Greinsler in early 1980s [102] on implanting absorbable PGA graft in rabbit aortas can be considered as the pioneer work in artificial absorbable vascular grafts. Greinsler's work seeded the idea of tissue-engineered vascular graft (TEVG), which has been pursued and developed by many researchers since then [103]. Niklason et al. [104]. used a PGA mesh as the scaffold and seeded autologous endothelial cells onto its luminal surface. They cultured the scaffold for three days and then implanted it into a pig. They found that the scaffold remained patent after four weeks. Shinoka et al. [105]. used a PLGA woven mesh sealed with nonwoven PGA mesh and observed change in diameter of the graft indicating the growth of endothelial lining. In addition to homopolymer grafts, TEVGs based on copolymers or polymer mixtures have received considerable attention as they show wider range of flexibility for controlling mechanical properties and degradation rate. Mid-term (four years) [106] and late-term (10 years) [107] in-vivo performance and patency of grafted PLA-co-PCL (50 : 50) TEVGs were examined in human body and no graft related mortality, evidence of aneurysm formation, graft rupture, graft infection, or ectopic were observed. These results indicate the potential of TEVG as promising solutions for patent vascular graft applications. He et al. [108]. also prepared electrospun tubular prosthesis of PLA-co-PCL (70 : 30) and seeded human coronary artery endothelial cells (HCAECs) on its luminal surface. The scaffold was cultured for 10 days and then implanted into rabbits to replace the inferior superficial epigastric veins. The obtained results indicate that the scaffold could retain its patency for seven weeks. Bouchet et al. [96]. prepared electrospun tubular prosthesis of polyurethane (PU)/PCL and showed that the scaffold was impermeable to blood for wall thicknesses above 50 μm. In addition, they showed that the prepared scaffold had higher compliance compared to the commercial ePTFE graft. Figure 8.10 compares the microstructure of their graft with the commercial Gore-Tex ePTFE graft.

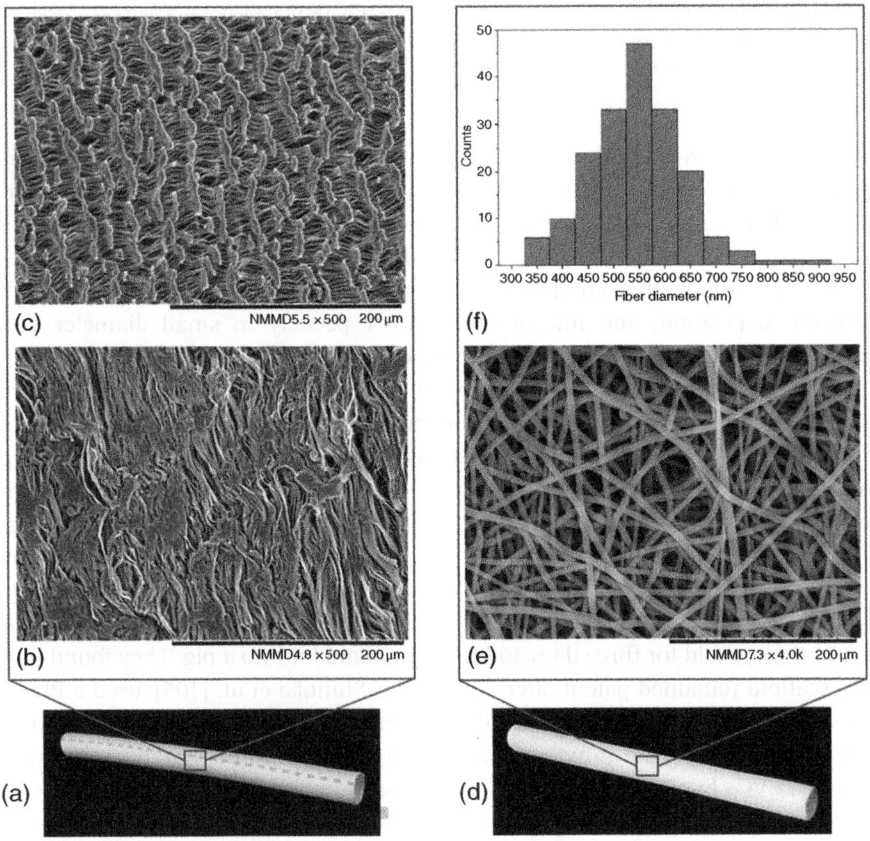


**Figure 8.10** Morphology of vascular prostheses: (a) commercial ePTFE, (b) surface morphology of ePTFE, (c) interior surface morphology of ePTFE, (d) PU/PCL prosthesis, (e) surface morphology of PU/PCL, (f) fiber diameter distribution of the electrospun PU/PCL. *Source:* Bouchet et al. [96]/with permission from Elsevier.

Sell et al. [109]. prepared vascular grafts from electrospun PDS/elastin blend and found that at 50 : 50 composition of PDS : elastin, mechanical properties mimic those of native femoral artery. They found that the presence of elastin led to migration and formation of cells through the thickness of the graft in seven days.

In addition to TEVG, application of polyester scaffolds for nerve regeneration has also been studied by many authors [89, 110–112]. As a suitable scaffold should have similar structure and biological function of nanoscale fibrous extracellular matrix (ECM), electrospinning has been used as the main approach to fabricate scaffolds for nerve regeneration. For example, Ghasemi-Mobarakeh et al. [112]. prepared scaffolds of PCL/gelatin and PCL/Gelatin/polyaniline (PANI) and examined the effect of electrical stimulation on nerve cells regeneration. As can be seen in Figure 8.11, they found that applying electrical stimulation led to considerable nerve cells proliferation and neurite outgrowth.

Electrospun tubular scaffolds of polyesters have also been used as the dural substitute for treatment of dural tears and defects caused by traumatic injury or during surgical operation [113–115]. Xie et al. [114]. showed that radial orientation of electrospun PCL nanofibers could induce migration of dural fibroblast cells along the nanofiber orientation direction and resulted in a complete coverage of the scaffold while the cells were not found in the center of the scaffold with random fiber orientation prepared by conventional electrospinning method. This indicates the importance of controlling fiber orientation in migration and

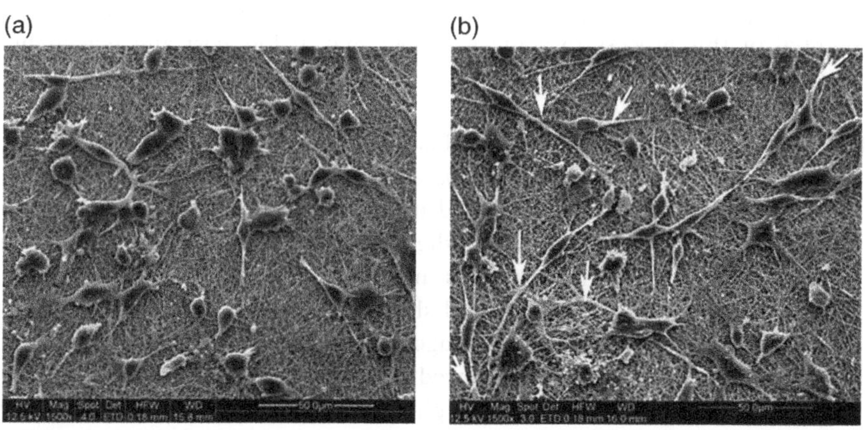


**Figure 8.11** Morphology of nerve cells (×1500 magnification) on PANI/PG nanofibrous scaffolds: (a) without electrical stimulation; (b) with electrical stimulation. Arrows show neurite extension of neurons. *Source:* Ghasemi et al. [112]/ with permission from John Wiley & Sons.

proliferation of dural fibroblast cells. In another interesting study, Kim et al. [116]. modified the collector drum and could produce a transparent hemispherical 3D scaffold of PCL/collagen nanofibers with radially oriented nanofibers in the middle and randomly oriented fibers around the center region. The produced 3D scaffold could mimic properties of the cornea including microscopic to macroscopic morphology and transparency.

Electrospun mats have also been used as scaffolds for skin tissue engineering and wound healing applications. Electrospun PLGA copolymers [117,118], PHBV [119–121], PCL [122,123], PLA-PCL copolymers [124,125], and PLLA [124,126] are among polyester scaffolds studied for skin tissue engineering. For example, Kumbar et al. [117]. used electrospun PLGA nanofibers and seeded the scaffold with human skin fibroblast. They reported that fibroblast proliferations rate depends significantly on the scaffold nanofiber diameter with the highest proliferation rate observed in the diameter range of 350–1100 nm.

Despite the unique characteristics and potentials of the electrospinning process, the high operation cost and difficulties in controlling this process restricted its application in tissue engineering [87,127]. Moreno et al. [95]. used the melt-blowing process, as a high throughout and low-cost technique, to produce nonwoven PET for vascular graft and examined cell adhesion and proliferation of the neat scaffold and gelatin-coated scaffold. Among different fiber diameters and pore sizes that they examined, the sample with the smallest fiber diameter and pore size showed the most promising results for the growth of human brain endothelial cells and aortic smooth muscle cells. Despite larger fiber sizes prepared by melt-blowing, these results indicate the potential of this process to achieve a fiber mat at low cost and reasonable reproducibility.

8) Scaffolds based on woven meshes: This approach comprises stacking sheets of woven bioresorbable sheets to prepare a 3D porous structure and has been used as a simple 3D model scaffold in some cartilage regeneration studies using PLA, PGA and PLGA meshes [63,128–131]. For example, Chen et al. [128]. stacked PLGA meshes and then immersed the structure in a bovine collagen acidic solution and showed that cell seeding and adhesion improved compared to the neat collagen sponge. Histological examination of the scaffold after implantation showed a homogeneous chondrocyte distribution and cartilaginous matrix throughout the implant. SEM images of their PLGA mesh and PLGA-collagen web are shown in Figure 8.12. Despite the simplicity of this method, it can only be used for large pore sizes scaffolds. In addition, inter-layer adhesion between mesh layers could affect the integrity of the scaffold.

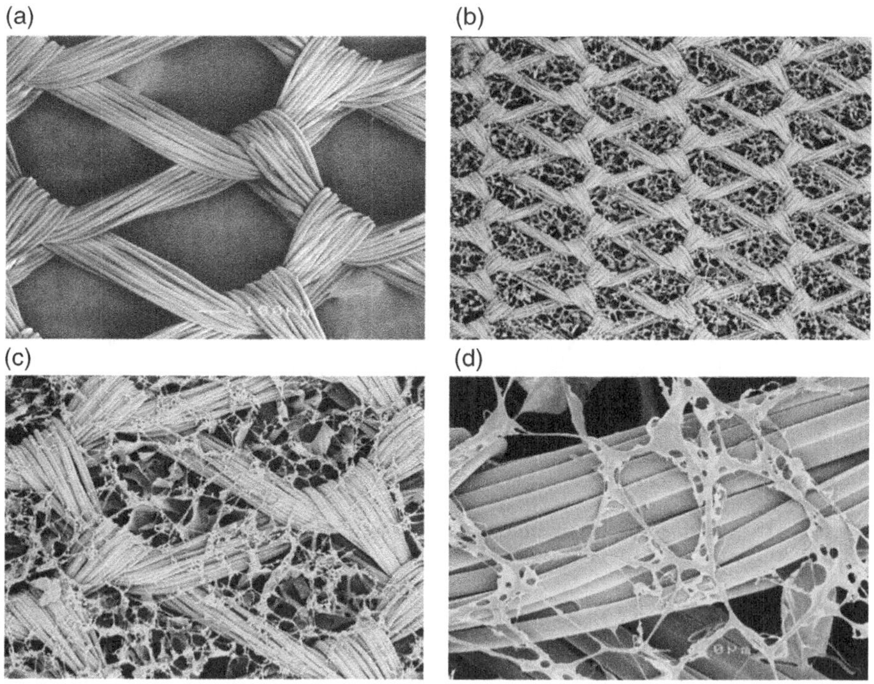


**Figure 8.12** SEM of (a) PLGA mesh, (b–d) PLGA–collagen web different magnifications. *Source:* Chen et al. [128]/ with permission from John Wiley & Sons.

## 8.3.2 Drug Delivery and Controlled Release Systems

Conventional delivery methods, also known as immediate-release systems, including tablets, pills, or injectables cause a prompt increase in drug level in the patient body which can reach the toxic concentration range, Figure 8.13. This is followed by a sudden decrease in drug concentration which results in a significant drug concentration fluctuation, and the drug concentration will be within the therapeutic window for a short period of time. Therefore, patients need to take the medication several times a day, which significantly influences the patient compliance and acceptance. The main purpose of a drug delivery system is to reduce the dosing frequency by at least a factor of two [132]. The ideal system would be a zero-order controlled release system, see Figure 8.13, in which release rate remains constant during the release process, which results in a constant drug concentration within the therapeutic concentration window during the treatment period. This eliminates the drug re-administration and reduces considerably the drug concentration fluctuation during the treatment period, resulting in shorter treatment period and reducing hospitalization and healthcare costs.

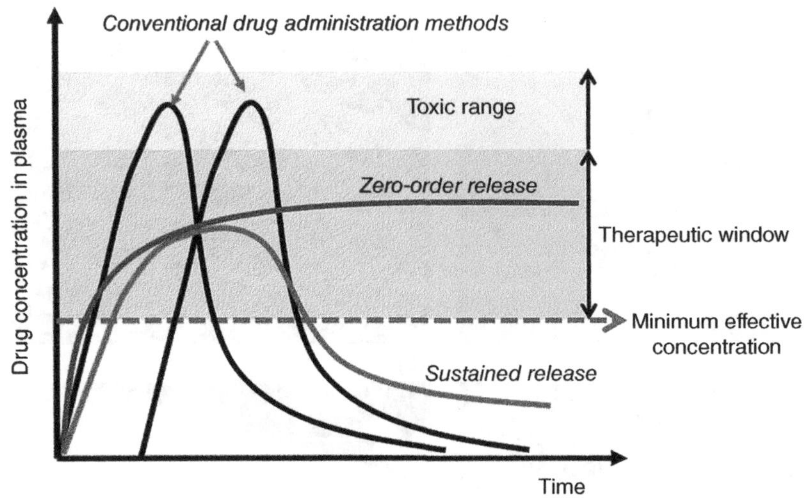


**Figure 8.13** Schematic representation of drug plasma levels after various dosing regimens.

However, complexity of manufacturing zero-order release systems and their high production cost have limited the commercial interest in zero-order drug release systems. The pre-programmed drug release systems are called modified release systems by Food and Drug Administration (FDA) [133] and world health organization (WHO) [134]. This category includes delayed-release systems and extended-release or sustained-release systems. The delayed systems are those with a time lag between drug administration time and drug concentration change in plasma. An extended release system results in extending and prolonging the release of the drug. It should be noted that both FDA and WHO consider extended release and sustained release as interchangeable terms. Targeted release systems are another interesting sector in drug delivery systems in which the drug is delivered and released only in the targeted area of the body. However, these systems have not been addressed in neither FDA nor WHO nomenclature.

Among different polyester materials, PLA [135–138], PLGA [139–143], PCL [144, 145], polyorthoesters [146, 147], and PDS [148, 149] have been used frequently in drug delivery applications. Different routes for administration of biomaterials are summarized in Figure 8.14.

Implantable drug delivery systems (IDDS) are among those delivery routes that were studied extensively and currently three IDDS products based on PLGA are available in the market. PLGA (50 : 50) subdermal implants with the commercial name of Zoladex have been used for monthly subcutaneous controlled delivery of goserelin acetate. PLGA (75 : 25) implants under the commercial name of Suprefact (also known as Profact Depot and Trigonist) have also been used to

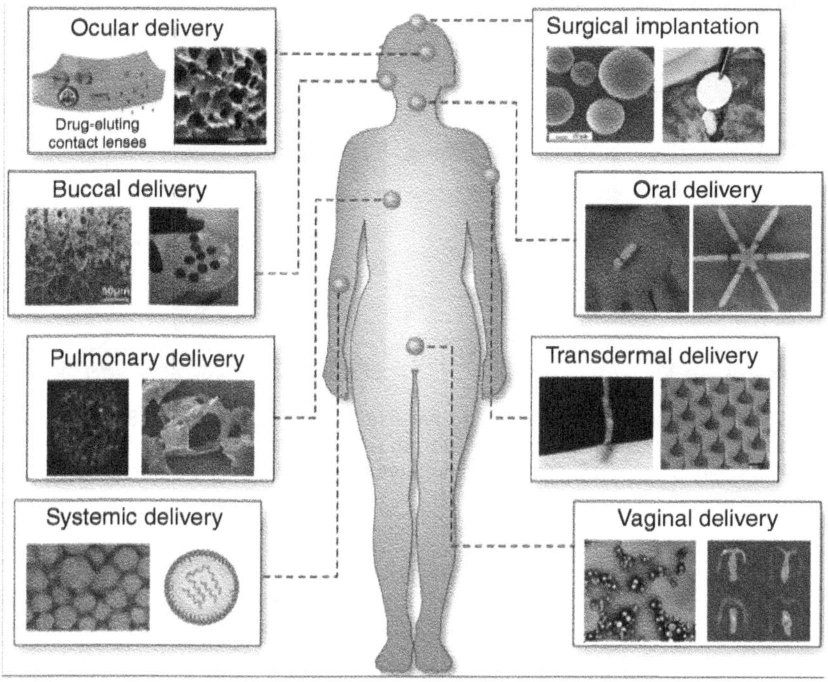


**Figure 8.14** Different administration strategies of biomaterials. *Source:* Li et al. [150]/ with permission from John Wiley & Sons.

release buserelin acetate in two to three months period. Another example of implantable drug delivery system available in the market is Ozurdex which is an intravitreal PLGA (50 : 50) implant for controlled-release of dexamethasone for the treatment of macular edema related to branch retinal vein occlusion (BRVO), central retinal vein occlusion (CRVO), and noninfectious uveitis [151].

In addition to IDDS, micro- and nanoparticles (or capsules in some cases) have also received considerable attention in drug delivery applications. Figure 8.15(a) shows SEM images of PLLA micro-particle loaded with chondroitin sulfate and PLGA nanoparticles.

Micro-spheres or micro-capsules can be manufactured by single or double emulsion, phase separation, and spray drying [154]. However, polydispersity of particles prepared by these methods reduces the control over release rate. Recently, production of micro-particles using microfluidic devices has attracted some attentions as a new route to prepare monodispersed micro-carriers. Xu et al. [152]. compared release profile of PLGA micro-particles prepared by single emulsion method and a microfluidic device as carriers for bupivacaine. As can be seen in Figure 8.16, they found that particles prepared by single emulsion method

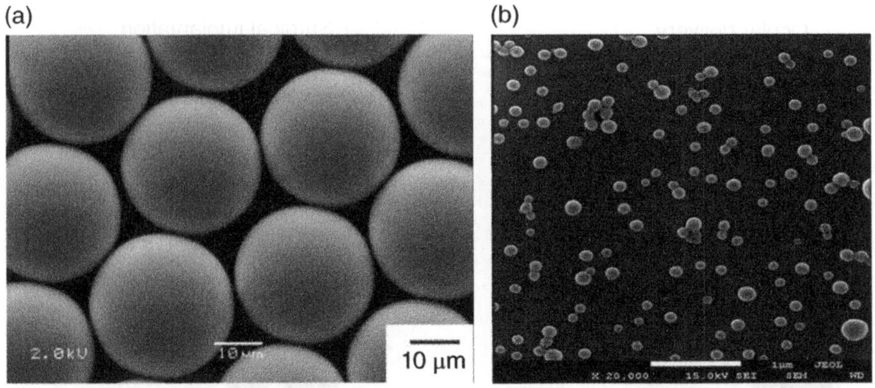


**Figure 8.15** SEM images of (a) PLGA micro-particles (average diameter = 28 μm) loaded with chondroitin sulfate. *Source:* Reproduced from [152]. Xu et al. 2009/John Wiley & Sons, (b) PLGA nanoparticles. *Source:* Carvalho Silva et al. [153] / with permission from John Wiley & Sons.

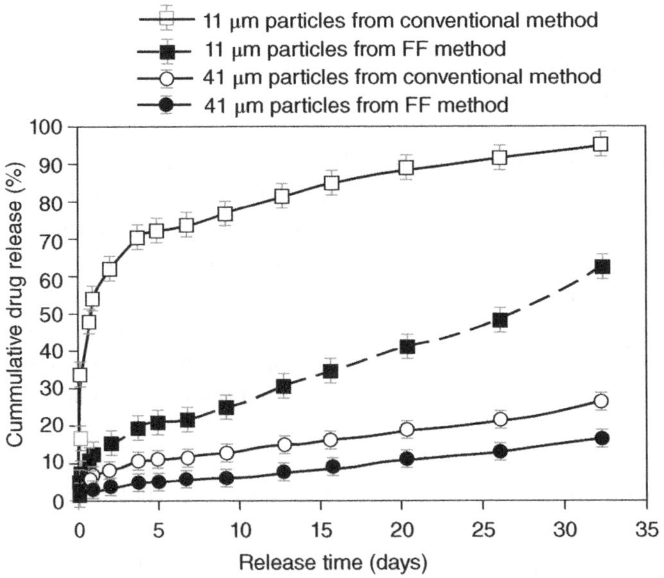


**Figure 8.16** Bupivacaine-release profile for PLGA microparticles prepared with microfluidic devices (FF method) and polydisperse microparticles prepared by single emulsion technique (conventional method). *Source:* Xu et al. [152]/with permission from John Wiley & Sons.

released bupivacaine at a very high rate in the first four days followed by a slower release period. On the other hand, bupivacaine was released at much uniform rate for the particles prepared by the microfluidic device. This indicates the significant importance of high level of control on particle size in achieving sustained release of drugs in drug delivery systems.

Some of the available commercial drug delivery products based on PLGA microcarriers are [155]: (i) microsphere loaded with leuprolide for treatment of prostate cancer with the commercial name of Lupron® Depot, (ii) microsphere loaded with somatropin for treatment of growth failure under commercial brand of Neutropin® Depot, and (iii) microsphere loaded with risperidone for treatment of certain mental disorder under commercial brand of Risperdal® Consta. It has been shown that particles greater than 500 nm cannot enter the systemic circulation or pass through mucosal membrane to deliver the drug to target tissue [156]. Therefore, preparing what is currently known as nanocarriers is essential in designing a drug delivery system that delivers drug using systemic circulation. This had led to extensive studies in the literature on nanocarriers based on polyester materials for drug delivery applications [137, 140–142, 157]. While nanocarriers can also be prepared by tuning processing condition in the manufacturing techniques for micro-particles [154], some new routes have also been proposed in the literature for producing nanocarriers for drug delivery [140, 158]. Some polyester-based nanocarrier are now in the clinical stage and expected to be introduced to the market soon. Among them, SEL-212 that is PLGA nanoparticles loaded with rapamycin for treatment of gout disease is now in the second phase of clinical trial [159]. PEG-PLA-PEG triblock nanoparticles loaded with docetaxel for treatment of solid tumors is in phase I clinical trials [160]. Also, PLA nanoparticles system loaded with Oxybenzone-Padimate O for sunscreen protection is now in phase I clinical trials [161].

### 8.3.3 Orthopedic Applications

Using biodegradable polyesters in orthopedic applications was limited in the past due to their much lower mechanical strength compared with metal implants. Despite their high mechanical strength, the lack of degradability and design flexibility of metallic implants and fixations have limited their applications and performance [6]. For example, due to the lack of degradability of metal implants, a second surgery might be required to remove them, which increases significantly their risk and implantation costs. The recent technological developments in the field of polymer materials and composites led to considerable improvements in their mechanical properties and introduced them as an interesting competitor for metal implants. Considering their lower processing cost and degradable nature, polyester materials provide a high design flexibility and allow tailoring final properties to a much higher extent compared to metals. Among polyesters, PLA, PGA,

PLGA copolymers, PCL, polydioxanone (PDS), and their composites [162–171] have been used in orthopedic implants and fixations. Many studies have been devoted to developments/modifications to enhance the design and implications of bioabsorbable fixation devices and, nowadays, these polymeric devices are commonly used for stabilization of fractures, reattachment of ligaments, tendons, and bone grafting [167]. Some examples of the commercial orthopedic devices based on polyesters are listed in Table 8.1.

The term self-reinforced polymer (SRP) is used frequently in orthopedic applications, which refers to a polymer system where matrix and reinforcing phases are made of the same polymer. The reinforcing phase is mainly produced by 1D, 2D, or 3D alignment of molecules in that phase. The importance of SRP systems lies in the fact that they provide higher mechanical properties at much lower density compared to polymer composites. In addition, an almost perfect interface between matrix and reinforcing phase ensures stress transfer from the matrix to the reinforcing phase [172]. Self-reinforced PLA (SR-PLA) and PGA (SR-PGA) were produced with this approach and have shown promising mechanical strength for orthopedic applications [173, 174].

**Table 8.1** Some available polyester-based commercial products for orthopedic applications.

| Application | Commercial name | Material |
|---|---|---|
| Interference screws for anterior cruciate ligament (ACL) graft fixation | Euroscrew | PLLA or PLLA/tricalcium phosphate (TCP) |
| | Biosteon | PLLA/Hydroxyapatite (HA) |
| | Sysorb | PDLA |
| | Matryx | PLA/TCP |
| | Smartscrew | PLLA |
| | MILAGRO® Advance | PLGA/TCP |
| | Biocomposite | PLLA/TCP |
| Maxillofacial and facial fracture fixation | LactoSorb | PLGA |
| | Fixsorb | PLLA |
| Suture anchors | Duet® Suture | PLLA |
| | Zimmer Bio-Statak | PLLA |
| | Panalok | PLGA |
| Pins | BioFix | SR-PGA or SR-PLLA |
| | Smartpins | PLLA |

### 8.3.4 Other Applications

Among different applications of polyester materials, polyester-based sutures and PET blood collection tubes (BCT) are two other important applications of polyesters in biomedical field that cannot be categorized under previously mentioned groups of applications and, therefore, are reviewed here in a separate section.

#### 8.3.4.1 Polyester-based Suture

Polyester-based suture are among the earlier applications investigated for polyester materials in biomedical field. The study on absorbable polyester suture (in this case PGA) was done by Frazza and Schimitt back in 1971 [175], and the first patent on using PLA-PGA in suture application dates back to 1973 [176]. Since then, many studies focused on the preparation and characterization of sutures based on different polyester materials [93, 148, 177, 178]. Currently, different commercial polyester-based sutures are available in the market such as Dexon Plus (PGA), Vicryl (polyglactin), Maxon (polyglyconate), and PDS (polydioxanone). Bourne et al. [178]. compared the in-vivo half-life tensile strength of these four products and found that half-life of Dexon Plus and Vicryl was two weeks while Maxon and PDS showed three and six weeks half-lives, respectively. Bourne et al. also found that, in addition to tensile strength, the knot security (indicated as the ability to hold the knot) is also very important when comparing performance and integrity of suture filaments. They found that among the four tested commercial sutures, Maxon showed superior knot security and performance.

#### 8.3.4.2 PET Blood Collection Tubes

Traditionally, BCTs were made from glass. However, the ease of processing, handling, and disposing of plastics and the lower production cost of plastic BCTs compared to glass increased significantly the application of plastic BCTs. Among different polymers, PET has been used commonly for BCTs due to its transparency and strength, which allows thin wall BCTs with acceptable mechanical properties. However, it has been shown that PET tubes can have complex interactions with blood serum and affect laboratory tests. Coating PET BCT walls with surfactant is the common method that tube manufacturer use to eliminate these interactions. Some researchers showed that these surfactants can leach out to the serum and interfere with tests results [14]. Plasma treatment has been presented as an approach to resolve the issue but, it was shown that plasma-treated PET could not retain surface hydrophilicity during the expected storage time of BCTs (18 months) [179]. In a recent study, Kim et al. [13]. grafted ethylene glycol on the surface of PET using their patented procedure [180]. They examined the performance of modified tubes in thyroid hormone tests in another study [16] and found negligible interference of modified PET BCT compared to commercial PET BCT samples.

## References


**1** Fitch Solutions (2018). *Global Medical Devices Report Q4*. Fitch Solutions.

**2** Lucintel (2019). *Medical Device Market Report: Trends, Forecast and Competitive Analysis*, 177. Lucintel.

**3** (2019). Medical plastic market. In: *Markets and Markets*, 155.

**4** Pawar, R.P. U Tekale S, U Shisodia S, T Totre J, and J Domb A. (2014). *Biomedical applications of poly (lactic acid). Recent Patents on Regenerative Medicine* 4 (1): 40–51.

**5** Malikmammadov, E., Tanir, T.E., Kiziltay, A. et al. (2018). *PCL and PCL-based materials in biomedical applications. Journal of Biomaterials Science, Polymer Edition* 29 (7–9): 863–893.

**6** Singhvi, M., Zinjarde, S., and Gokhale, D. (2019). *Poly-Lactic acid (PLA): synthesis and biomedical applications. Journal of Applied Microbiology* 127 (6): 1612–1626.

**7** Angelova, N. and Hunkeler, D. (1999). *Rationalizing the design of polymeric biomaterials. Trends in Biotechnology* 17 (10): 409–421.

**8** Manso, M., Valsesia A, Lejeune M, Gilliland D, Ceccone G, and Rossi F.(2005). *Tailoring surface properties of biomedical polymers by implantation of Ar and He ions. Acta Biomaterialia* 1 (4): 431–440.

**9** Ghavidel Mehr, N., Hoemann, C.D., and Favis, B.D. (2015). *Chitosan surface modification of fully interconnected 3D porous poly(ε-caprolactone) by the LbL approach. Polymer* 64: 112–121.

**10** Amiji, M. and Park, K. (1992). *Surface modification of polymeric biomaterials with poly(ethylene oxide), albumin, and heparin for reduced thrombogenicity*. In: *Polymer Biomaterials in Solution, as Interfaces and as Solids* (ed. S.L. Cooper, C.H. Bamford, and T. Tsuruta), 535–552. Utrech, Netherland: VSP.

**11** Ikada, Y. (1994). *Surface modification of polymers for medical applications. Biomaterials* 15 (10): 725–736.

**12** Chen, W. and McCarthy, T.J. (1998). *Chemical surface modification of poly(ethylene terephthalate). Macromolecules* 31 (11): 3648–3655.

**13** Kim, S., Bowen, R.A.R., and Zare, R.N. (2015). *Transforming plastic surfaces with electrophilic backbones from hydrophobic to hydrophilic. ACS Applied Materials & Interfaces* 7 (3): 1925–1931.

**14** Bowen, R.A.R., Chan Y, Ruddel ME, Hortin GL, Csako G, Demosky SJ, and Remaley AT. (2005). *Immunoassay interference by a commonly used blood collection tube additive, the organosilicone surfactant silwet L-720. Clinical Chemistry* 51 (10): 1874–1882.

**15** Bowen, R.A.R., Hortin GL, Csako G, Otañez OH, and Remaley AT. (2010). *Impact of blood collection devices on clinical chemistry assays. Clinical Biochemistry* 43 (1): 4–25.

**16** Jalali Dil, E., Kim SC, Saffar A, Ajji A, Zare RN, Sattayapiwat A, Esguerra V, and Bowen RAR. (2018). *Surface characterization and free thyroid hormones response of chemically modified poly(ethylene terephthalate) blood collection tubes. Applied Surface Science* 442: 602–612.

**17** Zhu, Y., Mao, Z., and Gao, C. (2013). *Aminolysis-based surface modification of polyesters for biomedical applications. RSC Advances* 3 (8): 2509–2519.

**18** Ghavidel Mehr, N. (2014). Highly structured and surface modified poly(ε-caprolactone) scaffolds derived from co-continuous polymer blends for bone tissue engineering. PhD. Thesis. Ecole Polytechnique de Montreal.

**19** Wales, S., Tan, G.K., and Cooper-White, J.J. (2015). *Controlling stem cell adhesion, proliferation, and differentiation with layer-by-layer films*. In: *Layer-by-Layer Films for Biomedical Applications*, 103–130. John Wiley & Son.

**20** Zhu, Y., Gao C, He T, Liu X, and Shen J. (2003). *Layer-by-layer assembly to modify poly (L-lactic acid) surface toward improving its cytocompatibility to human endothelial cells. Biomacromolecules* 4 (2): 446–452.

**21** Gong, Y., Zhu Y, Liu Y, Ma Z, Gao C, and Shen J. (2007). *Layer-by-layer assembly of chondroitin sulfate and collagen on aminolyzed poly (L-lactic acid) porous scaffolds to enhance their chondrogenesis. Acta Biomaterialia* 3 (5): 677–685.

**22** Ma, Z., Gao C, Yuan J, Ji J, Gong Y, and Shen J. (2002). *Surface modification of poly-L-lactide by photografting of hydrophilic polymers towards improving its hydrophilicity. Journal of Applied Polymer Science* 85 (10): 2163–2171.

**23** Zhu, Y., Gao C, Liu X, He T, and Shen J. (2004). *Immobilization of biomacromolecules onto aminolyzed poly (L-lactic acid) toward acceleration of endothelium regeneration. Tissue Engineering* 10 (1–2): 53–61.

**24** Sánchez, L.D., Brack N, Postma A, Pigram PJ, and Meagher L. (2016). *Surface modification of electrospun fibres for biomedical applications: a focus on radical polymerization methods. Biomaterials* 106: 24–45.

**25** Fadeev, A.Y. and McCarthy, T.J. (1998). *Surface modification of poly(ethylene terephthalate) to prepare surfaces with silica-like reactivity. Langmuir* 14 (19): 5586–5593.

**26** Inagaki, N., Narushim K, Tuchida N, and Miyazaki K. (2004). *Surface characterization of plasma-modified poly(ethylene terephthalate) film surfaces. Journal of Polymer Science Part B: Polymer Physics* 42 (20): 3727–3740.

**27** Kostov, K.G., Nishime TMC, Castro AHR, Toth A, and Hein LRO. (2014). *Surface modification of polymeric materials by cold atmospheric plasma jet. Applied Surface Science* 314: 367–375.

**28** Gonzalez, E., Barankin MD, Guschl PC, and Hicks RF. (2008). *remote atmospheric-pressure plasma activation of the surfaces of polyethylene terephthalate and polyethylene naphthalate. Langmuir* 24 (21): 12636–12643.

**29** Ping, X., Wang, M., and Xuewu, G. (2011). *Surface modification of poly(ethylene terephthalate) (PET) film by gamma-ray induced grafting of poly(acrylic acid) and its application in antibacterial hybrid film. Radiation Physics and Chemistry* 80 (4): 567–572.

**30** Rezaei, F., Dickey MD, Bourham M, and Hauser PJ. (2017). *Surface modification of PET film via a large area atmospheric pressure plasma: an optical analysis of the plasma and surface characterization of the polymer film. Surface and Coatings Technology* 309 (Supplement C): 371–381.

**31** Aziz, G., Thukkaram M, De Geyter N, and Morent R. (2017). *Plasma parameters effects on the properties, aging and stability behaviors of allylamine plasma coated ultra-high molecular weight polyethylene (UHMWPE) films. Applied Surface Science* 409: 381–395.

**32** Bacakova, L., Filova E, Parizek M, Ruml T, and Svorcik V. (2011). *Modulation of cell adhesion, proliferation and differentiation on materials designed for body implants. Biotechnology Advances* 29 (6): 739–767.

**33** Khang, D., Lu J, Yao C, Haberstroh KM, and Webster TJ. (2008). *The role of nanometer and sub-micron surface features on vascular and bone cell adhesion on titanium. Biomaterials* 29 (8): 970–983.

**34** Rezaei Kolahchi, A., Carreau, P.J., and Ajji, A. (2014). *Surface roughening of PET films through blend phase coarsening. ACS Applied Materials & Interfaces* 6 (9): 6415–6424.

**35** Persson, M., Lorite GS, Kokkonen HE, Cho S-W, Lehenkari PP, Skrifvars M, and Tuukkanen J. (2014). *Effect of bioactive extruded PLA/HA composite films on focal adhesion formation of preosteoblastic cells. Colloids and Surfaces B: Biointerfaces* 121: 409–416.

**36** Goddard, J.M. and Hotchkiss, J. (2007). *Polymer surface modification for the attachment of bioactive compounds. Progress in Polymer Science* 32 (7): 698–725.

**37** Li, S. and M. Vert, Biodegradation of aliphatic polyesters, in Degradable Polymers: Principles and Applications, G. Scott, Editor. 2002, Springer Netherlands: Dordrecht. p. 71–131.

**38** Gunatillake, P.A. and Adhikari, R. (2003). *Biodegradable synthetic polymers for tissue engineering. European Cells & Materials* 5 (1): 1–16.

**39** Middleton, J.C. and Tipton, A.J. (2000). *Synthetic biodegradable polymers as orthopedic devices. Biomaterials* 21 (23): 2335–2346.

**40** Rey-Vinolas, S., Engel, E., and Mateos-Timoneda, M.A. (2019). Polymers for bone repair. In: *Bone Repair Biomaterials*, 2e (ed. K.M. Pawelec and J.A. Planell), 179–197. Woodhead Publishing.

**41** Nishida, H. and Tokiwa, Y. (1993). *Effects of higher-order structure of poly (3-hydroxybutyrate) on its biodegradation. II. Effects of crystal structure on microbial degradation. Journal of Environmental Polymer Degradation* 1 (1): 65–80.

**42** Li, S. (1999). *Hydrolytic degradation characteristics of aliphatic polyesters derived from lactic and glycolic acids. Journal of Biomedical Materials Research: An Official Journal of the Society for Biomaterials, the Japanese Society for Biomaterials, and the Australian Society for Biomaterials* 48 (3): 342–353.

**43** Pierre, T.S. and Chiellini, E. (1986). *Review : biodegradability of synthetic polymers used for medical and pharmaceutical applications: part 1 – principles of hydrolysis mechanisms. Journal of Bioactive and Compatible Polymers* 1 (4): 467–497.

**44** Torres, A., Li S, Roussos S, and Vert M. (1996). *Degradation of L-and DL-lactic acid oligomers in the presence of Fusarium moniliforme and Pseudomonas putida. Journal of Environmental Polymer Degradation* 4 (4): 213–223.

**45** Chawla, A.S. and Chang, T. (1985). *In-vivo degradation of poly (lactic acid) of different molecular weights. Biomaterials, Medical Devices, and Artificial Organs* 13 (3–4): 153–162.

**46** Tracy, M.A., Ward KL, Firouzabadian L, Wang Y, Dong N, Qian R, and Zhang Y. (1999). *Factors affecting the degradation rate of poly(lactide-co-glycolide) microspheres in vivo and in vitro. Biomaterials* 20 (11): 1057–1062.

**47** Göpferich, A. (1996). *Mechanisms of polymer degradation and erosion. Biomaterials* 17 (2): 103–114.

**48** Huang, S.J. and P.G. Edelman, An overview of biodegradable polymers and biodegradation of polymers, in Degradable Polymers: Principles and applications, G. Scott and D. Gilead, Editors. 1995, Springer Netherlands: Dordrecht. p. 18–28.

**49** Chandra, P.K., Soker, S., and Atala, A. (2020). *Tissue engineering: current status and future perspectives.* In: *Principles of Tissue Engineering*, 5e (ed. R. Lanza et al.), 1–35. Academic Press.

**50** Langer, R. and Vacanti, J. (1993). *Tissue engineering. Science* 260 (5110): 920–926.

**51** Hutmacher, D.W. (2000). *Scaffolds in tissue engineering bone and cartilage. Biomaterials* 21 (24): 2529–2543.

**52** Schmitz, J.P. and Hollinger, J.O. (1988). *A preliminary study of the osteogenic potential of a biodegradable alloplastic-osteoinductive alloimplant. Clinical Orthopaedics and Related Research* 237: 245–255.

**53** Singhal, A., Agrawal, C., and Athanasiou, K. (1996). *Salient degradation features of a 50: 50 PLA/PGA scaffold for tissue engineering. Tissue Engineering* 2 (3): 197–207.

**54** Patrício, T., Domingos M, Gloria A, D'Amora U, Coelho J, and Bártolo P. (2014). *Fabrication and characterisation of PCL and PCL/PLA scaffolds for tissue engineering. Rapid Prototyping Journal* 92 (2): 366–376.

**55** Kramschuster, A. and Turng, L.S. (2010). *An injection molding process for manufacturing highly porous and interconnected biodegradable polymer matrices for use as tissue engineering scaffolds. Journal of Biomedical Materials Research Part B: Applied Biomaterials: An Official Journal of The Society for Biomaterials,*

*The Japanese Society for Biomaterials, and The Australian Society for Biomaterials and the Korean Society for Biomaterials* 92 (2): 366–376.

**56** Goldstein, A.S., Juarez TM, Helmke CD, Gustin MC, and Mikos AG. (2001). *Effect of convection on osteoblastic cell growth and function in biodegradable polymer foam scaffolds. Biomaterials* 22 (11): 1279–1288.

**57** Rambo, C., Recouvreux D, Carminatti C, Pitlovanciv A, Antônio R, and Porto L. (2008). *Template assisted synthesis of porous nanofibrous cellulose membranes for tissue engineering. Materials Science and Engineering: C* 28 (4): 549–554.

**58** Chen, G., Ushida, T., and Tateishi, T. (2000). *Hybrid biomaterials for tissue engineering: a preparative method for PLA or PLGA–collagen hybrid sponges. Advanced Materials* 12 (6): 455–457.

**59** Thomson, R.C., Yaszemski MJ, Powers JM, and Mikos AG. (1996). *Fabrication of biodegradable polymer scaffolds to engineer trabecular bone. Journal of Biomaterials Science, Polymer Edition* 7 (1): 23–38.

**60** Chuenjitkuntaworn, B., Inrung W, Damrongsri D, Mekaapiruk K, Supaphol P, and Pavasant P. (2010). *Polycaprolactone/hydroxyapatite composite scaffolds: preparation, characterization, and in vitro and in vivo biological responses of human primary bone cells. Journal of Biomedical Materials Research Part A* 94A (1): 241–251.

**61** Thadavirul, N., Pavasant, P., and Supaphol, P. (2014). *Development of polycaprolactone porous scaffolds by combining solvent casting, particulate leaching, and polymer leaching techniques for bone tissue engineering. Journal of Biomedical Materials Research Part A* 102 (10): 3379–3392.

**62** Liao, C.J., Chen CF, Chen JH, Chiang SF, Lin YJ, and Chang KY. (2002). *Fabrication of porous biodegradable polymer scaffolds using a solvent merging/particulate leaching method. Journal of Biomedical Materials Research: An Official Journal of The Society for Biomaterials, The Japanese Society for Biomaterials, and The Australian Society for Biomaterials and the Korean Society for Biomaterials* 59 (4): 676–681.

**63** Freed, L.E., Marquis J, Nohria A, Emmanual J, Mikos A, and Langer R. (1993). *Neocartilage formation in vitro and in vivo using cells cultured on synthetic biodegradable polymers. Journal of Biomedical Materials Research* 27 (1): 11–23.

**64** Nam, Y.S. and Park, T.G. (1999). *Porous biodegradable polymeric scaffolds prepared by thermally induced phase separation. Journal of Biomedical Materials Research* 47 (1): 8–17.

**65** Hua, F.J., Park, T.G., and Lee, D.S. (2003). *A facile preparation of highly interconnected macroporous poly(d,l-lactic acid-co-glycolic acid) (PLGA) scaffolds by liquid–liquid phase separation of a PLGA–dioxane–water ternary system. Polymer* 44 (6): 1911–1920.

**66** Huang, Y., Ren J, Chen C, Ren T, and Zhou X. (2008). *Preparation and properties of poly (lactide-co-glycolide)(PLGA)/nano-hydroxyapatite (NHA) scaffolds by* *thermally induced phase separation and rabbit MSCs culture on scaffolds. Journal of Biomaterials Applications* 22 (5): 409–432.

**67** Guarino, V., Guaccio A, Guarnieri D, Netti PA, and Ambrosio L. (2011). *Binary system thermodynamics to control pore architecture of PCL scaffold via temperature-driven phase separation process. Journal of Biomaterials Applications* 27 (3): 241–254.

**68** Salerno, A. and Domingo, C. (2015). *Pore structure properties of scaffolds constituted by aggregated microparticles of PCL and PCL-HA processed by phase separation. Journal of Porous Materials* 22 (2): 425–435.

**69** Ikeda, T., Ikeda K, Yamamoto K, Ishizaki H, Yoshizawa Y, Yanagiguchi K, Yamada S, and Hayashi Y. (2014). *Fabrication and characteristics of chitosan sponge as a tissue engineering scaffold. BioMed Research International* 2014.

**70** Hua, F.J., Kim GE, Lee JD, Son YK, and Lee DS. (2002). *Macroporous poly(L-lactide) scaffold 1. Preparation of a macroporous scaffold by liquid–liquid phase separation of a PLLA–dioxane–water system. Journal of Biomedical Materials Research* 63 (2): 161–167.

**71** Dounchis, J.S., Bae WC, Chen AC, Sah RL, Coutts RD, and Amiel D. (2000). *Cartilage repair with autogenic perichondrium cell and polylactic acid grafts. Clinical Orthopaedics and Related Research* 377: 248–264.

**72** Chu, C.R., Coutts RD, Yoshioka M, Harwood FL, Monosov AZ, and Amiel D. (1995). *Articular cartilage repair using allogeneic perichondrocyteseeded biodegradable porous polylactic acid (PLA): a tissue-engineering study. Journal of Biomedical Materials Research* 29 (9): 1147–1154.

**73** Schwartz, I., Robinson BP, Hollinger JO, Szachowicz EH, and Brekke J. (1995). *Calvarial bone repair with porous D, L-polylactide. Otolaryngology - Head and Neck Surgery* 112 (6): 707–713.

**74** Simpson, R.L., Wiria FE, Amis AA, Chua CK, Leong KF, Hansen UN, Chandrasekaran M, and Lee MW. (2008). *Development of a 95/5 poly(L-lactide-co-glycolide)/hydroxylapatite and β-tricalcium phosphate scaffold as bone replacement material via selective laser sintering. Journal of Biomedical Materials Research Part B: Applied Biomaterials* 84B (1): 17–25.

**75** Mehr, N.G., Hoemann CD, and Favis BD. (2014). *Poly (ε-caprolactone) scaffolds of highly controlled porosity and interconnectivity derived from co-continuous polymer blends: model bead and cell infiltration behavior. Journal of Materials Science: Materials in Medicine* 25 (9): 2083–2093.

**76** Mehr, N.G., Li X, Ariganello MB, Hoemann CD, and Favis BD. (2015). *Pore size and LbL chitosan coating influence mesenchymal stem cell in vitro fibrosis and biomineralization in 3D porous poly (epsilon-caprolactone) scaffolds. Journal of Biomedical Materials Research Part A* 103 (7): 2449–2459.

**77** Nam, Y.S., Yoon, J.J., and Park, T.G. (2000). *A novel fabrication method of macroporous biodegradable polymer scaffolds using gas foaming salt as a porogen* *additive. Journal of Biomedical Materials Research: An Official Journal of The Society for Biomaterials, The Japanese Society for Biomaterials, and The Australian Society for Biomaterials and the Korean Society for Biomaterials* 53 (1): 1–7.

**78** Harris, L.D., Kim, B.S., and Mooney, D.J. (1998). *Open pore biodegradable matrices formed with gas foaming. Journal of Biomedical Materials Research: An Official Journal of The Society for Biomaterials, The Japanese Society for Biomaterials, and the Australian Society for Biomaterials* 42 (3): 396–402.

**79** Yoon, J.J. and Park, T.G. (2001). *Degradation behaviors of biodegradable macroporous scaffolds prepared by gas foaming of effervescent salts. Journal of Biomedical Materials Research: An Official Journal of The Society for Biomaterials, The Japanese Society for Biomaterials, and The Australian Society for Biomaterials and the Korean Society for Biomaterials* 55 (3): 401–408.

**80** Kim, T.K., Yoon JJ, Lee DS, and Park TG. (2006). *Gas foamed open porous biodegradable polymeric microspheres. Biomaterials* 27 (2): 152–159.

**81** Mooney, D.J., Baldwin DF, Suh NP, Vacanti JP, and Langer R. (1996). *Novel approach to fabricate porous sponges of poly(d,l-lactic-co-glycolic acid) without the use of organic solvents. Biomaterials* 17 (14): 1417–1422.

**82** Hutmacher, D.W., Schantz T, Zein I, Ng KW, Teoh SH, and Tan KC. (2001). *Mechanical properties and cell cultural response of polycaprolactone scaffolds designed and fabricated via fused deposition modeling. Journal of Biomedical Materials Research: An Official Journal of the Society for Biomaterials, The Japanese Society for Biomaterials, and the Australian Society for Biomaterials and the Korean Society for Biomaterials* 55 (2): 203–216.

**83** Zein, I., Hutmacher DW, Tan KC, and Teoh SH. (2002). *Fused deposition modeling of novel scaffold architectures for tissue engineering applications. Biomaterials* 23 (4): 1169–1185.

**84** Korpela, J., Kokkari A, Korhonen H, Malin M, Närhi T, and Seppälä J. (2013). *Biodegradable and bioactive porous scaffold structures prepared using fused deposition modeling. Journal of Biomedical Materials Research Part B: Applied Biomaterials* 101B (4): 610–619.

**85** Bose, S., Vahabzadeh, S., and Bandyopadhyay, A. (2013). *Bone tissue engineering using 3D printing. Materials Today* 16 (12): 496–504.

**86** Yen, H.-J., Tseng C-S, Hsu S-H, and Tsai C-L. (2009). *Evaluation of chondrocyte growth in the highly porous scaffolds made by fused deposition manufacturing (FDM) filled with type II collagen. Biomedical Microdevices* 11 (3): 615–624.

**87** Liang, D., Hsiao, B.S., and Chu, B. (2007). *Functional electrospun nanofibrous scaffolds for biomedical applications. Advanced Drug Delivery Reviews* 59 (14): 1392–1412.

**88** Yoshimoto, H., Shin Y, Terai H, and Vacanti J. (2003). *A biodegradable nanofiber scaffold by electrospinning and its potential for bone tissue engineering. Biomaterials* 24 (12): 2077–2082.

**89** Ghasemi-Mobarakeh, L., Prabhakaran MP, Morshed M, Nasr-Esfahani M-H, and Ramakrishna S. (2008). *Electrospun poly (ε-caprolactone)/gelatin nanofibrous scaffolds for nerve tissue engineering. Biomaterials* 29 (34): 4532–4539.

**90** Zhou, C., Shi Q, Guo W, Terrell L, Qureshi AT, Hayes DJ, and Wu Q. (2013). *Electrospun bio-nanocomposite scaffolds for bone tissue engineering by cellulose nanocrystals reinforcing maleic anhydride grafted PLA. ACS Applied Materials & Interfaces* 5 (9): 3847–3854.

**91** Katsogiannis, K.A.G., Vladisavljević, G.T., and Georgiadou, S. (2015). *Porous electrospun polycaprolactone (PCL) fibres by phase separation. European Polymer Journal* 69: 284–295.

**92** Li, W.-J., Laurencin CT, Caterson EJ, Tuan RS, and Ko FK. (2002). *Electrospun nanofibrous structure: anovel scaffold for tissue engineering. Journal of Biomedical Materials Research* 60 (4): 613–621.

**93** Boland, E.D., Coleman BD, Barnes CP, Simpson DG, Wnek GE, and Bowlin GL. (2005). *Electrospinning polydioxanone for biomedical applications. Acta Biomaterialia* 1 (1): 115–123.

**94** Binan, L., Tendey C, De Crescenzo G, El Ayoubi R, Ajji A, and Jolicoeur M. (2014). *Differentiation of neuronal stem cells into motor neurons using electrospun poly-l-lactic acid/gelatin scaffold. Biomaterials* 35 (2): 664–674.

**95** Moreno, M.J., Ajji A, Mohebbi-Kalhori D, Rukhlova M, Hadjizadeh A, and Bureau MN. (2011). *Development of a compliant and cytocompatible micro-fibrous polyethylene terephthalate vascular scaffold. Journal of Biomedical Materials Research Part B: Applied Biomaterials* 97B (2): 201–214.

**96** Bouchet, M., Gauthier M, Maire M, Ajji A, and Lerouge S. (2019). *Towards compliant small-diameter vascular grafts: predictive analytical model and experiments. Materials Science and Engineering: C* 100: 715–723.

**97** Ong, C.S., Zhou X, Huang CY, Fukunishi T, Zhang H, and Hibino N. (2017). *Tissue engineered vascular grafts: current state of the field. Expert Review of Medical Devices* 14 (5): 383–392.

**98** Kurobe, H., Maxfield MW, Breuer CK, and Shinoka T. (2012). *Concise review: tissue-engineered vascular grafts for cardiac surgery: past, present, and future. Stem Cells Translational Medicine* 1 (7): 566–571.

**99** Giannico, S., Hammad F, Amodeo A, Michielon G, Drago F, Turchetta A, Di Donato R, and Sanders SP. (2006). *Clinical outcome of 193 extracardiac fontan patients: the first 15 years. Journal of the American College of Cardiology* 47 (10): 2065–2073.

**100** Shinoka, T. and Breuer, C. (2008). *Surgery issue: tissue-engineered blood vessels in pediatric cardiac surgery. The Yale Journal of Biology and Medicine* 81 (4): 161.

**101** McClure, M.J., Wolfe PS, Rodriguez IA, and Bowlin GL. (2011). *Bioengineered vascular grafts: improving vascular tissue engineering through scaffold design. Journal of Drug Delivery Science and Technology* 21 (3): 211–227.

**102** Greisler, H.P. (1982). *Arterial regeneration over absorbable prostheses. Archives of Surgery* 117 (11): 1425–1431.

**103** Naito, Y., Shinoka T, Duncan D, Hibino N, Solomon D, Cleary M, Rathore A, Fein C, Church S, and Breuer C. (2011). *Vascular tissue engineering: towards the next generation vascular grafts. Advanced Drug Delivery Reviews* 63 (4–5): 312–323.

**104** Schmedlen, R.H., Elbjeirami WM, Gobin AS, and West JL. (2003). *Tissue engineered small-diameter vascular grafts. Clinics in Plastic Surgery* 30 (4): 507–517.

**105** Shinoka, T., Shum-Tim D, Ma PX, Tanel RE, Isogai N, Langer R, Vacanti JP, and Mayer Jr JE. (1998). *Creation of viable pulmonary artery autografts through tissue engineering. The Journal of Thoracic and Cardiovascular Surgery* 115 (3): 536–546.

**106** Shin'oka, T., Matsumura G, Hibino N, Naito Y, Watanabe M, Konuma T, Sakamoto T, Nagatsu M, and Kurosawa H. (2005). *Midterm clinical result of tissue-engineered vascular autografts seeded with autologous bone marrow cells. The Journal of Thoracic and Cardiovascular Surgery* 129 (6): 1330–1338.

**107** Hibino, N., McGillicuddy E, Matsumura G, Ichihara Y, Naito Y, Breuer C, and Shinoka T. (2010). *Late-term results of tissue-engineered vascular grafts in humans. The Journal of Thoracic and Cardiovascular Surgery* 139 (2): 431–436.e2.

**108** He, W., Ma Z, Teo WE, Dong YX, Robless PA, Lim TC, and Ramakrishna S. (2009). *Tubular nanofiber scaffolds for tissue engineered small-diameter vascular grafts. Journal of Biomedical Materials Research Part A: An Official Journal of The Society for Biomaterials, The Japanese Society for Biomaterials, and The Australian Society for Biomaterials and the Korean Society for Biomaterials* 90 (1): 205–216.

**109** Sell, S., McClure MJ, Barnes CP, Knapp DC, Walpoth BH, Simpson DG, and Bowlin GL. (2006). *Electrospun polydioxanone–elastin blends: potential for bioresorbable vascular grafts. Biomedical Materials* 1 (2): 72.

**110** Gu, X., Ding, F., and Williams, D.F. (2014). *Neural tissue engineering options for peripheral nerve regeneration. Biomaterials* 35 (24): 6143–6156.

**111** Subramanian, A., Krishnan, U.M., and Sethuraman, S. (2009). *Development of biomaterial scaffold for nerve tissue engineering: biomaterial mediated neural regeneration. Journal of Biomedical Science* 16 (1): 108.

**112** Ghasemi-Mobarakeh, L., Prabhakaran MP, Morshed M, Nasr-Esfahani MH, Baharvand H, Kiani S, Al-Deyab SS, and Ramakrishna S. (2011). *Application of conductive polymers, scaffolds and electrical stimulation for nerve tissue engineering. Journal of Tissue Engineering and Regenerative Medicine* 5 (4): e17–e35.

**113** Wang, Y.-F., Guo, H.-F., and Ying, D.-J. (2013). *Multilayer scaffold of electrospun PLA–PCL–collagen nanofibers as a dural substitute. Journal of Biomedical Materials Research Part B: Applied Biomaterials* 101 (8): 1359–1366.

**114** Xie, J., MacEwan MR, Ray WZ, Liu W, Siewe DY, and Xia Y. (2010). *Radially aligned, electrospun nanofibers as dural substitutes for wound closure and tissue regeneration applications. ACS Nano* 4 (9): 5027–5036.

**115** Liu, W., Thomopoulos, S., and Xia, Y. (2012). *Electrospun nanofibers for regenerative medicine. Advanced Healthcare Materials* 1 (1): 10–25.

**116** Kim, J.I., Kim, J.Y., and Park, C.H. (2018). *Fabrication of transparent hemispherical 3D nanofibrous scaffolds with radially aligned patterns via a novel electrospinning method. Scientific Reports* 8 (1): 1–13.

**117** Kumbar, S.G., Nukavarapu SP, James R, Nair LS, and Laurencin CT. (2008). *Electrospun poly (lactic acid-co-glycolic acid) scaffolds for skin tissue engineering. Biomaterials* 29 (30): 4100–4107.

**118** Zhu, X., Cui W, Li X, and Jin Y. (2008). *Electrospun fibrous mats with high porosity as potential scaffolds for skin tissue engineering. Biomacromolecules* 9 (7): 1795–1801.

**119** Kuppan, P., Vasanthan KS, Sundaramurthi D, Krishnan UM, and Sethuraman S. (2011). *Development of poly (3-hydroxybutyrate-co-3-hydroxyvalerate) fibers for skin tissue engineering: effects of topography, mechanical, and chemical stimuli. Biomacromolecules* 12 (9): 3156–3165.

**120** Veleirinho, B., Coelho DS, Dias PF, Maraschin M, Ribeiro-do-Valle RM, and Lopes-da-Silva JA. (2012). *Nanofibrous poly (3-hydroxybutyrate-co-3-hydroxyvalerate)/chitosan scaffolds for skin regeneration. International Journal of Biological Macromolecules* 51 (4): 343–350.

**121** Keshel, S.H., Biazar E, Rezaei Tavirani M, Rahmati Roodsari M, Ronaghi A, Ebrahimi M, Rad H, Sahebalzamani A, Rakhshan A, and Afsordeh K. (2014). *The healing effect of unrestricted somatic stem cells loaded in collagen-modified nanofibrous PHBV scaffold on full-thickness skin defects. Artificial Cells, Nanomedicine, and Biotechnology* 42 (3): 210–216.

**122** Chong, E.J., Phan TT, Lim IJ, Zhang Y, Bay BH, Ramakrishna S, and Lim CT. (2007). *Evaluation of electrospun PCL/gelatin nanofibrous scaffold for wound healing and layered dermal reconstitution. Acta Biomaterialia* 3 (3): 321–330.

**123** Venugopal, J. and Ramakrishna, S. (2005). *Biocompatible nanofiber matrices for the engineering of a dermal substitute for skin regeneration. Tissue Engineering* 11 (5–6): 847–854.

**124** Nguyen, T.T.T., Ghosh C, Hwang S-G, Dai Tran L, and Park JS. (2013). *Characteristics of curcumin-loaded poly (lactic acid) nanofibers for wound healing. Journal of Materials Science* 48 (20): 7125–7133.

**125** Chandrasekaran, A.R., Venugopal J, Sundarrajan S, and Ramakrishna S. (2011). *Fabrication of a nanofibrous scaffold with improved bioactivity for culture of human dermal fibroblasts for skin regeneration. Biomedical Materials* 6 (1): 015001.

**126** Blackwood, K.A., McKean R, Canton I, Freeman CO, Franklin KL, Cole D, Brook I, Farthing P, Rimmer S, and Haycock JW. (2008). *Development of biodegradable electrospun scaffolds for dermal replacement. Biomaterials* 29 (21): 3091–3104.

**127** Lannutti, J., Reneker D, Ma T, Tomasko D, and Farson D. (2007). *Electrospinning for tissue engineering scaffolds. Materials Science and Engineering: C* 27 (3): 504–509.

**128** Chen, G., Sato T, Ushida T, Hirochika R, Shirasaki Y, Ochiai N, and Tateishi T. (2003). *The use of a novel PLGA fiber/collagen composite web as a scaffold for engineering of articular cartilage tissue with adjustable thickness. Journal of Biomedical Materials Research Part A* 67A (4): 1170–1180.

**129** Dai, W., Kawazoe N, Lin X, Dong J, and Chen G. (2010). *The influence of structural design of PLGA/collagen hybrid scaffolds in cartilage tissue engineering. Biomaterials* 31 (8): 2141–2152.

**130** Chung, C. and Burdick, J.A. (2008). *Engineering cartilage tissue. Advanced Drug Delivery Reviews* 60 (2): 243–262.

**131** Dunkelman, N.S., Zimber MP, LeBaron RG, Pavelec R, Kwan M, and Purchio A. (1995). *Cartilage production by rabbit articular chondrocytes on polyglycolic acid scaffolds in a closed bioreactor system. Biotechnology and Bioengineering* 46 (4): 299–305.

**132** Bruschi, M.L. (ed.) (2015). 2 - Modification of drug release. In: *Strategies to Modify the Drug Release from Pharmaceutical Systems*, 15–28. Woodhead Publishing.

**133** US Food Drug Administration Regulation Chapter 1, CFR-Code of Federal Regulations title 21. https://www.ecfr.gov/current/title-21/chapter-I

**134** World Health Organization (2011). *Revision of Monograph on Tablets: Final text for addition to The International Pharmacopoeia*, 10. QAS/09.324/Final.

**135** Dong, Y. and Feng, S.S. (2006). *Nanoparticles of poly (D, L-lactide)/methoxy poly (ethylene glycol)-poly (D, L-lactide) blends for controlled release of paclitaxel. Journal of Biomedical Materials Research Part A: An Official Journal of the Society for Biomaterials, the Japanese Society for Biomaterials, and the Australian Society for Biomaterials and the Korean Society for Biomaterials* 78 (1): 12–19.

**136** Tyler, B., Gullotti D, Mangraviti A, Utsuki T, and Brem H. (2016). *Polylactic acid (PLA) controlled delivery carriers for biomedical applications. Advanced Drug Delivery Reviews* 107: 163–175.

**137** Leroux, J.-C., Allémann E, De Jaeghere F, Doelker E, and Gurny R. (1996). *Biodegradable nanoparticles – from sustained release formulations to improved site specific drug delivery. Journal of Controlled Release* 39 (2–3): 339–350.

**138** Morita, Y., Saino, H., and Tojo, K. (1998). *Polymer blend implant for ocular delivery of fluorometholone. Biological and Pharmaceutical Bulletin* 21 (1): 72–75.

**139** Makadia, H.K. and Siegel, S.J. (2011). *Poly lactic-co-glycolic acid (PLGA) as biodegradable controlled drug delivery carrier. Polymers* 3 (3): 1377–1397.

**140** Govender, T., Stolnik S, Garnett MC, Illum L, and Davis SS. (1999). *PLGA nanoparticles prepared by nanoprecipitation: drug loading and release studies of a water soluble drug. Journal of Controlled Release* 57 (2): 171–185.

**141** Mu, L. and Feng, S. (2003). *A novel controlled release formulation for the anticancer drug paclitaxel (Taxol®): PLGA nanoparticles containing vitamin E TPGS. Journal of Controlled Release* 86 (1): 33–48.

**142** Danhier, F., Ansorena E, Silva JM, Coco R, Le Breton A, and Préat V. (2012). *PLGA-based nanoparticles: an overview of biomedical applications. Journal of Controlled Release* 161 (2): 505–522.

**143** Chereddy, K.K., Vandermeulen, G., and Préat, V. (2016). *PLGA based drug delivery systems: promising carriers for wound healing activity. Wound Repair and Regeneration* 24 (2): 223–236.

**144** Kim, S.Y. and Lee, Y.M. (2001). *Taxol-loaded block copolymer nanospheres composed of methoxy poly (ethylene glycol) and poly (ε-caprolactone) as novel anticancer drug carriers. Biomaterials* 22 (13): 1697–1704.

**145** Espuelas, M., Legrand P, Loiseau P, Bories C, Barratt G, and Irache J. (2002). *In vitro antileishmanial activity of amphotericin B loaded in poly (ε-caprolactone) nanospheres. Journal of Drug Targeting* 10 (8): 593–599.

**146** Heller, J. (2005). *Ocular delivery using poly(ortho esters). Advanced Drug Delivery Reviews* 57 (14): 2053–2062.

**147** Heller, J., Barr J, Ng SY, Abdellauoi KS, and Gurny R. (2002). *Poly(ortho esters): synthesis, characterization, properties and uses. Advanced Drug Delivery Reviews* 54 (7): 1015–1039.

**148** Goonoo, N., Jeetah R, Bhaw-Luximon A, and Jhurry D. (2015). *Polydioxanone-based bio-materials for tissue engineering and drug/gene delivery applications. European Journal of Pharmaceutics and Biopharmaceutics* 97: 371–391.

**149** Li, G., Chen Y, Hu J, Wu X, Hu J, He X, Li J, Zhao Z, Chen Z, and Li Y. (2013). *A 5-fluorouracil-loaded polydioxanone weft-knitted stent for the treatment of colorectal cancer. Biomaterials* 34 (37): 9451–9461.

**150** Fenton, O.S., Olafson KN, Pillai PS, Mitchell MJ, and Langer R. (2018). *Advances in biomaterials for drug delivery. Advanced Materials* 30 (29): 1705328.

**151** Sequeira, J.A.D., Santos AC, Serra J, Veiga F, and Ribeiro AJ. (2018). Poly(lactic-co-glycolic acid) (PLGA) matrix implants. In: *Nanostructures for the Engineering of Cells, Tissues and Organs* (ed. A.M. Grumezescu), 375–402. William Andrew Publishing.

**152** Xu, Q., Hashimoto M, Dang TT, Hoare T, Kohane DS, Whitesides GM, Langer R, and Anderson DG. (2009). *Preparation of monodisperse biodegradable polymer microparticles using a microfluidic flow-focusing device for controlled drug delivery. Small* 5 (13): 1575–1581.

**153** Carvalho Silva, R., Alexandre Muehlmann L, Rodrigues Da Silva J, de Bentes Azevedo R, and Madeira Lucci C. (2014). *Influence of nanostructure composition on its morphometric characterization by different techniques. Microscopy Research and Technique* 77 (9): 691–696.

**154** Jain, R.A. (2000). *The manufacturing techniques of various drug loaded biodegradable poly (lactide-co-glycolide)(PLGA) devices. Biomaterials* 21 (23): 2475–2490.

**155** Wright, J.C. and Burgess, D.J. (2011). *Long Acting Injections and Implants.* Springer.

**156** Brannon-Peppas, L. (1995). *Recent advances on the use of biodegradable microparticles and nanoparticles in controlled drug delivery. International Journal of Pharmaceutics* 116 (1): 1–9.

**157** Dinarvand, R., Sepehri N, Manoochehri S, Rouhani H, and Atyabi F. (2011). *Polylactide-co-glycolide nanoparticles for controlled delivery of anticancer agents. International Journal of Nanomedicine* 6: 877.

**158** Reis, C.P., Neufeld RJ, Ribeiro AJ, and Veiga F. (2006). Nanoencapsulation I. *Methods for preparation of drug-loaded polymeric nanoparticles. Nanomedicine: Nanotechnology, Biology and Medicine* 2 (1): 8–21.

**159** Azeem, R., Kivitz A, Plotkin H, Johnston L, Kishimoto TK, Park J, Smolinski S, and DeHaan W. (2019). *Phase 2 dose-ranging study of SEL-212 in symptomatic gout patients: selection of doses for further clinical development.* In: *Arthritis & Rheumatology*. Hoboken, NJ, USA: Wiley.

**160** Sim, T., Kim JE, Hoang NH, Kang JK, Lim C, Kim DS, Lee ES, Youn YS, Choi H-G, and Han H-K. (2018). *Development of a docetaxel micellar formulation using poly (ethylene glycol)–polylactide–poly (ethylene glycol)(PEG–PLA–PEG) with successful reconstitution for tumor targeted drug delivery. Drug Delivery* 25 (1): 1362–1371.

**161** Ragelle, H., Danhier F, Préat V, Langer R, and Anderson DG. (2017). *Nanoparticle-based drug delivery systems: a commercial and regulatory outlook as the field matures. Expert Opinion on Drug Delivery* 14 (7): 851–864.

**162** Hamad, K., Kaseem M, Yang H, Deri F, and Ko Y. (2015). *Properties and medical applications of polylactic acid: a review. Express Polymer Letters* 9 (5): 435–455.

**163** Parsons, A.J., Ahmed I, Haque P, Fitzpatrick B, Niazi MI, Walker GS, and Rudd CD. (2009). *Phosphate glass fibre composites for bone repair. Journal of Bionic Engineering* 6 (4): 318–323.

**164** Ahmed, I., Jones I, Parsons A, Bernard J, Farmer J, Scotchford C, Walker G, and Rudd C. (2011). *Composites for bone repair: phosphate glass fibre reinforced PLA with varying fibre architecture. Journal of Materials Science: Materials in Medicine* 22 (8): 1825–1834.

**165** Athanasiou, K.A., Niederauer, G.G., and Agrawal, C.M. (1996). *Sterilization, toxicity, biocompatibility and clinical applications of polylactic acid/polyglycolic acid copolymers. Biomaterials* 17 (2): 93–102.

**166** Agrawal, C., Athanasiou, K., and Heckman, J. (1997). Biodegradable PLA-PGA polymers for tissue engineering in orthopaedics. *Materials Science* 250: 115–128.

**167** Athanasiou, K.A., Agrawal CM, Barber FA, and Burkhart SS. (1998). *Orthopaedic applications for PLA-PGA biodegradable polymers. Arthroscopy: The Journal of Arthroscopic & Related Surgery* 14 (7): 726–737.

**168** Böstman, O., Hirvensalo E, Vainionpää S, Mäkelä A, Vihtonen K, Törmälä P, and Rokkanen P. (1989). *Ankle fractures treated using biodegradable internal fixation. Clinical Orthopaedics and Related Research* 238: 195–203.

**169** Bostman, O., Hirvensalo E, Makinen J, and Rokkanen P. (1990). *Foreign-body reactions to fracture fixation implants of biodegradable synthetic polymers. The Journal of Bone and Joint Surgery. British* 72 (4): 592–596.

**170** Böstman, O., Hirvensalo E, Vainionpää S, Vihtonen K, Törmälä P, and Rokkanen P. (1990). *Degradable polyglycolide rods for the internal fixation of displaced bimalleolar fractures. International Orthopaedics* 14 (1): 1–8.

**171** Hirvensalo, E. (1989). *Fracture fixation with biodegradable rods Forty-one cases of severe ankle fractures. Acta Orthopaedica Scandinavica* 60 (5): 601–606.

**172** Kmetty, Á., Bárány, T., and Karger-Kocsis, J. (2010). *Self-reinforced polymeric materials: a review. Progress in Polymer Science* 35 (10): 1288–1310.

**173** Majola, A., Vainionpää S, Rokkanen P, Mikkola H-M, and Törmälä P. (1992). *Absorbable self-reinforced polylactide (SR-PLA) composite rods for fracture fixation: strength and strength retention in the bone and subcutaneous tissue of rabbits. Journal of Materials Science: Materials in Medicine* 3 (1): 43–47.

**174** Törmälä, P., Vasenius J, Vainionpää S, Laiho J, Pohjonen T, and Rokkanen P. (1991). *Ultra-high-strength absorbable self-reinforced polyglycolide (SR-PGA) composite rods for internal fixation of bone fractures: in vitro and in vivo study. Journal of Biomedical Materials Research* 25 (1): 1–22.

**175** Frazza, E.J. and Schmitt, E.E. (1971). *A new absorbable suture. Journal of Biomedical Materials Research* 5 (2): 43–58.

**176** Wasserman, D. and Versfelt, C. (1974). Use of stannous octoate catalyst in the manufacture of L (-) lactide-glycolide copolymer sutures. Google Patents.

**177** Ray, J., Doddi N, Regula D, Williams J, and Melveger A. (1981). *Polydioxanone (PDS), a novel monofilament synthetic absorbable suture. Surgery, Gynecology & Obstetrics* 153 (4): 497–507.

**178** Bourne, R., Bitar H, Andreae P, Martin L, Finlay J, and Marquis F. (1988). *In-vivo comparison of four absorbable sutures: vicryl, dexon plus, maxon and PDS. Canadian Journal of Surgery.* 31 (1): 43–45.

**179** Vesel, A., Mozetic, M., and Zalar, A. (2007). *XPS study of oxygen plasma activated PET. Vacuum* 82 (2): 248–251.

**180** Zare, R.N., Kim, S., and Bowen, R.A.R. (2019). Methods for modifying a hydrophobic polymer surface and devices thereof, US Patent 10259922B2,2019.

# 9

# Upcycling of Polyesters

*Ion Olazabal[1], Elena Gabirondo[1], Coralie Jehanno[1,2], Kazuki Fukushima[3,4], and Haritz Sardon[1]*

[1] *POLYMAT, University of the Basque Country UPV/EHU, Donostia-San Sebastián, Spain*
[2] *POLYKEY, Joxe Mari Korta Center, Donostia-San Sebastian, Spain*
[3] *Department of Chemistry and Biotechnology, School of Engineering, The University of Tokyo, Tokyo, Japan*
[4] *Japan Science and Technology Agency (JST), Precursory Research for Embryonic Science and Technology (PRESTO), Saitama, Japan*

## 9.1 Introduction

Polyesters have a long history among synthetic polymers. Bischoff and Walden reported aliphatic polyesters such as polyglycolide and polylactide before 1900 [1, 2], and poly(alkylene alkanoate)s were systematically studied by Carothers in the 1930s [3, 4]. Although both ring-opening and step-growth polymerization were examined for these polymers, high molecular weight polymers were not obtained, resulting in poor mechanical and thermal properties and failed to practical applications such as textile fibers. Aromatic polyesters were then actively studied from the 1940s [5] after early studies were reported [3]. Commercial production of the representative aromatic polyester, poly(ethylene terephthalate) (PET), also started in this period [6]. Furthermore, from the late 1970s, liquid crystalline polyesters and all aromatic polyesters have been developed as high-performance engineering plastics [7, 8].

Ester bonds are more susceptible to nucleophilic attack than other bonds used in engineering condensation polymers such as amide, imide, urea, ketone, and ether. The wide range of reactivity of the ester bonds makes polyesters the leading player in the recycling/upcycling of polymers. Weak nucleophile water can react with the esters, providing carboxylic acids (or carboxylates) and alcohols under

*Polyester Films: Materials, Processes and Applications*, First Edition.
Edited by Miko Cakmak and Jehuda Greener.

acidic and alkaline conditions. However, in a polymer form, the ester bonds are not readily cleaved compared to small molecules because most of them are not soluble in water, and thus they form a barrier against water. Nevertheless, many polyesters can be decomposed by using the appropriate temperature, media, and catalysts. Accordingly, chemical recycling of polyesters is technically possible to close the material cycle loop, even though the process is not economically and environmentally feasible.

Upcycling was defined in the 1990s and means creating more valued materials from waste than virgin products and starting materials. As with recycling, there might be mechanical or material upcycling approaches where the waste materials are subjected to physical treatment, not a chemical process. This chapter, however, focuses on chemical upcycling. Theoretically, many polyesters can be recycled and upcycled. The development of the relevant technologies depends on the supply and social needs of the polyesters. PET is the most common polyester widely used in commodities such as films, fabrics, and bottles. Polylactides or poly(lactic acid)s (PLAs) are currently the most popular biodegradable polymer [9]. In this context, upcycling and recycling of these two polymers are the most intensely studied. Rational combinations of depolymerization and further transformations of the degraded byproducts, including repolymerization, lead to successful upcycling and recycling of polyesters. Alcoholysis, aminolysis, and transesterification are often employed in upcycling polyesters. Functional groups of the nucleophiles used in these reactions can add value to the degrades, e.g. monomers for more functional polymers than the original polyesters. Since various nucleophiles are available, there are many possibilities to produce functional polymers by upcycling polyesters. In addition, there are some polymer-to-polymer upcycling options, exploiting ring-chain equilibria inherent in aliphatic polyesters. This chapter describes recent progress and emerging techniques regarding the chemical upcycling of polyesters.

## 9.2 Upcycling of Poly(Ethylene Terephthalate)

PET is a well-known polyester widely used in many consumer products. In fact, PET is the fifth most produced polymer in the world, representing 9% of the world's polymers production [10]. The main applications of PET are the production of synthetic fibers and beverage bottles (40 and 60% in weight, respectively), along with packaging applications [11]. The industrial production of this material is divided in three steps, i.e. the esterification of terephthalic acid with ethylene glycol to create the monomer bis(2-hydroxyethyl) terephthalate (BHET), the formation of low molecular weight prepolymers through melt condensation and the solid state polymerization of the oligomers into high molecular weight PET. As for many polymeric materials, PET properties are key for the different applications with its crystallinity being one of the most important parameters. In the amorphous state, it is a colorless transparent material with good barrier properties to gases and moisture, which makes it a good candidate for beverage and food packaging. In contrast, highly crystalline PET is opaque under some conditions, and its mechanical properties such as tensile strength and impact resistance are improved compared with its amorphous version making it an excellent material for textiles. The industrial processes are tailored for the final product to achieve the desired properties. Thus, for recycling or upcycling, especially by depolymerization, the level and type of crystallinity are important factors that determine the conditions of the degradation process.

PET is the most recycled polymer in the world, according to European environmental agency (EEA), the PET recycled rate reached 57% in 2017 in Europe [12]. However, this huge amount of PET collected is mostly recycled mechanically, a method that despite of its economic viability and easy industrialization has important drawbacks [13]. The main being that the properties of recycled PET are deteriorated compared to virgin PET because of the decrease of molecular weight and the presence of contaminants in the matrix leading to a low-quality material. Therefore, mechanically recycled PET is mostly used by the fiber industry for carpets and low-quality textiles. In contrast, chemical recycling offers the possibility to recover monomers from PET that can subsequently be polymerized into a material of identical properties. Despite of being a promising alternative, depolymerization systems usually require such high temperatures that the energy cost restricts its wide use. As an alternative, PET upcycling allows synthesis of high-value-added molecules for a wide range of applications, from medicine to enhanced materials, overcoming the previously mentioned economic issues.

### 9.2.1 Hydrolysis

One of the most studied PET depolymerization pathway is hydrolysis, in which water is used as a reagent leading to terephthalic acid and ethylene glycol. Owing to the low nucleophilicity of the water this type of reaction is usually performed employing acidic (sulfuric or nitric acid), basic (strong bases such as NaOH or NaOMe), or neutral catalysis (organometallic catalysts), usually at high temperatures (200–270 °C) and pressures (1–2 MPa) [14, 15]. In the last decades, enzymatic catalyzed depolymerization has emerged as a promising alternative allowing the reaction to be carried out under milder conditions (25–80 °C, ambient pressure). However, the relatively low reaction rates remain an issue that must be solved. In an upcycling pathway, the terephthalic acid recovered from the hydrolysis of PET can be employed as a starting material leading to different molecules for high-added-value applications [16] (Figure 9.1).

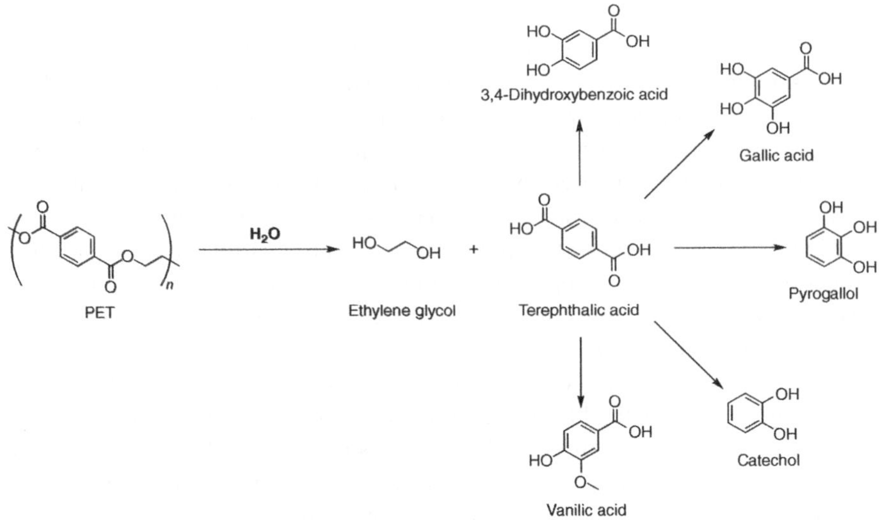


**Figure 9.1** Hydrolysis of PET into ethylene glycol and terephthalic acid and its upcycling into different organic synthons.

Kim et al. explored the biological valorization of waste PET in 2019 to create a wide range of molecules using terephthalic acid as starting material in mild condition using enzymes [16]. In this work, PET was depolymerized by hydrolysis into terephthalic acid and ethylene glycol dissolving 1 g PET in 13 ml of water at 230 °C under microwave irradiation for 50 minutes providing 99.9% of selectivity to terephthalic acid and 24.1% yield. The obtained terephthalic acid was treated by different enzymes to synthesize the high-value-added molecules including 3,4-dihydroxybenzoic acid, gallic acid, pyrogallol, catechol, or vanillic acid in yields ranging from 32.7 to 92.5%. Gallic acid and vanillic acid are used in the pharmaceutical industry in the production of antibacterial agents and analeptic drugs, respectively. Pyrogallol is currently employed as antioxidant in the oil industry. This example demonstrates how the degradation of PET by hydrolysis can lead to highly valuable molecules for a wide range of applications.

Sodium terephthalate salt has also been proven to be a useful upcycled product of PET and has shown interesting properties for energy-storing devices. Kumaresan et al. performed PET depolymerization under acidic conditions employing concentrated sulfuric acid ($H_2SO_4$) to obtain terephthalic acid at 120 °C for six hours [17]. The product was recovered by the removal of the colloidal suspension and washed over deionized water. Finally, disodium terephthalate salt ($Na_2TP$), was obtained by treating terephthalic acid with sodium carbonate ($Na_2CO_3$) and characterized by means of $^1H$, $^{13}C$ NMR, FT-IR, XRD, TGA, DSC, and its electrochemical properties were analyzed by cyclic

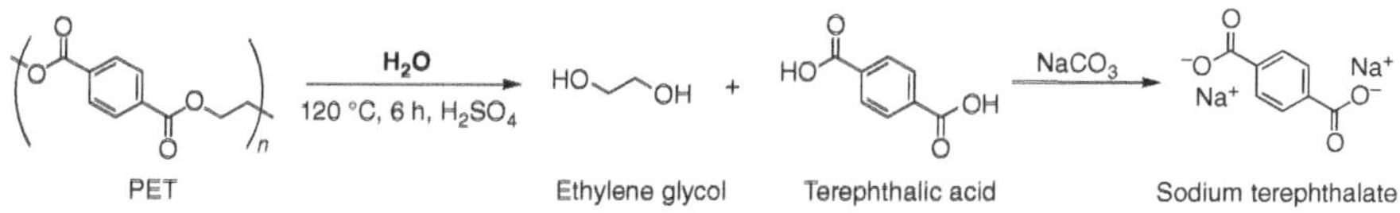


**Figure 9.2** Hydrolysis of PET for the synthesis of sodium terephthalate.

voltammetry (Figure 9.2). The results revealed that electrodes made of $Na_2TP$ in conjunction with conductive carbons such as single-walled carbon nanotubes (SWCNTs) and super P® black showed promising results. This work demonstrates that PET can be effectively upcycled to create promising materials with application in batteries, an emerging area that can take advantage of the upcycling of commodity polymers such as PET as a starting material.

In another study, the terephthalic acid obtained from PET was upcycled into polyhydroxyalkanoates and polyurea-urethanes [14]. Polyhydroxyalkanoates constitute a family of polymers well known for their biodegradability and good mechanical properties, which have been used in medicine and for rigid coatings. First, PET was enzymatically depolymerized into high-purity terephthalic acid and ethylene glycol, at 70 °C, by a thermostable polyester hydrolase. The obtained ethylene glycol and terephthalic acid can then be upcycled in two different ways. First, by conversion into a medium chain length poly(hydroxyalkanoate) (PHA), a type of polymers that are usually employed on medicine due to their interesting biodegradability and biocompatibility. Second, by conversion into a novel poly(amide-urethane) via hydroxyalkanoyloxy-alkanoates (HAA) followed by reacting with methylene diphenyl diisocyanate and butanediol (Figure 9.3). Overall, this study shows the potential of PET upcycling by means of enzymatic catalysis under mild conditions into new promising polymers.

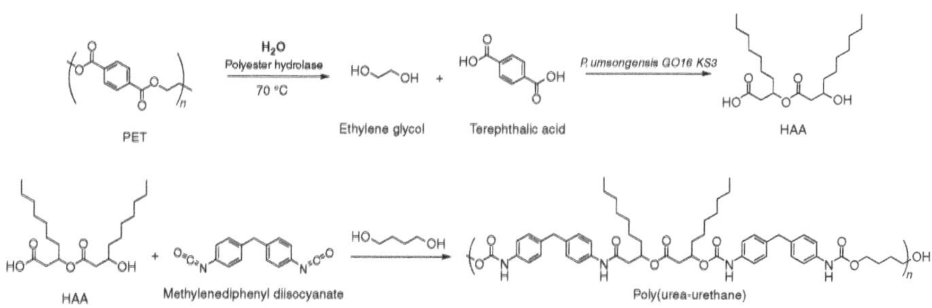


**Figure 9.3** Synthesis of HAA from the depolymerization of PET and upcycling into a novel poly(urea-urethane).

### 9.2.2 Alcoholysis

PET alcoholysis involves transesterification reaction between PET polymer chain and a molecule containing a hydroxyl group. Usually, this reaction requires strong Lewis acids, metallic catalyst such as zinc acetate, titanium butoxide, lead dioxide, cobalt acetate, or magnesium acetate, and relatively harsh reaction conditions to break the stable terephthalate ester bond. In addition, organocatalyst has also been proven as a valid alternative to the toxic metallic catalyst; organic bases such as 1,5,7-triazabicyclo[4.4.0]dec-5-ene (TBD) [18], 1,8-diazabicyclo[5.4.0]undec-7-ene (DBU) [19] or the thermally stable ionic salt TBD:MSA [20] to depolymerize PET by transesterification into BHET at 180 °C. Common transesterification includes methanolysis to obtain dimethyl terephthalate (DMT) or glycolysis to yield BHET [21,22]. Recycled DMT and BHET are more expensive than virgin ones, preventing the commercialization of the process. In contrast, upcycling by alcoholysis allows to obtain economically viable high-added-value molecules which contain the terephthalate moiety with varying pendant groups that usually contain functional end-groups that allow further reactions.

In 2012 Jaroslava and coworkers reported a method for the glycolysis of PET waste flakes into valuable monomers for the synthesis of polyurethane dispersion [23]. Polyurethanes (PUs) are commonly produced polymers widely used in a variety of applications. However, their traditional synthetic pathway raises several concerns, and in recent years, attempts were made to substitute the conventional polyols used in these processes with biobased or recycled sources. In this work, different molar ratios of triethylene glycol, propylene glycol, and poly(ethylene glycol) (PEG) of 400 g·mol$^{-1}$ were employed to assess the effect of the quantity and type of glycol used on the products obtained and their impact on the final polyurethane dispersion (Figure 9.4). As the molar ratio of glycol/PET increases the samples were found to possess better thermal

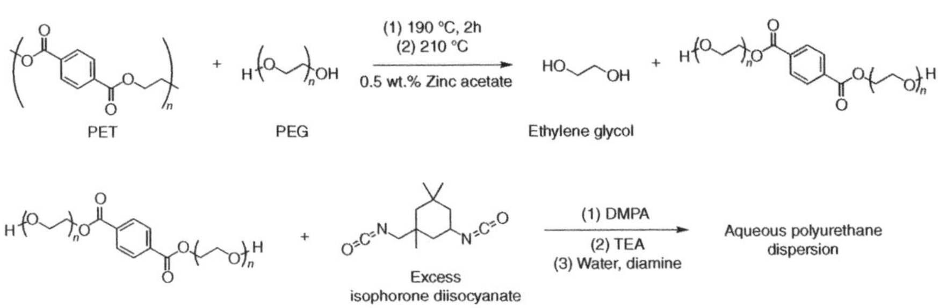


**Figure 9.4** Depolymerization of PET through alcoholysis to obtain aromatic polyols for the formulation of polyurethane dispersion.

properties and lower hardness due to the increase of the glycol chain, while when a low rate (2 : 1) was employed, two separate phases were observed each one corresponding to different hard segments. This work demonstrated the viability of depolymerizing PET into terephthalate-glycols moieties capable of being used for the synthesis of PUs with a relatively high content coming from the upcycled PET.

In a later study, Rajagopal and coworkers developed a similar approach for PET alcoholysis to obtain precursors of polyurethane foams [24]. The PET glycolysis was conducted between 180 and 190 °C using diethylene glycol and 0.5% weight of zinc acetate to increase the functionality of the glycosylate, glycerol was introduced to the reaction. Several PET: diethylene glycol (DEG) ratios were tested finding that 1 : 4 was the optimum one in terms of reaction rates (around seven hours), while 1 : 2 composition led to lower yields and higher viscosity. Later, the obtained glycosylate was reacted with sebacic acid and adipic acid demonstrating the higher reactivity of adipic acid, which led to a highly viscous product (Figure 9.5). Finally, the oligoesters previously obtained were further reacted with methyl diisocyanate (MDI) in different quantities to prepare aromatic polyurethane-polyisocyanurate foams. Different parameters important for their practical applications were measured such as cream and gel time and mechanical properties showing similar cream and gel times for all compositions, while the mechanical properties reveal that sample prepared using glycolyzed PET possesses much higher young modulus due to introduction of the terephthalate group. Thus, the foams synthesized prove to have comparable properties with industrially employed formulations and even surpassing some of the mechanical properties owing to the inclusion of the PET glycosylate.

In 2014 Gioia et al. developed a new strategy to synthesize polymers suitable for powder coating applications from recycled PET and monomers obtained from biomass such as isosorbide and succinic acid [25]. The depolymerization of PET was investigated at 230 °C with two equivalents of isosorbide as the

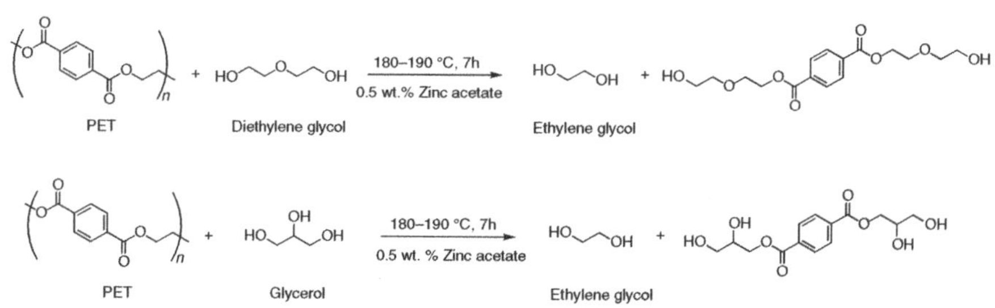


**Figure 9.5** Depolymerization of PET using diethylene glycol and glycerol.

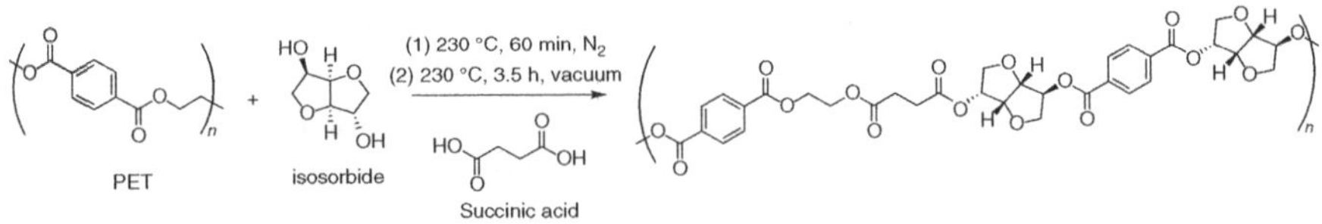


**Figure 9.6** Upcycling of PET into novel polyester through alcoholysis with isosorbide.

depolymerizing agent and different catalysts (Figure 9.6). Notably, the high catalytic activity of tin and titanium organometallic compounds was confirmed with conversion rates of 62 and 63%, respectively, while the other catalysts tested were not able to provide more than 5% of product in one hour. Subsequently, the oligomers obtained were polymerized with succinic acid and small quantities of glycerol as crosslinking agent to obtain a final formulation suitable for powder coating applications. The mechanical and thermal performances were tested demonstrating similar or even superior properties compared to commercial non-renewable-based coatings.

Recently Rorrer et al. published a paper in which PET was upcycled into two types of fiber-reinforced polymers [26]. In this work, the authors depolymerized PET by glycolysis employing excess of ethylene glycol or 1,4-butanediol as glycols using titanium butoxide as transesterification catalyst at 220 °C for four hour leading to the corresponding hydroxyl group terminated terephthalates in the range of 7000–50 000 $g{\cdot}mol^{-1}$ depending on the diol concentration [26] (Figure 9.7). After PET deconstruction, the oligomers were modified by transesterification reactions with malate, fumarate, and dimethyl muconate via melt blending to obtain unsaturated polyesters. Then the polymers were dissolved with a free radical initiator to form a resin, and finally, the resin was applied to a fiberglass mat obtaining fiber reinforced plastics (FRP). Comparing with previous mentioned studies, this work

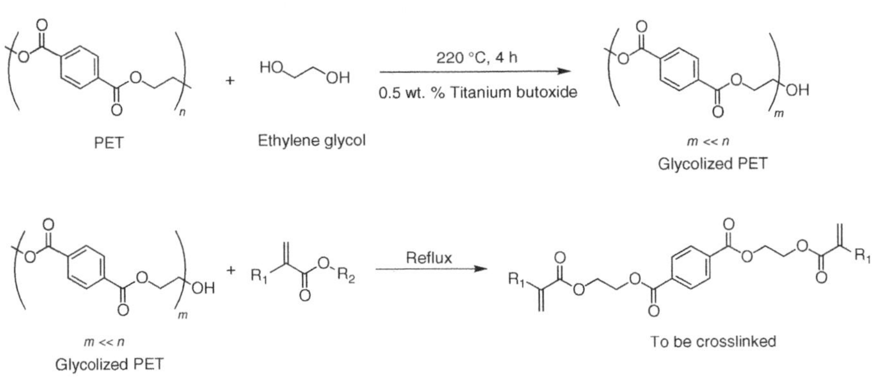


**Figure 9.7** Glycolysis of PET into novel material to be crosslinked.

was focused on obtaining a thermoset compatible with fiber reinforcements instead of the usual upcycling of thermoplastics demonstrating their potential for the creation of other types of materials.

### 9.2.3 Aminolysis

The process of PET degradation into small molecules by primary or secondary amines is called aminolysis. These reactions are faster than alcoholysis due to higher nucleophilicity of amines leading to reaction that can be carried out even in the absence of catalyst at milder reaction conditions (20–200 °C) and shorter reaction times to produce molecules containing terephthalamide moieties. Commonly, diamines or aminoalcohols have been employed in conjunction with metallic catalysts like sodium acetate, dibutyl tin oxide or potassium sulfate as well as organocatalysts or deep eutectic solvents [27–29]. The obtained products have shown a great potential to be employed in a wide range of applications. When they are polymerized, the presence of the terephthalamide groups in the chain's backbone confers an extraordinary rigidness to the material. In addition, some of the molecules obtained by aminolysis can be employed in medical applications [30–32] or as adhesives while further applications [33] of these compounds are currently a subject of intensive research.

One of the first studies to investigate an application from PET upcycling through aminolysis was by Shamsi et al. employing the recovered *N,N′*-bis(2-hydroxyethyl) terephthalamide (BHETA) for the synthesis of novel PUs [34]. PET was reacted with six equivalents of ethanolamine under reflux using 1% of sodium acetate. At the end of the reaction distilled water was added to the crude to precipitate the BHETA, and for further purification, it was recrystallized from boiling water and dried at 80 °C (Figure 9.8). After its characterization polyurethanes were synthesized by mixing the obtained BHETA with a polyol, butanediol and dibutiltin dilauretate (DBTDL) as catalyst in DMSO before the addition

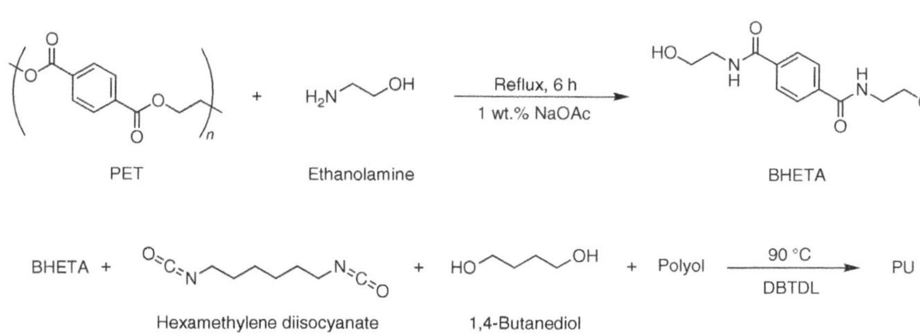


**Figure 9.8** Aminolysis with ethanolamine for the synthesis of polyurethane.

of the desired quantity of hexamethylene diisocyanate. The reaction was carried out at 90 °C employing different proportions of BHETA to study the effect of these molecules on PU properties. Tensile shear strength confirmed that the addition of BHETA lead to a higher maximum load elongation. In addition, the thermal analysis by means of DSC showed a shift of the thermal transition to higher temperatures due to the presence of terephthalamide moieties inside the polymer chains. Moreover, the presence of BHETA in the polymer structure resulted in a shift in the onset of degradation temperature from 140 to 235 °C because of the aromatic ring of BHETA. These PUs were a good example of upcycling since they possessed some improved properties compared to traditional formulations owing to the presence of BHETA coming from waste PET sources, making them a promising alternative to traditional materials for coatings and adhesives.

In 2016 Sharma et al. used molecular derivate from PET and cardanol for the synthesis of adhesives [35]. In this work, PET was depolymerized by aminolysis of ethylene diamine under microwave irradiation. The study revealed the high dependence of reaction rates on temperature showing that complete depolymerization was only achieved at 250 °C close to the melting temperature of PET (260 °C) (Figure 9.9). In addition, it was determined that the 1 : 6 ratio of PET and ethylenediamine was the optimum for which no oligomer was formed, and further dilution did not provide any advantage. The recovered bis(2-aminoethyl) terephthalamide (BAET) was reacted with cardanol and paraformaldehyde in a Mannich-type condensation at 80 and 90 °C for one and two hours, respectively, and after a liquid extraction with chloroform and water, cardanol-BAET (C-BAET) resin was obtained. The same procedure was carried out for the terephthalamide oligomers (AOET) to obtain the cardanol-AOET (C-AOET) material which led to an increase of 24 °C for the curing temperature up to 239 °C due to the longer terephthalamide chain. The lap shear strength tests show remarkable properties for both materials with the C-BAET material providing the highest value, probably because of the higher content of cardanol promoting the formation of hydrogen bonds in the material. Both materials demonstrated to be equally effective for the formation of high-performance adhesive resins from discarded PET leading the way for further studies in this field.

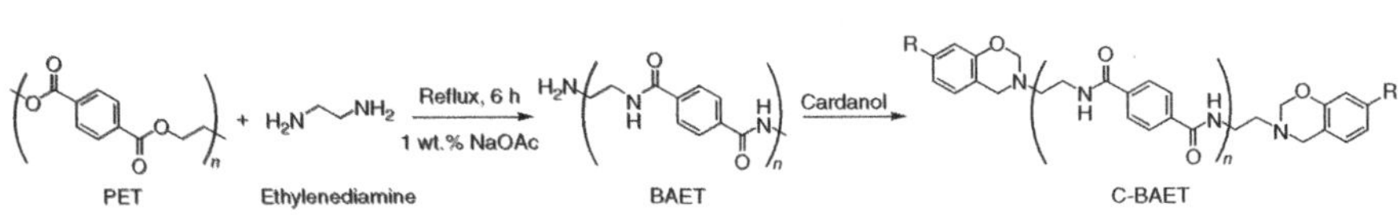


**Figure 9.9** Aminolysis of PET for the synthesis of cardanol terminated terephthalamide chains.

Nevertheless, PET can be degraded into high-value molecules suitable for medical applications. As an example, Tan et al. developed a strategy for the synthesis of antimicrobial polyionene materials including upcycled PET terephthalamides in the polymer backbone [36]. PET was depolymerized into a wide range of secondary terephthalamides containing tertiary amine groups that were subsequently polymerized by polyaddition employing 1,4-bis(chloromethyl) benzene leading this way to the corresponding ionenes (Figure 9.10). All materials were evaluated for antibacterial activity against common bacteria such as *S. aureus*, *E. coli*, *P. aeruginosa*, and *C. albicans*. The results revealed that hydrophilic Ionenes possess similar antimicrobial potency while keeping lower hemolytic values which was negligible. In addition, they were able to attack *Mycobacterium tuberculosis* (Mtb; a type of bacteria known for its ability to develop drug resistance) strains; thus, demonstrating the high effectiveness and applicability of these polymeric materials.

Recently Demarteau et al. reported the organocatalytic upcycling of PET employing different alcohols to create novel polymers. Bottle grade PET was depolymerized employing a wide range of organocatalysts comparing their performance at 180 °C [27]. All reactions were carried out in short times, between 10 and 15 minutes, but of all of them the protic ionic salt TBD : MSA provided fast depolymerization rates while being able to be recycled due to its thermal resistance (Figure 9.11). Different mesh sizes were degraded revealing the low impact of this parameter on the reactions but, by contrast, high crystallinity commercial pellets depolymerized much slower and provided lower yields compared with the amorphous counterparts. Finally, the obtained monomers were polymerized by polycondensation to create novel poly(ester-amides) with three different dimethyl esters at 200 °C using the metal-free catalyst DBU : BA opening the avenue for new PET upcycled materials.

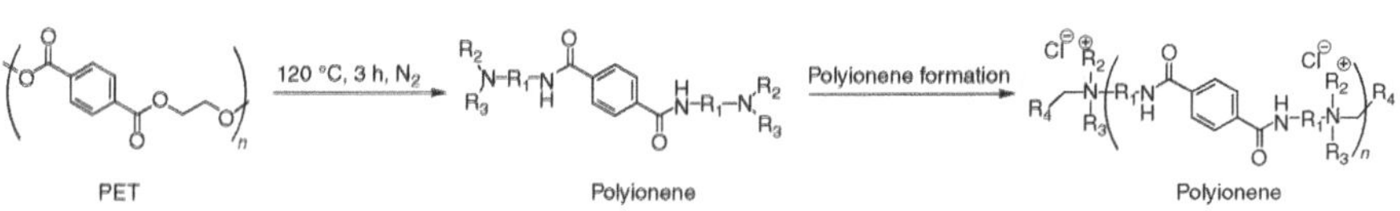


**Figure 9.10** Upcycling of PET into ionene.

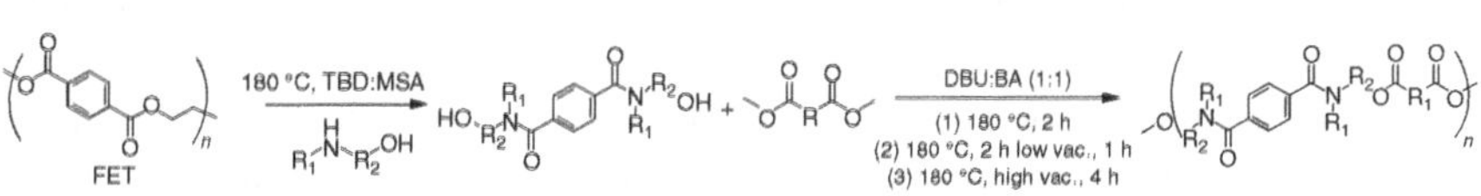


**Figure 9.11** Organocatalysed depolymerization of PET for upcycling into novel poly(ester-amide)s.

## 9.3 Upcycling of Poly(Lactic Acid)

PLA is a common aliphatic polyester well-known for its biocompatibility and natural origin, which could be synthesized by microbial fermentation of starch [37]. However, it is not the only way to produce it; it is also widely synthesized industrially by the ring-opening polymerization (ROP) of lactide. The industrial process is the main synthetic route, having the ability to control molecular weight and crystallinity of the polymer. The control of the synthetic process and properties of PLA has made it a highly sought material for a wide variety of applications in the fields of medicine, packaging, and 3D printing [38].

PLA is a biodegradable and compostable polymer under specific conditions. In this process, the polymer is decomposed to lactic acid and its oligomers and to other small compounds, such as $CO_2$ and $H_2O$ [38]. However, the drawback of this process is that the polymer is absorbed by the soil, and the material is lost. To take advantage of the material at the end of its useful life, many end-life assessments have been proposed, including recycling and upcycling. The latter is a very promising option for PLA due to significant potential opportunities.

### 9.3.1 Hydrolysis of PLA

To obtain value-added monomers, the first step to be undertaken is the depolymerization process by hydrolysis. In this step, the PLA is depolymerized into lactic acid monomer in the presence of water by ester bond breakage. The recovery of the lactic acid enables the resynthesis of PLA, closing the recycling loop. However, many other value-added monomers could also be synthesized starting from the hydroxy acid.

### 9.3.2 Alcoholysis of PLA

Lately, these monomers are attracting considerable attention as organic biobased solvents and as potential alternative for petroleum-based ones. Since they are esters, they are biodegradable. In addition, their toxicity and vapor pressure are low, making them easy to handle. The synthesis of alkyl lactates could be performed by two different routes: one is the direct depolymerization, which could be carried out by alcoholysis, using a short alcohol, such as ethanol or methanol, to break the ester bond of the PLA polymer and to synthesize a new low molecular weight ester monomer (Figure 9.12). The other synthetic route would be the depolymerization of PLA in water followed by esterification reaction [39].

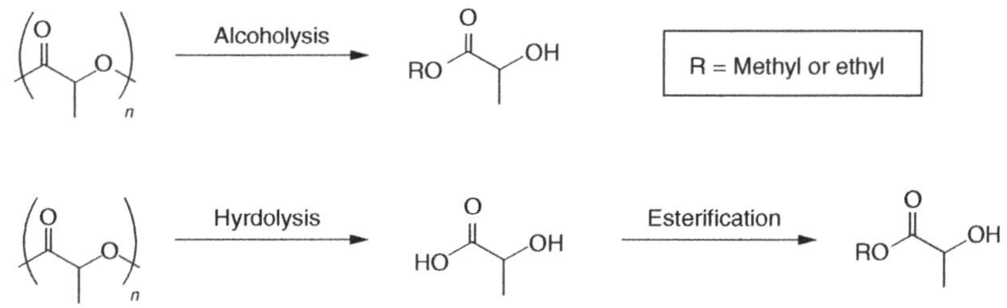


**Figure 9.12** Alkyl lactate synthesis from the depolymerization of PLA.

Taking into account the efficiency of the process, the first synthetic route is the most commonly used to obtain alkyl lactates from PLA. For that process, different catalysts have been studied including metal catalysts and organocatalysts. Metal catalysts, such as, Zn and Fe, have been used for alcoholysis in mild conditions and at relatively short reaction times. However, their toxicity and metal presence point to the need to find a more sustainable catalyst. To address this issue, acid catalysts such as sulfuric acid were proposed, although the problem of toxicity was not solved [40]. In order to help solve the toxicity problem, the use of Ionic liquids (IL) have been proposed. Song et al. tried the methanolysis of PLA by [Bmim][OAc], obtaining methyl lactate at 93% yield at 115 °C in three hours [41]. Even though they could be good candidates for the depolymerization process of PLA they are not suitable industrially due to their high cost.

Organic bases have also been studied for alcoholysis, TBD, for instance, was studied by Leibfarth et al., and the results obtained were extremely promising, achieving up to 90% of conversion to ethyl lactate in three minutes at room temperature [42]. In addition, they reported that the complex TBD is forming with ethyl lactate helps to preserve the stereochemistry of the product in more than 95%. 4-Dimethylaminopyridine (DMAP) (organic base) was also tested for the controlled depolymerization of PLA by Hedrick and coworkers. The aim of their project was to obtain a method that could enable the depolymerization into desired molecular weights [43]. However, it has been seen that full depolymerization into methyl lactate could be obtained by the use of the same catalysts in 10 minutes at 180 °C by microwave irradiation [44].

### 9.3.3 Hydrogenation of PLA

Most common depolymerization methods give monomers that contain an ester or an acid group. However, a high value-added diol such as propylene glycol could be obtained by reductive polymerization.

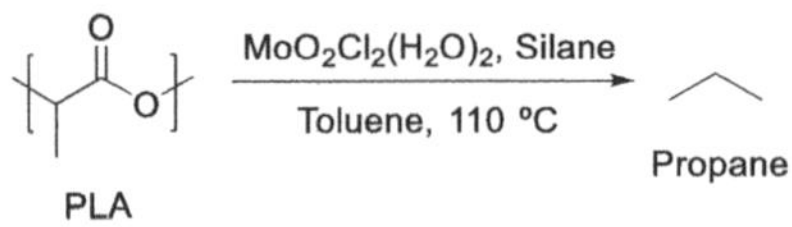


**Figure 9.13** Reductive depolymerization of PLA into propane.

This monomer is prepared by reductive depolymerization of PLA polymer by hydrogenation reaction with the catalysts employed being mainly metallic. Klankermayer et al. employed a Ruthenium complex for the depolymerization at 14 hours and 100 bar $H_2$ [45]. Fernandes et al. have reported the use of $Zn(OAc)_2 \cdot 2H_2O$ in toluene at 110 °C getting propylene glycol in 65% yield after 48 hours of reaction. The yield was enhanced up to 71% adding $PhSiH_3$ to the reaction mixture [46]. Nunes et al. reported the use of $MoO_2Cl_2(H_2O)_2$ complex and silane for the reductive depolymerization obtaining propane as final product [47] (Figure 9.13).

## 9.4 Other Polyesters

Poly(hydroxybutyric acid) (PHB) is a naturally occurring aliphatic polyester that is attracting interest due to its excellent barrier properties making it a promising material for packaging applications [48]. It is synthesized by microorganisms using sugars and oils as precursors, and the process is totally microorganism dependent [49]. In this process, it is still challenging to control molecular weight or crystallinity at the same level as chemical polymerization approach [50]. The good barrier properties of this material could not be fully exploited due to its poor mechanical and processing properties.

PHB polymer can be biodegraded by microorganisms secreting enzymes that break down the polymer into its molecular building blocks, called hydroxy acids, which are oxidized to $CO_2$ or methane. For that process, the PHB is exposed to soil, compost, or marine sediment, being biodegraded in both aerobic and anaerobic environments, without forming any toxic products. Aerobically, PHB is completely degraded to water and carbon dioxide, but in anaerobic conditions, it is degraded to methane and carbon dioxide. The degradation rate of PHB typically ranges from few months in aerobic sewage to years in seawater. The recycling and upcycling of the polymer is an interesting alternative in order to find a useful end-of-life assessment for PHB together with a synthetic method that could enable control over the polymerization. In addition, the possibility of synthesizing PHB chemically could offer the opportunity to enhance its poor mechanical properties and make it easier to process [51].

Melchiors et al. tried the depolymerization of PHB polymer for the first time in 1996 [52]. They tried to depolymerize in melt and in solution, obtaining different results for each depolymerization method. In melt, the temperature employed was always 220 °C, and volatile products were collected and characterized. The product of this depolymerization method was mainly crotonic acid although hydroxybutyrate was also obtained in specific cases. Observing these poor results of the depolymerization of PHB was also performed in solution. Common organic solvents were used including tetrahydrofuran, chloroform, 1,2-chloroethane, and toluene. Only toluene promotes the depolymerization at 111 °C. Each catalyst results in different product. The use of $Ti(OiPr)_4$ led to linear products (short-chain oligomers) and cyclic products (of three repeating unit mainly). $Bu_2Sn(OMe)_2$ and *p*-TsOH. $H_2O$ only provided cyclic ester after 17–20 hours. The cyclic product was isolated and characterized observing mostly the formation of trimer cyclic (TBL). Once the depolymerization product was obtained, the repolymerization was investigated by ROP in melt. The yields obtained were quite low, with a maximum of 60%, depending on the initiator used [52]. This method enables the chemical recycling of PHB, even though the molecular weights obtained are low and not comparable with the PHB synthesized by microorganisms (Figure 9.14).

Polyesters constitute a large polymer class, but only a few are comercially produced. However, there are some specialty polyesters used in niche areas such as in biomedical applications, an area that requires very specific characteristics that cannot always be found in common polymers. In addition, some biobased or biodegradable polyesters have a great potential for large production in the future because they are not dependent of petroleum-based raw materials. Thus, the interest in upcycling other engineering polyesters has been emerging over the last decade, and upcycling exploration of these materials is expected to grow over the time.

In 2019 Liu et al. performed the upcycling of poly(glycolic acid) (PGA) a polymer used for suture production by novel transesterification strategy with the bioderived γ-butyrolactone (BL) [53] (Figure 9.15). In this process, γ-butyrolactone was

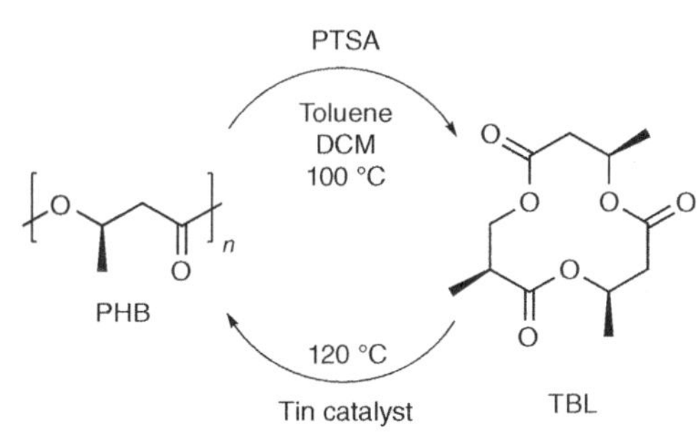


**Figure 9.14** Depolymerization of PHB into macrocycles.

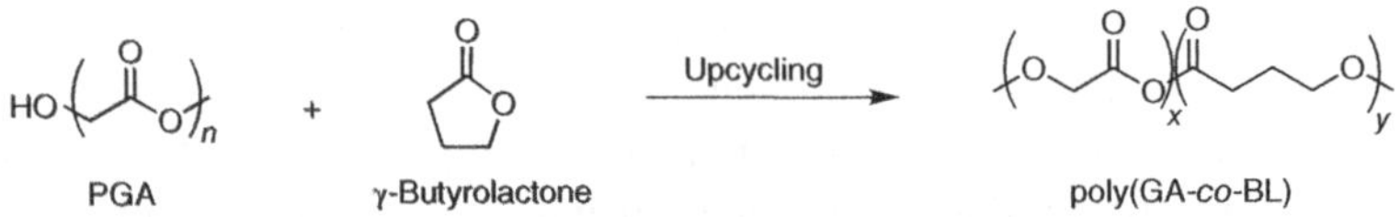


**Figure 9.15** Upcycling of poly(glycolic acid) for the formation of poly(GA-*co*-BL).

employed both as reactant and solvent avoiding the use of hazardous solvents. This way a well-defined poly(GA-*co*-BL) copolymer could be obtained, which exhibited superior thermal stability compared with both homopolymers. A computational study was able to explain the sequence–thermostability relationship, which according to the results depends on the presence of glycolate sequences, especially when they are at the chain ends. Finally, the chemical recyclability of this new copolymer was tested by thermolysis at 300 °C and chemical recycling in the presence of Lewis acid catalyst such as zinc chloride at 120 °C. The glycolide was obtained during the thermolysis but in conjunction with glycolic acid and remaining PGA oligomer impurities. Both acidic and neutral media did not provide the expected results, but the basic reaction carried out in 1M sodium hydroxide water solution drastically improved the degradation rate, leading to sodium glycolate and sodium y-hydroxybutyrate which can be easily precipitated in ethanol.

Another interesting approach would be the hydrogenation of common terephthalic polymers to recover an aliphatic cyclohexane dicarboxylic acid and the corresponding diol. Recently Lende et al. reported the hydrogenation over Rh-Pt catalyst supported on carbon black of Polybutylene terephthalate using 1,1,1,3,3, 3-hexafluro-2-propanol (HFIP) as solvent to synthesize the environmentally friendly polyester poly(butylene-1,4-cycxlohexanedicarboxylate) (PBC) [54] (Figure 9.16). Notably, the whole process was carried out at mild temperatures (50 °C) for one hour and thanks to its solvation capacity there was no need to reach melting point of PET, achieving 100% conversion. Among the wide range of catalysts tested, the bimetallic ruthenium-platinum-based catalyst provided the best results compared to the monometallic ruthenium ones. In addition, the effect of $H_2$ pressure, temperature, and catalyst support was studied for the complete optimization of the system. Interestingly, compressed $CO_2$ was used as antisolvent to precipitate the polymer, a green alternative to traditional precipitation techniques that involve large quantities of volatile organic compounds.

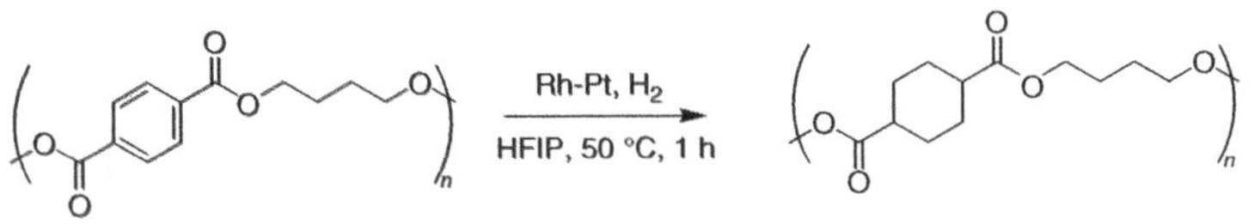


**Figure 9.16** Hydrogenation of poly(butylene terephthalate).

## 9.5 Conclusion

The global environmental situation associated with plastics, such as the urgent need for zero carbon emissions, gives impetus to the circular plastic economy, with special emphasis on upcycling. Upcycling of polyesters is more promising and flexible than recycling and more interesting from an economic standpoint. Nonetheless, chemical upcycling involves similar issues to chemical recycling due to waste-based processes: secure source and values and costs of the products. Success or failure in upcycling the most common polyester, PET, can be a benchmark for chemical upcycling of other polyesters. Most PET degradations proceed in a heterogeneous system. Therefore, progress in process chemistry and catalysts will be critical for the industrialization of chemical upcycling. Successful chemical upcycling of aromatic polyesters, including PET, can serve as an aromatic feedstock, which will be important for realizing zero carbon emissions in the future. However, most upcycled materials are currently not designed for future applications and should exclude landfills. The success of upcycling will be closing the material loop, including reuse and recycling of the upcycled products.

## Acknowledgment

K.F. thanks financial supports by JSPS KAKENHI Grants (JP19H05716) and JST PRESTO Grant Number JPMJPR21N7. The authors from UPV/EHU thank the Ministry of Spain for the EUR2020-112080 financial support. C.J. acknowledges the financial support from el Ministerio de ciencia e innovación from the Juan de la Cierva Program (FJC2020045872-I).

## References

1 Bischoff, C.A. and Walden, P. (1893). Ueber das glycolid und seine homologen. *Berichte der deutschen chemischen Gesellschaft* 26 (1): 262–265. https://doi.org/10.1002/cber.18930260158.

2 Bischoff, C.A. and Walden, P. II. (1894). Ueber derivate der Milchsäure. *Justus Liebigs Annalen der Chemie* 279 (1–2): 71–99. https://doi.org/10.1002/jlac.18942790107.

3 Carothers, W.H. and Arvin, J.A. (1929). Studies on polymerization and ring formation. II. Poly-esters. *J. Am. Chem. Soc.* 51 (8): 2560–2570. https://doi.org/10.1021/ja01383a042.

4 Carothers, W.H., Dorough, G.L., and van Natta, F.J. (1932). Studies of polymerization and ring formation. X. The reversible polymerization of

6-membered cyclic esters. *J. Am. Chem. Soc.* 54 (2): 761–772. https://doi.org/10.1021/ja01341a046.

5 Whinfield, J.R. (1946). Chemistry of 'Terylene'. *Nature* 158: 930–931. https://www.nature.com/articles/158930a0.

6 (1992). The early years of polyester. *J. Soc. Dyers Colour.* 108 (9): 363–364. https://doi.org/10.1111/j.1478-4408.1992.tb01478.x.

7 Volksen, W., Lyerla, J.R. Jr., Economy, J., and Dawson, B. (1983). Liquid-crystalline copolyesters based on poly(*p*-oxybenzoate) and poly(*p*,*p*-biphenylene terephthalate). *J. Polym. Sci.: Polym. Chem. Ed.* 21 (8): 2249–2259. https://doi.org/10.1002/pol.1983.170210810.

8 Schwarz, G. and Kricheldorf, H.R. (1995). Whiskers. 12. Whisker-like crystals of poly(4'-hydroxybiphenyl-4-carboxylic acid). *Macromolecules* 28 (11): 3911–3917. https://doi.org/10.1021/ma00115a024.

9 Fukushima, K. and Kimura, Y. (2006). Stereocomplexed polylactides (Neo-PLA) as high-performance bio-based polymers: their formation, properties, and application. *Polym. Int.* 55 (6): 626–642. https://doi.org/10.1002/pi.2010.

10 Geyer, R., Jambeck, J.R., and Law, K.L. (2017). Production, use, and fate of all plastics ever made. *Sci. Adv.* 3 (7): e1700782. https://doi.org/10.1126/sciadv.1700782.

11 Barnard, E., Arias, J.J.R., and Thielemans, W. (2021). Chemolytic depolymerisation of PET: a review. *Green Chem.* 23 (11): 3765–3789. https://doi.org/10.1039/D1GC00887K.

12 (2020). *Plastics - The Facts 2020*. Plastics Europe.

13 Al-Salem, S.M., Lettieri, P., and Baeyens, J. (2009). Recycling and recovery routes of Plastic Solid Waste (PSW): a review. *Waste Manage.* 29 (10): 2625–2643. https://doi.org/10.1016/j.wasman.2009.06.004.

14 Tiso, T., Narancic, T., Wei, R. et al. (2021). Towards bio-upcycling of polyethylene terephthalate. *Metab. Eng.* 66: 167–178. https://doi.org/10.1016/j.ymben.2021.03.011.

15 Ikenaga, K., Inoue, T., and Kusakabe, K. (2016). Hydrolysis of PET by combining direct microwave heating with high pressure. *Procedia Eng.* 148: 314–318. https://doi.org/10.1016/j.proeng.2016.06.442.

16 Kim, H.T., Kim, J.K., Cha, H.G. et al. (2019). Biological valorization of poly(ethylene terephthalate) monomers for upcycling waste PET. *ACS Sustainable Chem. Eng.* 7 (24): 19396–19406. https://doi.org/10.1021/acssuschemeng.9b03908.

17 Kumaresan, L., Kirubakaran, K.P., Priyadarshini, M. et al. (2021). Sustainable-inspired design of efficient organic electrodes for rechargeable sodium-ion batteries: conversion of P-waste into E-wealth device. *Sustainable Mater. Technol.* 28: e00247. https://doi.org/10.1016/j.susmat.2021.e00247.

18 Fukushima, K., Coulembier, O., Lecuyer, J.M. et al. (2011). Organocatalytic depolymerization of poly(ethylene terephthalate). *J. Polym. Sci. A Polym. Chem.* 49 (5): 1273–1281. https://doi.org/10.1002/pola.24551.

**19** Fukushima, K., Coady, D.J., Jones, G.O. et al. (2013). Unexpected efficiency of cyclic amidine catalysts in depolymerizing poly(ethylene terephthalate). *J. Polym. Sci. A Polym. Chem.* 51 (7): 1606–1611. https://doi.org/10.1002/pola.26530.

**20** Jehanno, C., Flores, I., Dove, A.P. et al. (2018). Organocatalysed depolymerisation of PET in a fully sustainable cycle using thermally stable protic ionic salt. *Green Chem.* 20 (6): 1205–1212. https://doi.org/10.1039/C7GC03396F.

**21** Ren, J., Dyosiba, X., Musyoka, N.M. et al. (2016). Green synthesis of Chromium-Based Metal-Organic Framework (Cr-MOF) from waste Polyethylene Terephthalate (PET) bottles for hydrogen storage applications. *Int. J. Hydrogen Energy* 41 (40): 18141–18146. https://doi.org/10.1016/j.ijhydene.2016.08.040.

**22** Hundertmark, T., Mayer, M., McNally, C. and Simons, T. Recycling and the future of the plastics industry | McKinsey. https://www.mckinsey.com/industries/chemicals/our-insights/how-plastics-waste-recycling-could-transform-the-chemical-industry (accessed 18 november 2021).

**23** Cakić, S.M., Ristić, I.S., Cincović, M. et al. (2012). Glycolyzed products from PET waste and their application in synthesis of polyurethane dispersions. *Prog. Org. Coat.* 74 (1): 115–124. https://doi.org/10.1016/j.porgcoat.2011.11.024.

**24** Roy, P.K., Mathur, R., Kumar, D., and Rajagopal, C. (2013). Tertiary recycling of poly(ethylene terephthalate) wastes for production of polyurethane-polyisocyanurate foams. *J. Environ. Chem. Eng.* 1 (4): 1062–1069. https://doi.org/10.1016/j.jece.2013.08.019.

**25** Gioia, C., Vannini, M., Marchese, P. et al. (2014). Sustainable polyesters for powder coating applications from recycled PET, isosorbide and succinic acid. *Green Chem.* 16 (4): 1807–1815. https://doi.org/10.1039/c3gc42122h.

**26** Rorrer, N. A.; Nicholson, S.; Carpenter, A.; Biddy, M. J.; Grundl, N. J.; Beckham, G. T. Combining reclaimed PET with bio-based monomers enables plastics upcycling. *Joule* 2019, 3 (4), 1–22https://doi.org/10.1016/j.joule.2019.01.018.

**27** Demarteau, J., Olazabal, I., Jehanno, C., and Sardon, H. (2020). Aminolytic upcycling of poly(ethylene terephthalate) wastes using a thermally-stable organocatalyst. *Polym. Chem.* 11 (30): 4875–4882. https://doi.org/10.1039/d0py00067a.

**28** Fukushima, K., Lecuyer, J.M., Wei, D.S. et al. (2013). Advanced chemical recycling of poly(ethylene terephthalate) through organocatalytic aminolysis. *Polym. Chem.* 4 (5): 1610–1616. https://doi.org/10.1039/C2PY20793A.

**29** Soni, R.K. and Singh, S. (2005). Synthesis and characterization of terephthalamides from poly(ethylene terephthalate) waste. *J. Appl. Polym. Sci.* 96 (5): 1515–1528. https://doi.org/10.1002/app.21502.

**30** Fukushima, K., Liu, S., Wu, H. et al. (2013). Supramolecular high-aspect ratio assemblies with strong antifungal activity. *Nat. Commun.* 4: 2861. https://doi.org/10.1038/ncomms3861.

**31** Liu, S., Fukushima, K., Venkataraman, S. et al. (2018). Supramolecular nanofibers self-assembled from cationic small molecules derived from repurposed poly(ethylene teraphthalate) for antibiotic delivery. *Nanomedicine* 14 (1): 165–172. https://doi.org/10.1016/j.nano.2017.09.007.

**32** Fukushima, K., Tan, J.P.K., Korevaar, P.A. et al. (2012). Broad-spectrum antimicrobial supramolecular assemblies with distinctive size and shape. *ACS Nano* 6 (10): 9191–9199. https://doi.org/10.1021/nn3035217.

**33** Fukushima, K., Jones, G.O., Horn, H.W. et al. (2020). Formation of bis-benzimidazole and bis-benzoxazole through organocatalytic depolymerization of poly(ethylene terephthalate) and its mechanism. *Polym. Chem.* 11 (30): 4904–4913. https://doi.org/10.1039/D0PY00436G.

**34** Shamsi, R., Abdouss, M., Sadeghi, G.M.M., and Taromi, F.A. (2009). Synthesis and characterization of novel polyurethanes based on aminolysis of poly(ethylene terephthalate) wastes, and evaluation of their thermal and mechanical properties. *Polym. Int.* 58 (1): 22–30. https://doi.org/10.1002/pi.2488.

**35** Sharma, P., Lochab, B., Kumar, D., and Roy, P.K. (2016). Sustainable bis-benzoxazines from cardanol and PET-derived terephthalamides. *ACS Sustainable Chem. Eng.* 4 (3): 1085–1093. https://doi.org/10.1021/acssuschemeng.5b01153.

**36** Tan, J.P.K., Tan, J., Park, N. et al. (2019). Upcycling poly(ethylene terephthalate) refuse to advanced therapeutics for the treatment of nosocomial and mycobacterial infections. *Macromolecules* 52 (20): 7878–7885. https://doi.org/10.1021/acs.macromol.9b01333.

**37** Payne, J., McKeown, P., and Jones, M.D. (2019). A circular economy approach to plastic waste. *Polym. Degrad. Stab.* 165: 170–181. https://doi.org/10.1016/j.polymdegradstab.2019.05.014.

**38** Dechy-Cabaret, O., Martin-Vaca, B., and Bourissou, D. (2004). Controlled ring-opening polymerization of lactide and glycolide. *Chem. Rev.* 104 (12): 6147–6176. https://doi.org/10.1021/cr040002s.

**39** McKeown, P. and Jones, M.D. (2020). The chemical recycling of PLA: a review. *Sustainable Chem.* 1 (1): 1–22. https://doi.org/10.3390/suschem1010001.

**40** Brake, L. D. (1991). Preparation of alkyl esters by depolymerization background of the invention 1. 5,264,617 1. Field of the Invention. 5264617.

**41** Song, X., Zhang, X., Wang, H. et al. (2013). Methanolysis of poly(lactic acid) (PLA) catalyzed by ionic liquids. *Polym. Degrad. Stab.* 98 (12): 2760–2764. https://doi.org/10.1016/j.polymdegradstab.2013.10.012.

**42** Leibfarth, F.A., Moreno, N., Hawker, A.P., and Shand, J.D. (2012). Transforming polylactide into value-added materials. *J. Polym. Sci. A Polym. Chem.* 50 (23): 4814–4822. https://doi.org/10.1002/pola.26303.

**43** Nederberg, F., Connor, E.F., Glausser, T., and Hedrick, J.L. (2001). Organocatalytic chain scission of poly(lactides): a general route to controlled molecular weight, functionality and macromolecular architecture. *Chem. Commun.* 2066–2067. https://doi.org/10.1039/B106125A.

**44** Alberti, C., Damps, N., Meißner, R.R.R., and Enthaler, S. (2019). Depolymerization of end-of-life poly(lactide) via 4-dimethylaminopyridine-catalyzed methanolysis. *ChemistrySelect* 4 (23): 6845–6848. https://doi.org/10.1002/slct.201901316.

**45** Westhues, S.; Idel, J.; Klankermayer, J. Molecular catalyst systems as key enablers for tailored polyesters and polycarbonate recycling concepts. *Sci. Adv.* 2018, 4 (8), eaat966. https://www.science.org/doi/epdf/10.1126/sciadv.aat9669.

**46** Fernandes, A.C. (2021). Reductive depolymerization of plastic waste catalyzed by $Zn(OAc)_2$ $2H_2O$. *ChemSusChem* 14 (19), cssc.202100130. https://doi.org/10.1002/cssc.202100130.

**47** Nunes, B.F.S., Oliveira, M.C., and Fernandes, A.C. (2020). Dioxomolybdenum complex as an efficient and cheap catalyst for the reductive depolymerization of plastic waste into value-added compounds and fuels. *Green Chem.* 22 (8): 2419–2425. https://doi.org/10.1039/c9gc04206g.

**48** Meereboer, K.W., Misra, M., and Mohanty, A.K. (2020). Review of recent advances in the biodegradability of Polyhydroxyalkanoate (PHA) bioplastics and their composites. *Green Chem.* 22 (17): 5519–5558. https://doi.org/10.1039/D0GC01647K.

**49** Taniguchi, I., Kagotani, K., and Kimura, Y. (2003). Microbial production of poly(hydroxyalkanoate)s from waste edible oils. *Green Chem.* 5 (5): 545–548. https://doi.org/10.1039/B304800B.

**50** Phan, H.T., Hosoe, Y., Guex, M. et al. (2022). Directed evolution of sequence-regulating polyhydroxyalkanoate synthase to synthesize a Medium-Chain-Length–Short-Chain-Length (MCL–SCL) block copolymer. *Biomacromolecules* 23 (3): 1221–1231. https://doi.org/10.1021/acs.biomac.1c01480.

**51** Tang, X. and Chen, E.Y.-X. (2018). Chemical synthesis of perfectly isotactic and high melting bacterial poly(3-hydroxybutyrate) from bio-sourced racemic cyclic diolide. *Nat. Commun.* 9 (1): 2345. https://doi.org/10.1038/s41467-018-04734-3.

**52** Melchiors, M., Keul, H., and Höcker, H. (1996). Depolymerization of poly[(R)-3-hydroxybutyrate] to cyclic oligomers and polymerization of the cyclic trimer: an example of thermodynamic recycling. *Macromolecules* 29 (20): 6442–6451. https://doi.org/10.1021/ma9604350.

**53** Liu, X., Hong, M., Falivene, L. et al. (2019). Closed-loop polymer upcycling by installing property-enhancing comonomer sequences and recyclability. *Macromolecules* 52 (12): 4570–4578. https://doi.org/10.1021/acs.macromol.9b00817.

**54** Lende, A.B., Bhattacharjee, S., and Tan, C.S. (2020). Production of environmentally friendly polyester by hydrogenation of poly(butylene terephthalate) over Rh-Pt catalysts supported on carbon black and recovery by a compressed $CO_2$ antisolvent technique. *Ind. Eng. Chem. Res.* 59 (49): 21333–21346. https://doi.org/10.1021/acs.iecr.0c04378.

# Index



*Polyester Films: Materials, Processes and Applications*, First Edition.
Edited by Miko Cakmak and Jehuda Greener.

## *q*



## *r*



## *s*

Printed and bound by CPI Group (UK) Ltd, Croydon, CR0 4YY

17/09/2026

14964318-0003